76th Annual Meeting of Sugar Industry Technologists 2017

Sugar Industry Technologists, Inc. Volume LXXVI

Tainan, Taiwan
7 - 10 May 2017

ISBN: 978-1-5108-5388-1

Printed from e-media with permission by:

Curran Associates, Inc.
57 Morehouse Lane
Red Hook, NY 12571

Some format issues inherent in the e-media version may also appear in this print version.

Printed by Curran Associates, Inc. (xxxx)

Sugar Industry Technologists Inc. (SIT)
4032 Captiva Row
Myrtle Beach, South Carolina 29579
USA

Phone: +1 863 983 3637
Fax: +1 843 796 4743

SITcontact@sucrose.com

Additional copies of this publication are available from:

Curran Associates, Inc.
57 Morehouse Lane
Red Hook, NY 12571 USA
Phone: 845-758-0400
Fax: 845-758-2633
Email: curran@proceedings.com
Web: www.proceedings.com

TABLE OF CONTENTS

TECHNICAL PRESENTATIONS

POSTERS:

PURPOSE OF SUGAR INDUSTRY TECHNOLOGISTS, INC.

Sugar Industry Technologists, Inc. was established in 1941 to serve the professional interests of its members by providing a forum for the exchange of scientific information and other technical aspects of sugar refining, by encouraging original research in sugar technology, by promoting a broader acquaintance among its members and by cooperating with other engineering, technical and scientific societies.

Past Meeting Sites

1941 - 1967, New York
1968 - Montreal, Canada
1969 - New York
1970 - New Orleans
1971 - New York
1972 – Houston
1973 - Miami
1974 - Vancouver, Canada
1975 - Savannah
1976 - Toronto, Canada
1977 - San Francisco
1978 - London, England
1979 - Boston
1980 - Chicago
1981 - Fort Lauderdale
1982 - Atlanta
1983 - New York
1984 - Houston
1985 - Saint John, Canada
1986 - Baltimore
1987 – Sydney, Australia
1988 – Savannah, GA, USA
1989 - New Orleans, LA, USA
1990 - Vancouver, Canada
1991 - New York
1992 – Brussels, Belgium
1993 – Toronto, Canada
1994 - Honolulu, Hawaii
1995 - La Romana, Dominican Republic
1996 - Durban, South Africa
1997 - Montreal, Canada
1998 - Marseille, France
1999 - Estoril, Portugal
2000 - New Orleans
2001 –Taipei, Taiwan, ROC
2002 - Delray Beach, Florida
2003 – Hamilton Island, Australia
2004 – Vancouver, Canada
2005 – Dubai, United Arab Emirates
2006 – La Baule, France
2007 – Baltimore, Maryland, USA
2008 – Leipzig, Germany
2009 – New Orleans, LA, USA
2010 – Savannah, GA, USA
2011 – Montreal, Canada
2012 – Auckland, New Zealand
2013 - Guangzhou, China
2014 - Toronto, Canada
2015 – Osaka, Japan
2016 - New York, NY, USA
2017 – Tainan, Taiwan, R.O.C.

APPROVED FUTURE MEETING SITES

2018 – March 25-28, Naples, Fl, USA
2019 – May 5-8, Durban, South Africa

FOREWORD

The Seventy-Sixth Annual Technical Meeting of the Sugar Industry Technologists, Inc. was held in Tainan, Taiwan from May 7th through May 10th, 2017. Twenty four countries were represented and included 147 delegates and 14 guests for a total of 161 attendees.

Our meeting kicked off with a performance by the Ten Drum Percussion Group which enthralled delegates and guests alike. Our guest speaker was the Taiwan Minister for Economic Development, Dr. Chih-Kung Lee, who discussed the historical role of sugar in Taiwan.

Four technical sessions were presided over by David Pelham (U.S. Sugar Corporation), Paul Schorn (Tongaat Hulett Sugar South Africa Limited), Trevor Ohlson (Lantic Inc.) and Michael Burchell (ASR Group – C&H Sugar). Eighteen technical papers were presented and all were well received with numerous questions asked and lively discussion. Nine Technical posters were on display and generated much discussion between sessions. Thanks to David Pelham for organising the technical program.

The symposium topic was “Energy Improvements in Sugar Refineries”, a topic of great interest to all members. The panel represented a wealth of knowledge and perspectives and their presentations were well received. Thanks to Paul Schorn for organizing the symposium.

On May 10th delegates and guests visited the Tainan refinery. The tour began with a Chinese Dragon performance that captivated all who saw it. Thanks to Tainan Sugar Corporation for their hospitality during the refinery tour and the banquet luncheon.

A special thank you goes to Edgar Aguirre, Executive Director, for all his efforts to make this another smooth, successful and professional technical conference. As the incoming S.I.T. President I would like to personally invite you to the 77th S.I.T. Technical Meeting in Bonita Springs, Florida on March 25th – 28th, 2018.

Trevor Ohlson
2017-2018 S.I.T. President

SUGAR INDUSTRY TECHNOLOGISTS, INC.

2016-2017

OFFICERS

President	Michael Burchell	ASR Group – C&H Sugar Company, Inc.
1st Vice President	Trevor Ohlson	Lantic Inc. - Vancouver
2nd Vice President	Paul Schorn	Tongaat Hulett Sugar South Africa Ltd.
Vice President – Finance	David Demone	ASR Group - American Sugar Refining, Inc.
Immediate Past President	David Pelham	United States Sugar Corporation
Executive Director	Edgar Aguirre	Sugar Industry Technologists, Inc.

EXECUTIVE COMMITTEE

Sune Berghall	Finnsugar Ltd.
Bob Copeland	Lantic Inc.
David Meadows	Tongaat Hulett Ltd.
Thomas Rathke	Imperial Sugar Company
Frank van Noord	Suiker Unie
Katarina Silfversparre	Nordic Sugar AB
Peter Simpson	New Zealand Sugar Co. Ltd.
Ahmed Vawda	United Sugar Company

COMMITTEE CHAIRMAN

By-Laws	Thomas Rathke	Imperial Sugar Company
Finance	David Demone	ASR Group - American Sugar Refining, Inc.
Crystal Award	Charley Richard	Sugar Processing Research Institute
Meade Award	Trevor Ohlson	Lantic Inc. - Vancouver
Chapman Memorial Award	Trevor Ohlson	Lantic Inc. - Vancouver.
Nominating	Bob Copeland	Lantic Inc.
Membership	Edgar Aguirre	Sugar Industry Technologists, Inc.
Symposium	Paul Schorn	Tongaat Hulett Sugar South Africa Ltd.
Technical Meeting Program	David Pelham	United States Sugar Corporation
Publications	Edgar Aguirre	Sugar Industry Technologists, Inc.

LOCAL ARRANGEMENTS

2017 – Tainan, Taiwan, R.O.C.	Hsi-Chun Tso	Taiwan Sugar Corporation
2018 – Bonita Springs, FL, U.S.A.	David Pelham	United States Sugar Corporation

S.I.T. OFFICERS

2016 – 2017

Michael Burchell, Trevor Ohlson, Paul Schorn, David Demone, David Pelham and Edgar Aguirre

TREVOR OHLSON
President-Elect

PAUL SCHORN
1st Vice President-Elect

SUGAR INDUSTRY TECHNOLOGISTS, INC.

2017 – 2018

OFFICERS

President	Trevor Ohlson	Lantic Inc. – Vancouver
1st Vice President	Paul Schorn	Tongaat Hulett Sugar South Africa Ltd.
2nd Vice President	Jack Thompson........................	Louisiana Sugar Refining, LLC
Vice President – Finance	David Demone	ASR Group - American Sugar Refining, Inc.
Immediate Past President	Michael Burchell....................	ASR Group – C&H Sugar Company, Inc.
Executive Director....................	Edgar Aguirre	Sugar Industry Technologists, Inc.

EXECUTIVE COMMITTEE

Sune Berghall..	Finnsugar Ltd.
David Pelham	United States Sugar Corporation.
Dave Meadows..	Tongaat Hulett Ltd.
Frank van Noord	Suiker Unie
Thomas Rathke	Imperial Sugar Company
Katarina Silfversparre..	Nordic Sugar AB
Peter Simpson..	New Zealand Sugar Co., Ltd.
Ahmed Vawda..	United Sugar Company

COMMITTEE CHAIRMAN

By-Laws....................................	Thomas Rathke..............................	Imperial Sugar Company
Finance	David Demone	ASR Group - American Sugar Refining, Inc.
Crystal Award	Peter Maraia	ASR Group - American Sugar Refining, Inc.
Meade Award	Paul Schorn	Tongaat Hulett Sugar South Africa Ltd.
Chapman Memorial Award	Paul Schorn	Tongaat Hulett Sugar South Africa Ltd.
Nominating	Ahmed Vawda	United Sugar Company
Membership.........................	Edgar Aguirre.........................	Sugar Industry Technologists, Inc.
Symposium	Jack Thompson	Louisiana Sugar Refining, LLC
Technical Meeting Program.....	Michael Burchell....................	ASR Group – C&H Sugar Company, Inc.
Publications	Edgar Aguirre..............................	Sugar Industry Technologists, Inc.

LOCAL ARRANGEMENTS

2018 – Naples, Florida, USA	David Pelham	United States Sugar Corporation
2019 – Durban, South Africa	Paul Schorn	Tongaat Hulett Sugar South Africa Ltd.

SUGAR INDUSTRY TECHNOLOGISTS, INC.

DIRECTORS

2017-2018

PAST PRESIDENTS

of

SUGAR INDUSTRY TECHNOLOGISTS, INC.

1941 **ROBERT S. KENT***
Robert Sayre Kent, Inc.

1942 **ROBERT S. KENT***
Robert Sayre Kent, Inc.

1943 **R. A. RING**
Atlantic Sugar Refineries Co. Ltd.

1944 **ALEX M. ORMOND**
Savannah Sugar Refining Corporation

1945 **FRED M. GIBSON**
American Sugar Refining Company

1946 **FRANK C. STAPLES**
American Molasses Company

1947 **FRANK C. STAPLES**
American Molasses Company

1948 **FRANK C. STAPLES**
American Molasses Company

1949 **FRANK M. EXLEY**
Savannah Sugar Refining Company

1950 **CURTIS L. TAGGART**
Revere Sugar Refinery

1951 **ANSON J. FOWLER**
National Sugar Refining Company

1952 **JAMES M. BROWN**
Revere Sugar Refinery

1953 **JAMES M. BROWN**
Revere Sugar Refinery

1954 **FRANK M. CHAPMAN**
Tate & Lyle Refineries, Ltd.

1955 **PETER H. PETRI**
American Molasses Company

1956 **GEOFFREY CONSTABLE**
St. Lawrence Sugar Refineries, Ltd.

1957 **FRANK A. MONROE, JR.**
Revere Sugar Refinery

1958 **W. AUSTIN CLARNO**
Savannah Sugar Refining Company

1959 **ELMER J. CULP**
American Sugar Refining Company

1960 **G. ERNEST WARING**
Atlantic Sugar Refining Co., Ltd.

1961 **A. D. HINRICHS**
Refined Syrups & Sugar Inc.

1962 **H. J. JACOBS**
South Coast Corporation

1963 **FREDERICK BRUDER**
SuCrest Corporation

1964 **WATSON A. BEMIS**
Revere Sugar Refinery

1965 **FREDERICK W. SCHWER**
National Sugar Refining Company

1966 **E. D. GILLETTE**
Refined Syrups & Sugars, Inc

1967 **RUFUS J. HERRING**
Savannah Sugar Refining Corporation

1968 **H. JAMES REICHE**
Refined Syrups & Sugars Inc.

1969 **HENRY G. GERSTNER**
Colonial Sugars Company

1970 **DORR E. TIPPENS**
Amstar Corporation

1971 **PHILIP F. MEADS**
California & Hawaiian Sugar Co.

1972 **JOSEPH A. METZLER**
Goudchaux-Henderson Sugar Co.,

**Chairman of Executive Committee*

PAST PRESIDENTS

of

SUGAR INDUSTRY TECHNLOLGISTS, INC.

1973	**WARREN L. REED** *Revere Sugar Refinery*	1991	**JOSEPH W. ALBERINO** *Savannah Sugar Refinery*
1974	**STEPHEN STACHENKO** *Redpath Sugars Limited*	1992	**THOMAS F. CHANDLER** *Redpath Sugars, Limited*
1975	**GEORGE FAWCETT** *Savannah Sugar Refinery*	1993	**DONALD E. WEBSTER** *B. C. Sugar*
1976	**A. M. BARTOLO** *Imperial Sugar Company*	1994	**BRIAN T. HARRISON** *Imperial Holly Corporation*
1977	**K. R. HANSON** *Amstar Corp. American Sugar Div.*	1995	**C. RICHARD DONNELLY** *Everglades Sugar Refinery, Inc.*
1978	**STANLEY E. GEORGE** *B. C. Sugar*	1996	**KEITH SINCLAIR** *Tate & Lyle Sugars*
1979	**R. STUART PATTERSON** *California & Hawaiian Sugar Co.*	1997	**MARCEL BRAECKMAN** *Raffinerie Tirlemontoise, S. A.*
1980	**JOHN V. LOPEZ-ONA** *National Sugar Refining Company*	1998	**LUIS SAN MIGUEL BENTO** *Refinarias de Acucar Reunidas*
1981	**JOSEPH A. HARRISON, SR.** *Supreme Sugar Company*	1999	**RAY BURKE** *Savannah Sugar Refinery*
1982	**MICHAEL C. BENNETT** *Tate & Lyle, Ltd.*	1999	**LUIS SAN MIGUEL BENTO** *Refinarias de Acucar Reunidas*
1983	**BRUCE A. FOSTER** *St. Lawrence Sugar, Ltd.*	2000	**CHUNG CHI CHOU** *Dr. Chou Technologies, Inc.*
1984	**LOUIS E. MAHONEY III** *Revere Sugar Corporation*	2001	**STEPHEN CLARKE** *Florida Crystals Corporation.*
1985	**EDWARD D. STEPHENSON** *Lantic Sugar Limited*	2002	**DENIS BOURÉE** *Saint Louis Sucre*
1986	**DAVID M. HUMM** *California & Hawaiian Sugar Co.*	2003	**JEAN-PAUL MERLE** *C & H Sugar Company*
1987	**LEON A. ANHAISER** *Imperial Holly Sugar Company*	2004 2005	**EDGAR AGUIRRE** *United States Sugar Corp*
1988	**C. FRANK STOWE** *Amstar Sugar Corporation*	2006	**JEFFREY ROBINSON** *American Sugar Refining Inc.*
1989	**LEON SUE** *Raffinerie Tirlemontoise S. A...*	2007	**PHILIP ROWLAND** *Imperial Sugar Company*
1990	**CHRISTIAN LAUR** **Tate & Lyle France**	2008	**DONALD RYAN** *Lantic Inc. – Vancouver*

PAST PRESIDENTS

of

SUGAR INDUSTRY TECHNOLOGISTS, INC.

2009 **AMELIA FERRAZ**
Rar-Refinarias de Azucar Reunides

2010 **AHMED VAWDA**
United Sugar Company

2011 **DAVID DEMONE**
ASR Group - American Sugar Refining Inc.

2012 **THOMAS RATHKE**
Imperial Sugar Company

2013 **MICHAEL BURCHELL**
ASR Group – C&H Sugar Company

2014 **DAVE MEADOWS**
Tongaat Hulett Limited

2015 **DAVID PELHAM**
United States Sugar Corporation

2016 **MICHAEL BURCHELL**
ASR Group – C&H Sugar Company

THE JOE HARRISON MEMORIAL SCHOLARSHIP AWARD - 2016

Awarded to Ramiro Rafael Cifuentes de Leon, Ingenio La Union Sugar Mill, Guatemala, for achieving the highest score at the Cane Sugar Refiner's Institute at Nicholls State University in 2016

Dr. Darcey Wayment, Director, Sugar Institute, Nicholls State University, Ramiro Rafael Cifuentes de Leon, Ingenio La Union Sugar Mill, Guatemala, Leon A. Anhaiser, Sugar Industry Technologists, Inc.

FRANK CHAPMAN MEMORIAL AWARD

RECIPIENTS

2007 **M. LAFRANCE AND J. LEBLANC**
Lantic Sugar Co. Ltd.
F. GULA AND X. LANCRENON
Applexion/Novasep

2008 **KARL-HEINZ THEISEN**
and I. GEYER
pro-M-tec

2009 **GILLIAN EGGLESTON**
USDA-ARS-SRRC
BELISARIO MONTES
Alma Plantation LLC

2010 **LUIS SAN MIGUEL BENTO**
Porto

2011 **NIKLAS BRAUN**
Nordic Sugar

2012 **I. KADDOUR, A. FERREIRA,**
L. BENTO and F. ROCHA
University of Porto

2013 **G. EGGLESTON, et. al.**
USDA-ARS-SRRC
GILBERT, I. ROSE
AgroParisTech

2014 **JUAN J MENDOZA, REGINA VISO,**
Prodek Inc. .
HERNANDO AFANADOR, *Ingenio Riopaila, Colombia*, **and DOMINGO SERPA,** *LIPESA*

2015 **ITSUKI SHIMASATO**
Toyo Jidoki Co., Ltd.

2016 **JOSE M. CHORAO**
RAR-Rafinarias de A□úcares Reunidas
CARLOS A. ALVES and ARNALDO O. ARAÚJO
Air Liquide Portugal

JOE HARRISON MEMORIAL SCHOLARSHIP

RECIPIENTS

2004 **TERRY TAHAL**
Guyana Sugar Corporation, Guyana

2005 **RAFAEL ZAUN**
Compañía Azucarera Salvadoreña

2006 **ALBERT KATRYAN**
Guyana Sugar Corporation, Guyana

2007 **NORMA SANCHEZ FUENTES**
Sugar Saenz Group

2008 **JUAN RAMIRO SAYAGO**
Ledesma S.A.A.I., Argentina

2009 **BENJAMIN LARA**
Central Izalco, El Salvador

2010 **ADAN ERNESTO ANAYA**
Compañía Azucarera Salvadoreña

2011 **ALVARO MOISES**
Compañía Azucarera Salvadoreña

2012 **GIOVANNI MOLINA**
Compañía Azucarera Salvadoreña

2013 **DAVID ARMANDO FLAMENCO**
C.A.S.S.A. Group, El Salvador

2014 **BYRON LOPEZ**
CENGICAÑA, Guatemala

2015 **AMILCAR EDUARDO AYALA**
Ingenio La Union, Guatemala

2016 **RAMIRO R. CIFUENTES DE LEON**
Ingenio La Union, Guatemala

THE FRANK CHAPMAN MEMORIAL AWARD - 2016

Poster No. 1149 - Pre-Carbonatation with pure CO_2 in a Cane Sugar Refinery

By

Jose M. Chorao, RAR-Refinarias de A□úcares Reunidas, Carlos A. Alves and Arnaldo O. Araújo, Air Liquide Portugal

Paul Schorn presenting the plaque to
Jose M. Chorao, also accepting for Carlos A. Alves and Arnoldo O. Araújo

THE GEORGE AND ELEANORE MEADE AWARD - 2016

Paper No. 1140 – Operational Improvements of Granular Activated Carbon Station through Continuous Improvement

By

Jesse Williams, American Sugar Refining, Inc., Yonkers, NY, U.S.A.

David Demone presenting the George and Eleanore Meade Award

To

Jesse Williams, also accepting for Elizabeth Mendonca

THE GEORGE AND ELEANORE MEADE AWARD

RECIPIENTS

1965 **WARREN L. REED & JAMES VLADYKA**
Revere Sugar Refinery

1966 **W. A. BENNETT; E.W.P. CUNNEEN & DR. A. M. HERTZBERG**
Colonial Sugar Refining Co. Ltd.

1967 **DR. JAMES C. P. CHEN**
Empresa Agricola Chicana, Ltd.

1968 **JOSEPH F. DOWLING**
Refined Syrups & Sugars, Inc.

1969 **A. M. MOULT**
Tate & Lyle Refineries, Ltd.

1970 **M. A. COHEN; O. G. DIONISIO & S. J. DRESHER**
American Sugar Company
DR. C. M. BENNETT & J. C. ABRAM
Tate & Lyle Technical Services

1971 **DR. CHUNG CHI CHOU & K.R. HANSON**
Amstar Corporation

1972 **P. J. LANGLEY & C. J. BOHLIG**
California & Hawaiian Sugar Co.

1973 **A. E. NORCOTT; C. V. RICH & J. T. RUNDELL**
Tate & Lyle Enterprises, Ltd.

1974 **R. S. PATTERSON**
California & Hawaiian Sugar Co.

1975 **PETER H. PETRI**
Goudchaux-Henderson Sugar Co. Ltd.

1976 **JOHN V. LOPEZ-ONA & EVERETT G. JONES**
The National Sugar Refining Co.

1977 **WILLIAM F. BARTON**
Atlantic Sugar Limited

1978 **DORR E. TIPPENS; LEIGH C. HAYES & DR. CHUNG CHI CHOU**
Amstar Sugar Corporation

1979 **WARREN L. REED**
Revere Sugar Corporation

1980 **JOHN V. LOPEZ-ONA**
The National Sugar Refining Co.

1981 **D. J. BARDWELL; H. R. DELANEY & D. T. HAWKINS**
CSR, Ltd. Sugar Division
H. C. WILSON & JOHN A. RICHMOND
Holly Sugar Corporation

1982 **W. J. BARTON**
Atlantic Sugar Limited, N.B.
W. J. KNEBEL
Calgon Corporation, Pittsburgh, PA

1983 **J. CUEL & C. LONGUE EPEE**
Begin-Say

1984 **LIEF RAMM-SCHMIDT & GORAN HYOKY**
Finnish Sugar Company, Ltd.

1985 **DIETRICH BOSSE**
BMA/Silver Engineering

1986 **DIETER FRANK; JOHN PARK & LINCOLN D. METCALFE**
Kazoo Chemise America

1987 **D. J. RADFORD; D. J. TAYFIELD & M. G. S. COX**
Hulett Refineries Limited

1988 **M. KAWAKAMI; H. IWABE & T. NAKAMURA**
Mitsui Sugar Company, Ltd.

1989 **LUIS ROCHA SAN MIGUEL BENTO**
RAR - Refinarias de Acucar Reunidas, S.A.R.L.

1990 **SIMON J. GIBBONS & RICHARD B. RATCLIFFE**
Tate & Lyle Sugars

1991 **PETER J. BILLING & MICHAEL LOW**
Tate & Lyle Sugars
GEORGE RAPOS
Redpath Sugars Limited

1992 **MARGARET A. CLARKE; LESLIE A. EDYE & NGHIA TO**
Sugar Processing Research Inst.
CYNTHIA McDONALD-LEWIS
N.I.R. Systems

THE GEORGE AND ELEANORE MEADE AWARD

RECIPIENTS

1993 **SEBASTIAO FEYO de AZEVEDO**
JOSE CHORAO, MARIA JOAO GONCALVES
University of Porto, Portugal &
LUIS SAN MIGUEL BENTO
RAR Refinarias de Acucar Reunidas S.A.

1994 **DAVE M. MEADOWS**
Tongaat-Hulett Sugar Limited

1995 **BRYAN C. TUNGLAND, GENE SNOOK,**
ANNETTE CAMPBELL & JANIS KARDATZKE
Imperial Holly Corporation

1996 **GAVIN B. DALGLEISH & GRANT M. HUBBARD**
Tongaat-Hulett Sugar Limited

1997 **DERRALD H. HOUSTON & TED D. MILNER**
Silver Weibull

1998 **PETER BURNS & PETER FIELDS**
CSR Sugar

1999 **D. E. REARICK, M. KEARNEY & D. COSTESSO**
Amalgamated Research Inc.

2000 **PETER J. DIBELLA**
Manildra Harwood Sugars

2001 **MARY AN GODSHALL & KHALID IQBAL**
Sugar Processing Research Institute, Inc.

2002 **JAMAL AL GHURAIR, G. C. SINGH, &**
B. K. PATIL
Al Khaleej Sugar Company

2003 **P. SIMPSON, A. CARREON & S. SPILIOPOULOS**
New Zealand Sugar Company & Sugar Australia

2004 **PHILIP ROWLAND**
Imperial Sugar Company – Savannah

2005 **REINHOLD HEMPELMANN**
BMA

2006 **LES EDYE**
Queensland University
GAVIN KHAN
Carbochem Inc.

2007 **ARNO JANSEN**
Schaffer and Associates
MALCOLM WADE
United States Sugar Corporation

2008 **GEORGE CARTER**
Redpath Sugar

2009 **BARBARA MUIR, SMRI, and**
GILLIAN EGGLESTON, USDA-ARS-SRRC

2010 **AHMED VAWDA AND G.N. KASSIM**
Savola Sugar Middle East, Cairo, Egypt

2011 **GEORGE CARTER**
Redpath Sugar

2012 **JEFFREY C. ROBINSON**
American Sugar Refining, Inc.

2013 **PETER SIMPSON**
New Zealand Sugar Co. Ltd.

2014 **MARSHA R. COLE, GILLIAN EGGLESTON**
YOO JIN CHUNG, ***USDA-ARS-SRRC and***
AUDREY GILBERT, ***AgroParisTech***

2015 **CRAIG JENSEN & DAVID LOVE**
Tongaat Hulett Limited

2016 **JESSE WILLIAMS & ELIZABETH MENDONCA**
ASR Group, American Sugar Refining, Inc.

S.I.T. CRYSTAL AWARD 2017

Peter Maraia

Born and raised in Brooklyn New York. Graduate from the University of Louisville BA Chemistry.

Upon graduation, worked at the Indiana Army Ammunition Plant in Charlestown Indiana as a bench chemist in the plant's modernization program in the production of black gun powder from 1970-1974

Join Amstar Corp in 1974 as a Process Development Engineer in the development of a Continuous Boiling Process utilizing two conventional batch vacuum pans at the Philadelphia Refinery. Held a number of Refinery Staff positions from Energy Conservation to Area Supervisor learning and understanding the various unit operations of cane sugar refining.

Upon the closing of Philadelphia Refinery in 1982 was transferred to the company's Chalmette Refinery as a staff engineer. In 1986 became the Process Manager until 1991 when I assume the Engineering & Maintenance Manager role at Chalmette

In Oct. of 2005 became Plant Manager of The Chalmette Refinery until May 2016 where I now assumed the position of Vice President of Sugar Technology.

From 1995 to present instructor at The Cane Sugar Refiner's Institute on Low & High Grade Crystallization at Nicholls State University. Attended CSRI in 1988

Was a member on the Foundation Member & Process Technologists Board of Directors at Nunez Community College in St. Bernard LA. from May 2005 to May 2016. Was one of the founding members of Nunez's Process Technologists curriculum.

For the last 42 years and going forward. I am very proud, honored and privileged to be in the industry of Cane Sugar Refining.

RECIPIENTS OF THE S.I.T. CRYSTAL AWARD**

(S.I.T. Achievement Award in Sugar Technology)

1961 **DR. VICTOR DIETZ**
National Bureau of Standards

1961 **LOHN LOWE**
Revere Sugar Refinery

1962 **HAROLD E. C. POWERS**
Tate & Lyle Refineries, Ltd.

1962 **EUGENE C. GILLETT**
C & H Sugar Refining Corporation

1963 **DR. GEORGE P. MEADE**
Colonial Sugars Company

1963 **ALFRED L. WEBRE**
Jackson Industries, Inc.

1964 **FRANK M. EXLEY**
Savannah Sugar Refining Corp.

1964 **HAROLD J. JACOBS. SR.**
The South Coast Corp.

1965 **FRANK M. CHAPMAN**
Tate & Lyle Refineries, Ltd.

1965 **DR. PIETER HONIG**
Consultant

1966 **WILLIAM H LOUVIERE**
Imperial Sugar Company

1966 **PAUL P. WOEHRLE**
National Sugar Refining Co.

1967 **DR. ELLIOT P. BARRETT**
Kerr-McGee Chemical Corp.

1967 **WATSON A. BEMIS**
Revere Sugar Refinery

1968 **E. DWIGHT GILLETTE**
Refined Syrups & Sugars, Inc.

1969 **DR. PHILIP F. MEADS**
California & Hawaiian Sugar Co.

1969 **ELMER J. CULP**
American Sugar Company

1970 **HENRY G. GERSTNER**
Colonial Sugars Co.

1971 **STEPHEN STACHENKO**
Canada & Dominion Sugar Co. Ltd.

1972 **THOMAS R. GILLETT**
California & Hawaiian Sugar Co.

1973 **WILLIAM W. BLANKENBACH**
B.C. Sugar Refining Co. Ltd.

1974 **W. AUSTIN CLARNO**
Savannah Sugar Refinery

1975 **DR. FRANK G. CARPENTER**
Cane Sugar Ref. Research Project

1976 **FRANK C. STAPLES**
SuCrest Corporation

1977 **R. STUART PATTERSON**
California & Hawaiian Sugar Co.

1978 **DR. JAMES C. P. CHEN**
Cane Tech Sugar Consultants

1978 **DR. MICHAEL C. BENNETT**
Tate & Lyle Engineering Ltd.

1979 **KENNETH R. HANSON**
Amstar Sugar Corporation

1980 **FRANKLYN M. WILLIAMS**
Calgon Corp. Activated Carbon Div.

1981 **ALBERT BERGE**
Raffinerie Tirlemontoise

1981 **WARREN L. REED**
Revere Sugar Refinery

1982 **JOHN V. LOPEZ-ONA**
The National Sugar Refinery Co.

1983 **GEORGE W. MULLER, JR.**
Executive Director, S.I.T.

1984 **JOHN B. ALEXANDER**
Tongaat-Hulett Sugar Ltd.

RECIPIENTS OF THE S.I.T. CRYSTAL AWARD**

(S.I.T. Achievement Award in Sugar Technology)

1984 **Dr. MARGARET A. CLARKE**
Sugar Processing Research Inc.

1985 **A. MARK HERTZBERG**
CSR Lmited

1985 **STANLEY GEORGE**
B. C. Sugar

1986 **DONALD K. LUKE**
Sugar y Azucar

1986 **NEIL L. PENNINGTON**
California & Hawaiian Sugar Co.

1987 **GEORGE FAWCETT**
Savannah Sugar Refinery

1987 **HEIKKI J. HONGISTO**
Finnish Sugar Company, Ltd.

1988 **ADOLPH M. BARTOLO**
Imperial Sugar Company

1989 **DR. CHUNG CHI CHOU**
Amstar Sugar Corporation

1989 **DR. R. A. MCGINNIS**
Beet Sugar Institute (Retired)

1990 **JOSEPH A. HARRISON, SR.**
Supreme Sugar Company, Inc.

1990 **EARLE J. ROBERTS**
Sugar Processing Research, Inc.

1991 **ENRIQUE R. ARIAS**
Sugar Cane Growers Coop of Florida

1992 **GEORGE C. CONRAD**
Western States Machine Co.

1993 **DAVID M. HUMM**
California & Hawaian Sugar Co.

1994 **LEON A. ANHAISER**
Imperial Holly Corporation

1995 **LUIS SAN MIGUEL BENTO**
RAR Refinarias de Acucar Reunidas

1996 **MARY AN GODSHALL**
Sugar Processing Research Institute

1997 **PETER REIN**
Tongaat-Hulett Sugar Limited

1997 **RICHARD RIFFER**
California & Hawaiian Sugar Co.

1998 **KEITH SINCLAIR**
Tate & Lyle Sugars

1999 **PETER FIELD**
CSR Sugar

2000 **MURRAY R. PLAYER**
CSR Sugar

2001 **STEPHEN J. CLARKE**
Florida Crystals Corporation

2002 **MICHAEL SASKA**
Audubon Sugar Institute, LSU

2003 **PETER G. WRIGHT**
Sugar Research Limited

2004 **MICHAEL DONOVAN**
Tate and Lyle Sugars

2005 **G. C. SINGH**
Al Khaleej Sugar Co. (L.L.C.)

2006 **DENIS BOURÉE**
Saint Louis Sucre

2006 **KLAUS E. AUSTMEYER**
Horn GmbH &Co. KG

2007 **JEAN-PAUL MERLE**
C & H Sugar Co.

2008 **JEFFREY ROBINSON**
American Sugar Refining Co.

2009 **DAVID MEADOWS**
Tongaat-Hulett Sugar Ltd.

2010 **BRIAN HARRISON**
Imperial Sugar Company

RECIPIENTS OF THE S.I.T. CRYSTAL AWARD**

(S.I.T. Achievement Award in Sugar Technology)

2011 **MIKE INKSON**
Sugar Knowledge International Ltd.

2012 **DONALD A. RYAN**
Lantic Inc. - Vancouver

2013 **AHMED VAWDA**
United Sugar Company

2014 **ROBERT FALGOUT**
Nicholls State University

2015 **PETER SIMPSON**
New Zealand Sugar C., Ltd.

2016 **CHARLEY RICHARD**
Sugar Processing Research Institute, Inc.

2017 **PETER MARAIA**
ASR Group – American Sugar Refining, Inc.

**** Formerly – Honorary Award**

S.I.T. HONORARY MEMBERSHIP

1963 – Dan Gutleben
1964 – Edgar W. Harris
1968 – Curtis L. Taggart
1970 – George P. Meade
1998 - Earl Roberts
1999 – Bruce Foster
1999 – Mario de Campos Vidal
2002 – Marcel Braeckman
2008 – G. C. Singh
2009 – Leon A. Anhaiser
2010 - Robert Falgout
2016 - Chung Chi Chou

MICHAEL BURCHELL

President

PRESENTATION OF THE GAVEL

President Michael Burchell presents the gavel to

President elect Trevor Ohlson

GROUP OF GUESTS AT THE 2017 S.I.T. TAINAN CONFERENCE, TAIWAN, R.O.C.

SUGAR INDUSTRY TECHNOLOGISTS INCORPORATED

HISTORY

Sugar Industry Technologists was founded in the United States in 1941 by several North American sugar refineries. It has since grown to an international association with a membership of about fifty corporate sugar refining companies, and about eighty corporate allied companies with over three hundred individual members from around the world.

ANNUAL TECHNICAL MEETING

The association holds an annual convention, usually in May. At this convention, technical sessions are held where members present technical papers based on research or development in cane sugar processing technology, quality control, environmental issues, and other topics of interest to the cane sugar refining industry.

Non-members may attend this conference on payment of a non-member fee equal to a membership fee.

A spousal program, social functions and an awards banquet are also held at the convention where the members have the opportunity to get acquainted and discuss mutual technical concerns.

PUBLICATIONS

A Proceedings of the papers presented at the technical sessions is published. Each Corporate member receives a copy along with an updated membership directory. Extra copies or past publications may be purchased from the S.I.T. office for $60.00 USD per copy

AWARDS

Sugar Industry Technologists presents three awards and one scholarship annually.

- The Crystal Award is presented to individuals who have contributed notably to the technological advancement of the Sugar Refining Industry.
- The George and Eleanore Meade Award is presented in recognition of the best paper presented at the Annual Technical Sessions.
- The Frank Chapman Memorial Award is presented for the best poster presentation at the Annual Technical Sessions.
- The Joe Harrison Sr. Memorial Scholarship is awarded to a student attending the Cane Sugar Refiners Institute Course at Nicholls State University.

MEMBERSHIP

A Corporate membership is $450.00 USD with individual Associate membership at $40.00 USD. Affiliate memberships are available to academic and scientific persons at $60.00 USD. All fees are payable in U.S. funds. Annual fees cover the period from July 1 to June 30.

Charles Huang

Chairman, Taiwan Sugar Corporation

Chairman, Taiwan Circular Economy Network

Prior to his appointment as chairman of Taiwan Sugar Corporation in September 2016, Mr. Huang has served as Director on the Board of a number of listed companies such as Continental Holdings Corp., CTCI Advanced Systems Inc. etc., and has been the Chairman and CEO of TSRC Corp. He was also board member of non-profit organizations such as New Frontier Foundation and Hao Ran Foundation. In March 2015, Charles founded Taiwan Circular Economy Network, an organization advocating and promoting resource efficiency and a circular economic model.

Charles graduated from University of Pennsylvania in the US as well as an MBA degree from The Wharton School, U of Pennsylvania.

Minister Lee, President Michael Burchell, Dr. Aguirre and all our distinguished guests. Good morning!

First off, I like to extend my personal appreciation to Minister Lee for making this special trip just to be with us this morning. Your presence symbolizes S.I.T.'s importance in this part of the world and in particular, Taiwan.

Secondly, on behalf of everyone from Taiwan Sugar Corporation (Taiwan Sugar), I like to extend our warmest welcome to all delegates for joining us here in Taiwan. We are deeply honored to have this opportunity to play host to S.I.T. this year.

In the early 20th century, Taiwan's sugar industry was modernized with significant capital investments in R&D and production facilities, and employed highly qualified professionals to manage these modern sites, Taiwan Sugar ascertained a string leadership position in global sugar market. Incidentally, sugar industry is Taiwan's first flagship industry in the history of Taiwan's economic development, preceding Taiwan's petrochemical and IT industries, which had emerged after the sugar industry only in the second half of the last century.

Taiwan Sugar's History: Taiwan and East Asia's LEGACY

Taiwan, with annual production of 1.4 million tons of sugar, ranked number three in sugar export before World War II. Harvesting and production of sugar was Taiwan's economic lifeline for decades. On the social side, with 42 sugar mills and a sugar-rail network of nearly 3,000 KM, of which, more than 600 KM are dedicated to passenger transport, Taiwan's social livelihood is highly interwoven with Taiwan's Sugar's activities.

Taiwan Sugar's Future: ACCELERATOR of Circular Economy

There exists greater urgency for us to move from the highly wasteful and "***degenerative***" Linear Economy to that of a more "***regenerative***" Circular Economy. As supply for natural resources becomes more scarce, and demand for resources continue to increase, this imbalance coupled with the so-called "wastage", which often ended up as pollutants in the air we breathe, water we drink and the soil we grow our food. Time awaits no one!

In recent years, Taiwan Sugar has initiated numerous programs to accelerate the company's commitment to move toward Circular Economy, including, redesigning of our sugar operations with greater circularity; aiming for zero waste and zero emission. To begin with, Taiwan Sugar is redesigning our sugar plantations for minimum water and carbon footprint and full biodiversity. Taiwan Sugar will also introduce and reinvest with more innovative circular economy solutions along the transportation and production line.

At Taiwan Sugar, we are committed to translate good IDEAS into good BUSINESSES. We will protect our soil, water, air and biodiversity and generate diverse source of revenues beyond sales of sugar, to include sales of feed, fuel, fertilizer and bio-materials. Henceforth, I like to take this unique opportunity to invite SIT to collaborate with us, as we move toward circular economy. Only through collaboration, new and innovative circular economy solutions can readily be researched and developed. Perhaps, SIT would consider bringing our community together, by forming a Circular Economy Committee.

Once again, we are thankful and deeply honored to have all of you here in Taiwan. I hope all of you will enjoy southern Taiwan in the next few days, please do have a safe and pleasant stay here. Finally, let me wish the conference great success and hope to see you all again soon.

S.I.T. Paper #1153

Development of a Linear Pan Growth Approach

Chris Mayhew, Carlton Haynes, Robert Howe, Craig Parker and Chris Rhoten
British Sugar and Escon America Inc.

The fine control of the mother liquor saturation is key to efficient crystal production. Over many years the industry has developed several methods to approximate the mother liquor conditions across the total cycle of the batch pan. Both total massecuite brix and mother liquor brix have been used individually to control the degree of supersaturation indirectly, but neither method can accurately guarantee control of crystal growth in isolation. It is the interaction between both these brix measurements that can be used to portray the bigger picture of how the crystal is growing.

This study, over the 2016-17 beet campaign at our Wissington Factory, will report on the investigation into how combining the output from a microwave brix meter and a refractometer can be used effectively to improve crystal growth within a batch pan.

Introduction

There are many factors that influence crystal quality in white batch pans; some are plant related, some due to production requirements, whilst others are driven by process chemistry. One key factor is the crystal growth rate. The better the control achieved here, the more likely the final product quality will improve along with associated benefits in centrifugal performance and syrup purity profiles.

It is well documented [1] [2] that the degree of supersaturation in the mother liquor drives the crystal growth rate. At the same time, maintaining the supersaturation within the metastable range is crucial in controlling the quality of the crystal produced. However, measuring the degree of supersaturation in the mother liquor directly is currently not possible. The majority of control methods imply mother liquor supersaturation, either through direct or indirect measurement of the mother liquor brix. In recent years British Sugar has used Boiling Point Elevation, stirrer power and total massecuite brix (microwave meters) to assimilate mother liquor brix and hence infer degree of saturation. Although reasonable results have been achieved with these control strategies, ultimately the repeatability of the final crystal quality is dependent on the accuracy of supersaturation control achieved within the pan cycle. This is often where the process falters.

Accurate measurement alone will not deliver the optimum crystallisation performance; this requires the appropriate control program that can drive the batch pan to meet quality, throughput, extraction and energy requirements batch after batch. Most batch pan programs focus on a profile to deliver a supersaturation profile.

After a review of current available pan control strategies employed within the sugar industry, British Sugar opted to utilise one that focussed on the primary elements of supersaturation control and crystal growth rate. Linear Pan Growth has been successfully employed in a number of high purity syrup environments, where it defines a mother liquor brix profile to deliver a defined rate of crystal growth over the pan cycle. Following the introduction of seeding crystals at the desired seeding point (either microwave or refractometric measurement of syrup brix), Linear Pan Growth utilises the difference between massecuite and mother liquor brix, a "delta brix", to control the rate of crystal growth. This mechanism balances the rate of crystal growth in the massecuite against the replenishment of sucrose in the mother liquor, at the ideal degree of supersaturation.

Linear Growth Modelling

Numerous authors [3] [4] [5] have published growth rate predictive models based on parameters covering mother liquor supersaturation, temperature and purity and crystal morphology. However, these models would require a range of live process data points, many of which are still not available via online instrumentation. Therefore, most pan boiling programs rely on controlling the key growth driver, the degree of supersaturation in the mother liquor, and approximate its value based on direct, or indirect, methods of measuring the brix of the mother liquor. The challenge here is how to select the ideal profile to control the chosen measurement to.

The principle of linear pan growth modelling relates to the growth rate of a crystal in any single (linear) direction. In high purity solutions, like those in white pan operation, this growth rate factor does not vary greatly for a chosen degree of supersaturation. At the same time, in white batch pans, the operating conditions (feed, vacuum, steam, etc.) remain significantly repeatable from batch to batch. Under this degree of repeatability, these complex interrelationships can be simplified considerably for the purpose of modelling into a suitable process control algorithm. For modelling purposes, the complex sucrose crystal shape also can also be simplified to that of a cube, which can then be used to calculate the mass gain of the crystals in the solution.

To complete the simplified model, the volume of seed required for the batch to achieve the final product size is fixed based on the seed quality (MA and crystal content) and the target final product requirements. This provides the model with the number of seed sites required for full seeding and therefore the initial surface area of the seed when added.

The model developed for the trial was originally designed and created by Chris Rhoten for slurry seeded batch white pans. This model was modified to incorporate the larger Brunswick seed used at Wissington, with steam and vacuum profiles to match the campaign setting for Wissington.

The basic operation of the model follows a pre-determined mother liquor supersaturation profile over the growth phase of the batch cycle. The rate of evaporation is driven by the chosen steam and vacuum profiles over the cycle of the pan, which leads to an increased massecuite brix and mother liquor brix within the pan. At the same time, the linear growth rate calculates the masse of crystal grown from the sucrose in the mother liquor. This growth removes sugar from the mother liquor, causing the mother liquor brix to reduce. The model balances the change in both massecuite and mother liquor brixes and the resultant mother liquor supersaturation against the chosen supersaturation profile, using the syrup feed rate as the adjusting factor. The added syrup increases the level of the pan and is used by the model, in combination with the difference (delta brix) between the massecuite and mother liquor brixes to generate a target delta brix to pan level control profile.

Process modifications for the trial

At Wissington the 6 white batch pans are arranged into two streams of 3 pans, with A-side being pans 1, 2 & 3 and B-side consisting of pans 4, 5 & 6. Each stream runs separately through a dedicated receiver, continuous vacuum crystalliser (CVC), associated batch centrifuges and granulator before merging together prior to the sugar cooler. All 6 white pans operate from a common feed syrup system and also share a common steam supply and vacuum system.

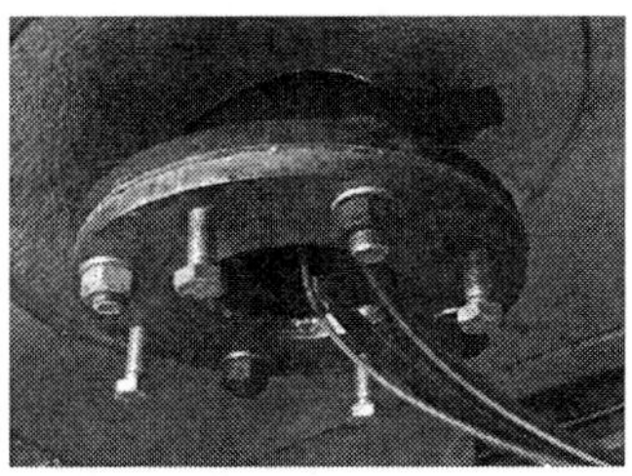

Each pan is fitted with a proMtec microwave brix meter, installed below the calandria. The microwave meters carry on-board temperature measurement for process compensation calculations. The pans are also fitted with three temperature probes as a secondary control back-up.

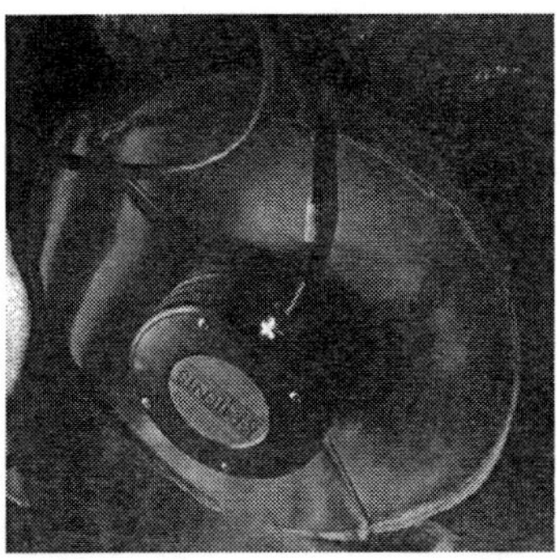

In November 2016, pans 4, 5 and 6 were fitted with K-Patents PR-23 refractometers, installed below the calandria at the same level as the microwave meters, to ensure both devices measured the same massecuite conditions. They were installed with an automated lens flushing facility, connected to the pan floor water flushing system, as per the manufacturer's requirements.

A further K-Patents refractometer was also installed into the cooling crystalliser, which produces seed massecuite for the white pans. Insulation collars were locally fabricated to provide both heat protection for the refractometer's sensor electronics (mounted locally to the sensor) and also prevent any localised point cooling around the probe as it entered the pan.

The signals from both the microwave meters and refractometers were linked back to the factory distributed control systems, with their outputs linked into the pan program controller. The pan program for pans 4, 5 and 6 used both values to calculate the difference between the massecuite and mother liquor brix ("delta brix") and used the "delta brix" as the process variable for the feed syrup controller. The set point for the feed syrup controller was calculated based on the control algorithm from the linear growth profile. The output of this algorithm projects the required "delta brix" for a given pan level and creates a moving set point for the feed controller.

In the linear growth model, the rate of evaporation over the pan cycle was set to drive the desired saturation profile. In the pan this evaporation profile was driven by controlling the cycle to a predetermined steam and vacuum profile, with the syrup feed valve controlling the mother liquor brix required to achieve the desired saturation profile.

The original program was left in control of pans 1, 2 and 3, running the feed profile to deliver a microwave brix profile relating to the massecuite level in the pan. The steam and vacuum profiles followed identical profiles and control mechanisms as described earlier for pans 4, 5 and 6.

At the start of this trial all 6 white pans were seeded on a microwave brix target, set to deliver a degree of saturation (Dsat) of 1.15. Once seeded, pans 1, 2 and 3 would then feed to the predefined microwave brix profile vs massecuite level, whereas pans 4, 5 and 6 would feed to the calculated "delta brix" profile vs massecuite level.

Performance Measurements

During the long campaigns, now common place within British Sugar, juice quality deterioration over the duration of the campaign is the norm. Therefore, the benefit of running two parallel streams on the same juice at the same time eliminates concerns that any performance changes could be related solely to juice quality variation.

For the performance comparison, massecuite samples were taken from both A and B-side batch pans and from the corresponding CVC's discharge lines prior to the BMA batch centrifuges. These samples were analysed in the factory laboratory for brix, purity, mother liquor purity and crystal quality (MA and CV). At the same time, samples of wash syrup and high green were also collected from the batch centrifuges and analysed for brix and purity. To ensure that the syrup samples were representative of the centrifuge performance over the complete cycle, a consistent trickle flow of syrup from both the high green and wash syrup lines was piped into two separate 20 litre containers. The contents of these containers were then sealed and agitated to ensure good mixing in the containers, prior to their testing in the laboratory.

Samples of dry sugar were also collected from the outlets of both A and B-side granulators and these were screened in the laboratory for MA and CV comparisons.

Results and Discussion

The 2016-17 beet campaign at Wissington started in October with all 6 white pans running on their standard microwave brix profile, from seeding through to the high level feed cut off. The syrup feed flow being regulated to control the massecuite brix to the pre-determined profile. The feed controller would switch over to a stirrer power profile during the feed phase, should the stirrer power exceed a pre-determined setting for each pan. If the microwave brix meter fails, the program automatically switches the control of feed syrup addition to meet a boiling point elevation control profile, which was based on achieving the same massecuite brix profile.

Data, collected from both A and B-side systems during this period (Table 1), showed a difference in both crystal particle size (MA) and CV between the two sides, with A side delivering an average particle size (MA) 14μm larger than B side. The average CV of A side was lower than B side, by an average of 0.7 points. There was also a difference in high green and wash syrup purities, as measured from the white centrifuges on A-side and B-side. The purities from the A-side centrifuges were lower, with high green purity an average of 0.98 points lower and wash syrup purity an average 0.5 points lower than on B-side.

Table 1 Syrup and crystal average analysis pre-trial and during the trial

	Stream	Pre-Trial Standard Juice	Linear Growth Standard Juice
	Syrup Streams		
High Green Purity	A Side	84.17	83.43
	B Side	85.15	83.48
	Difference	*0.98*	*0.05*
Wash Syrup Purity	A Side	87.39	86.14
	B Side	87.89	85.65
	Difference	*0.50*	*-0.48*
Feed Syrup	*To pans*	94.54	94.02
	MA of Sugar produced (μm)		
Ex Granulators	A Side	564.0	562.3
	B Side	549.9	547.1
	Difference	*-14.1*	*-15.3*
	% Through 224 Screen		
Ex Granulators	A Side	2.9	3.1
	B Side	4.0	3.3
	Difference	*1.1*	*0.2*
	CV of Sugar produced		
Ex Granulators	A Side	34.2	34.4
	B Side	34.9	34.1
	Difference	*0.7*	*-0.3*

During the first week in November 2016, the K-patents refractometers were installed and commissioned on the B side pans 4, 5 and 6 and the cooling crystalliser. The feed programs for these pans were then switched to operate on the linear growth profiles provided by the model and adjusted to suit the pan cycles.

Once the revised programs were commissioned and proven to be repeatable, the sampling regime was re-instated and the results analysed. Under these revised brix profiles, developed by Chris Rhoten for the Wissington pans, the B side pan performance improved, both in crystal size and syrup purities. The B-side system reduced the high green purity difference between sides by 0.93 points, from the initial 0.98 delta to a far more meaningful 0.05 points differential (well within laboratory purity analysis repeatable standards for such systems). A similar change for the wash syrup purity was seen, switching the B-side system from being 0.5 purity points higher to 0.48 points lower when operating the linear growth program profiles.

Over the same period, the wash water timers to the B side centrifuges automatically reduced, typically by 2-3 seconds, cutting the wash water flows on these centrifuges by around 10% compared with the A side machines. The wash water timers are automatically controlled by ITECA in-line sugar colour cameras to maintain a target sugar colour to silo.

Analysis of the final product granulated sugar from both A-side and B-side granulators showed a similar small change in performance over the trial period (Table 1). The average CV from A-Side was 34.2 prior to the trial, compared with 34.9 from B-Side (a difference of 0.7 in favour of A-side). During the linear growth control trial, the CV of A-side averaged 34.4, whilst B-side reduced to 34.1 (a difference of 0.3 in favour of B-Side). Thus a swing of around 1 point of CV was noted in favour of the B-side pans and the linear pan growth control strategy.

Over the total trial period the average MA from both A-Side and B-Side remained consistent, with A-side averaging 564µm pre-trial and 562µm during the trial and B-side averaging 550µm pre-trial and 547µm during the trial. This level of consistency was maintained by the factory by varying the seed volumes to the pans to meet production targets to silo.

Calibration challenges

The K-Patents refractometers came pre-calibrated, with the manufacturer's claim that they remain accurate during normal operation and require no "in-use" calibration checks whilst in use over the campaign.

Once installed, the process team crossed checked the output from the refractometers by comparing them with laboratory analysis on process juice, sampled during the boil down phase of the pans. At the same time the output from the microwave brix meters were also noted and compared with both the laboratory result and the output of the refractometer. Across pans 4, 5 and 6 the refractometer outputs were within 0.2 brix of the laboratory results, whilst the microwave meters varied by up to 0.6 brix for the same juice.

Sampling of the massecuite post seeding was also undertaken at various points during the pan cycle to compare the readings from both the microwave meters and refractometers with the laboratory results. The massecuite samples were split in to two equal portions in the laboratory, with one portion of the sample tested for total massecuite brix using the 50:50

dilution method on the laboratory refractometer. The mother liquor was separated from the second massecuite portion, using the laboratory vacuum chamber, and this liquor was also analysed on the laboratory refractometer, again using the 50:50 dilution method. The results of these tests were also compared with the process readings from the microwave meters and refractometers.

From the results it became evident that the microwave meters performance drifted over time and they required regular online recalibration to maintain their accuracy. A monthly program consisting of a four point massecuite brix calibration check was then instigated, to individually recalibrate the microwave meters across all 6 white pans.

During the boil down and initial feed phases of the pan cycle, it was observed that the microwave meters regularly registered brix values below those of the refractometers. This created problems using the "delta brix" method to feed the pans over this period as the control algorithm was not able to handle a negative "delta brix". This resulted in the pan program shutting off the feed syrup flow and false grain establishing as the pans over brixed. Regular recalibration of the microwave meters did restore the "delta brix" output for the control profile for short periods, but the output would eventually drift again impacting on the "delta brix" profile and the associated feed control.

In contrast to the variability of the microwave meters, the refractometers performed well over the early phases of the pan cycle, maintaining repeatability from batch to batch. It was therefore decided not only to seed the pans using the refractometers, but also to start feeding the pans on a refractometer brix profile until a positive "delta brix" could be measured. The program would then switch over to delta brix control for the remainder of the feed phase.

Surface area and its effect on saturation control

With the B-side trial pans established on the refractometer and delta brix profile, it was noted that the feed syrup control profiles were varying from strike to strike during the "delta brix" phase. Initially, this was associated with the issues in maintaining the calibration of the microwave meters (as discussed earlier). However, when the microwave meters had been re-calibrated and a positive "delta brix" output restored, the feeding variability continued.

From the linear growth model, the mother liquor brix was predicted to rise to the seeding point and then follow a very shallow curve, rising slightly during the brix-up phase (figure 1).

In the pans it was observed that the mother liquor brix would rarely follow this profile from start to finish. More often, the mother liquor brix would decline after seeding, either after a short period of stability (figure 2), or more rapidly following the seed addition (figure 3). During some cycles, the mother liquor brix would continue to rise for a short period after the seed addition before falling back (figure 4). In this final scenario, the formation of false grain could be observed on the pan microscopes, as the mother liquor brix started to fall. In all scenarios, once the pan had reached high level and the feed valve had closed, the mother liquor Brix rose during the brix up phase, in line with the model's prediction. In the cycles where the mother liquor brix continued to rise or fall after seeding, the feed syrup value output chased the "delta brix" target profile, leading to periods of false grain formation occurring later in the pan cycle.

Figure 1 - Model Brix Profiles vs. Pan Level

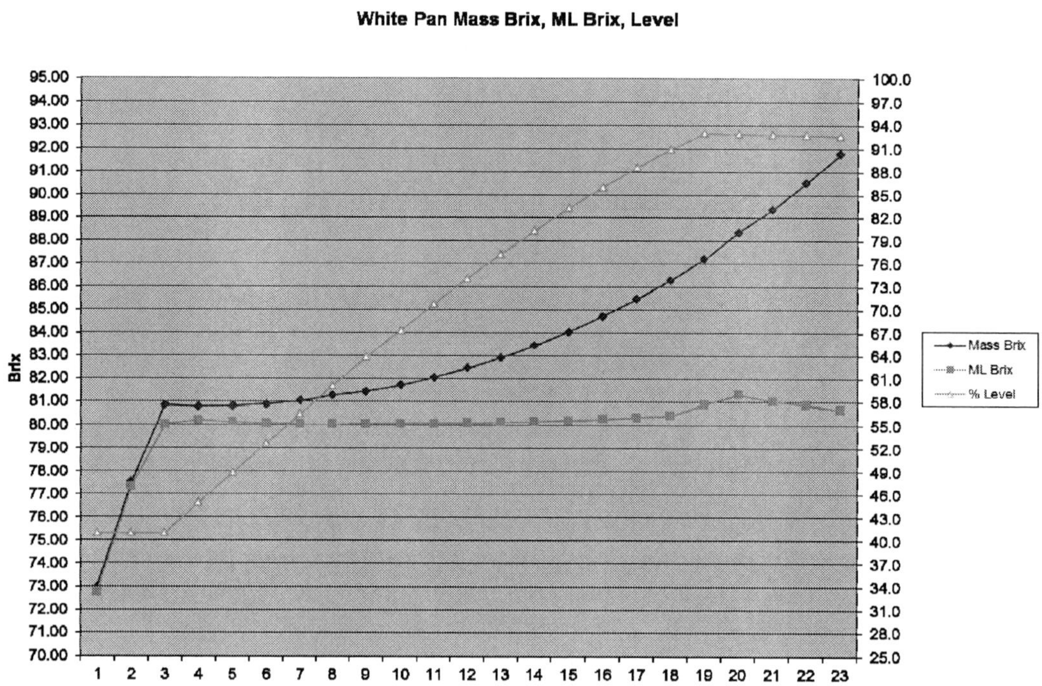

Figure 2 Shallow mother liquor brix profile (Thicker purple line) in Delta Brix control mode

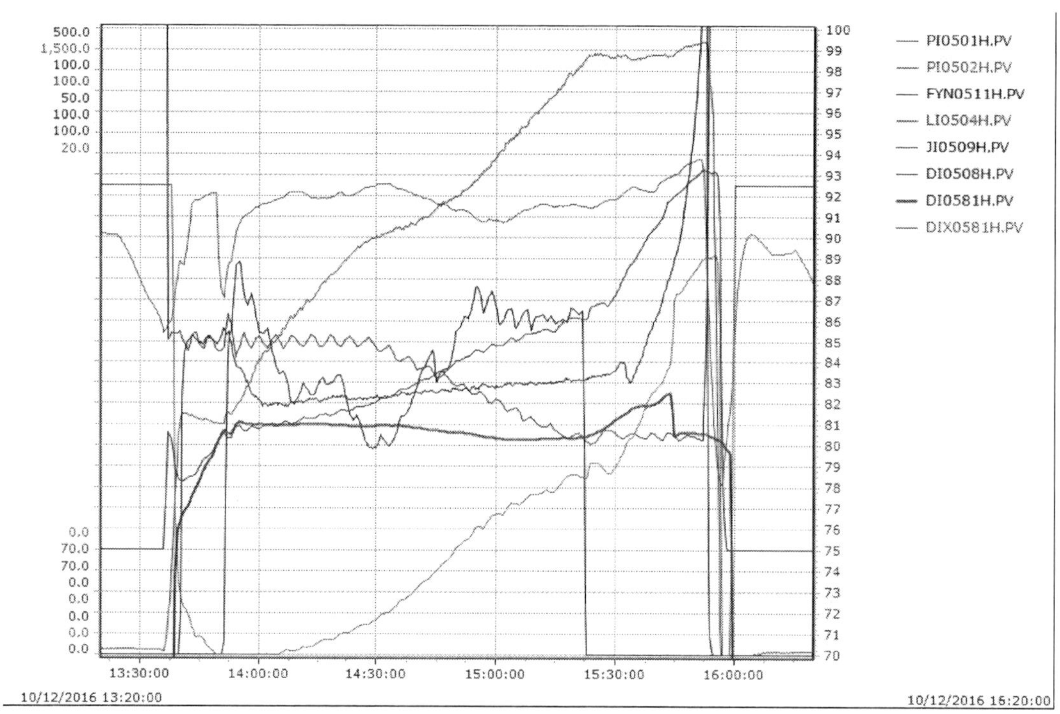

Figure 3 - Rapid decline in mother liquor brix post seeding (Thicker purple line)

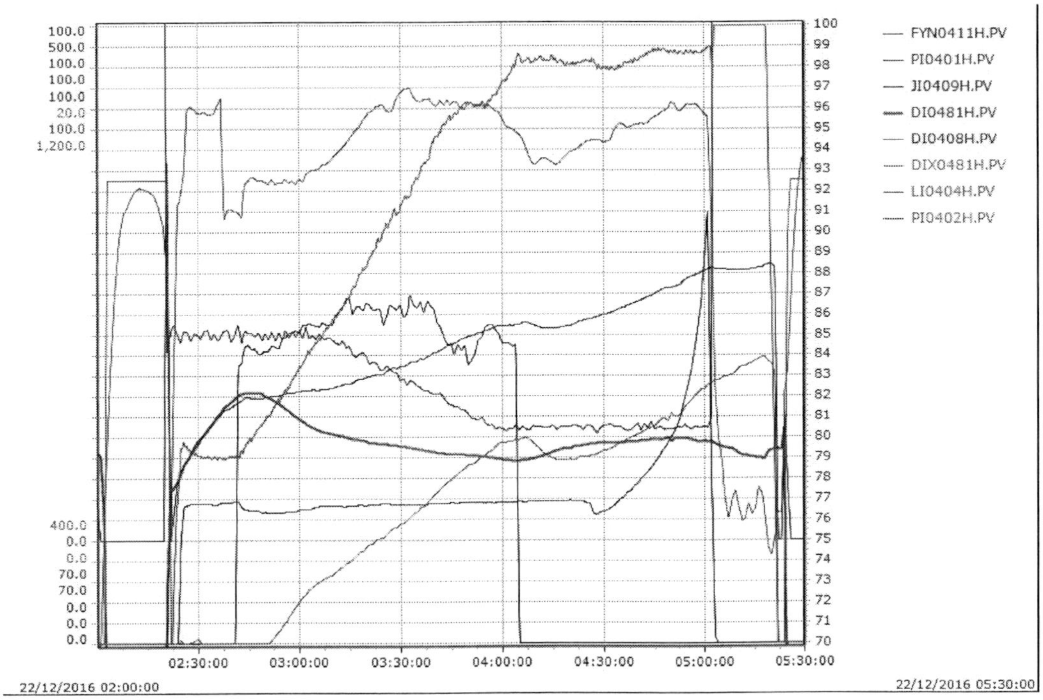

Figure 4 - Continued rise in mother liquor brix post seeding (Thicker purple line)

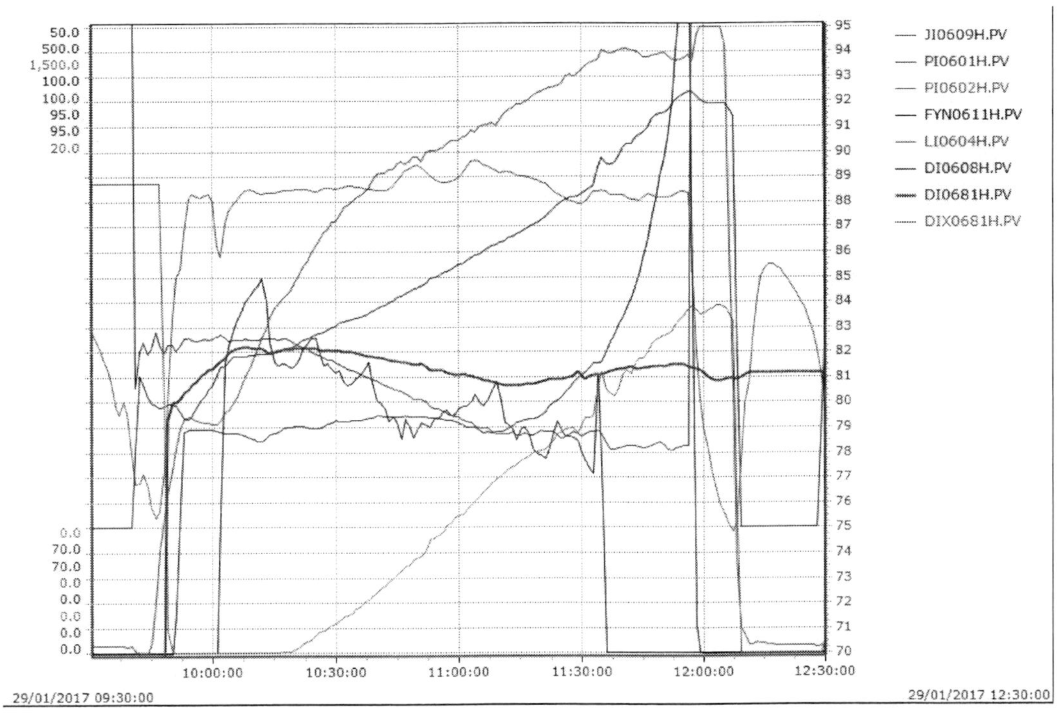

Initially, it was believed that the instability, described in the paragraph above, occurred due to the drift seen in the microwave brix meter's performance leading to a false "delta brix" measurement. At this point a decision was made to switch the pan control to use the output from the refractometers to control the mother liquor brix to the linear growth profile. This action resulted in a significantly smoother syrup feed profile over the early and middle feed sections of each pan cycle. However, in almost all cases under this control, the system lost its way during the final quarter of the pan (pre-brix up), with the mother liquor brix falling away from the desired profile and the feed valve responding by slowly closing. Ultimately, the pan stirrer power would rise to the point where the control system would switch to stirrer feed control in the final feed phase of the pan.

Whilst running the trial, the factory continued to operate its standard method to control the white sugar MA to silo, by varying the Brunswick seed volumes added to the pans (Figure 5). At the start of the trial, the seed volume to all 6 white pans was as low as 17% of the head pot volume (equivalent to approximately 0.8m³ of Brunswick seed per pan) and rose to as high as 56% of the seed pot (2.8m³ of Brunswick seed per pan) before falling back down to as low as 8% (0.4m³ of Brunswick seed per pan).

Figure 5 - Variation of seed volumes to the white pans over the trial period

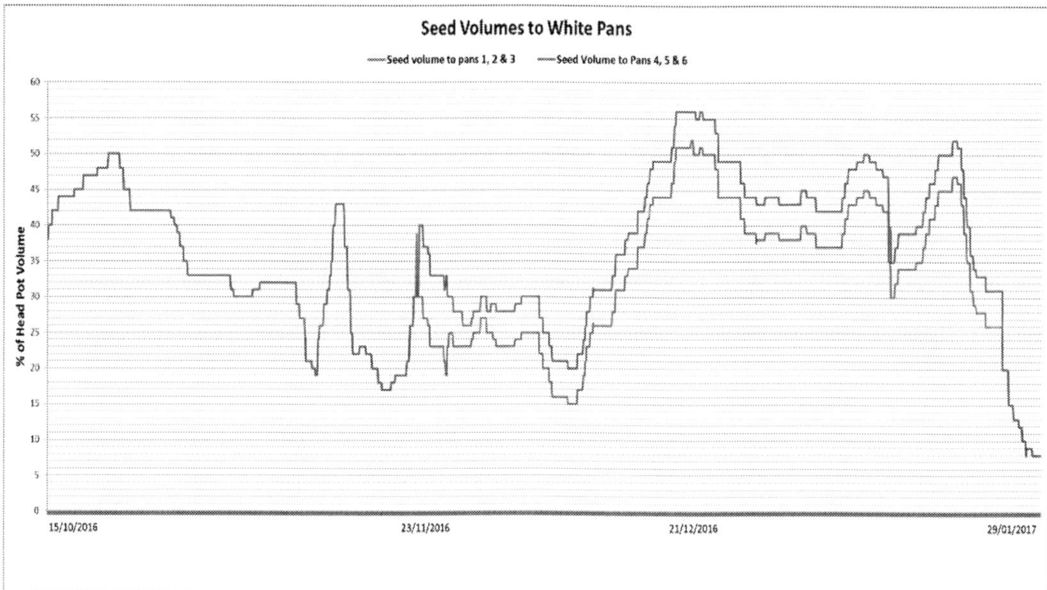

Based on a target 130µm seed size at 28% crystal content, this would equate to a variance in the number of seed sites introduced into a pan from 4.5 x 10^{10} at a 8% seed pot volume, through to 3.2 x 10^{11} at a 56% seed pot volume. In terms of the initial surface area variability, over the same range of seed pot volumes this equates to 4,607m² at 8% rising to 32,250m² at 56% (figure 6). This is a 7 fold increase in surface area added at the beginning of the growth phase from the minimum to maximum seed volumes experienced.

Figure 6 - Impact of variable seed volumes on surface area and crystal number

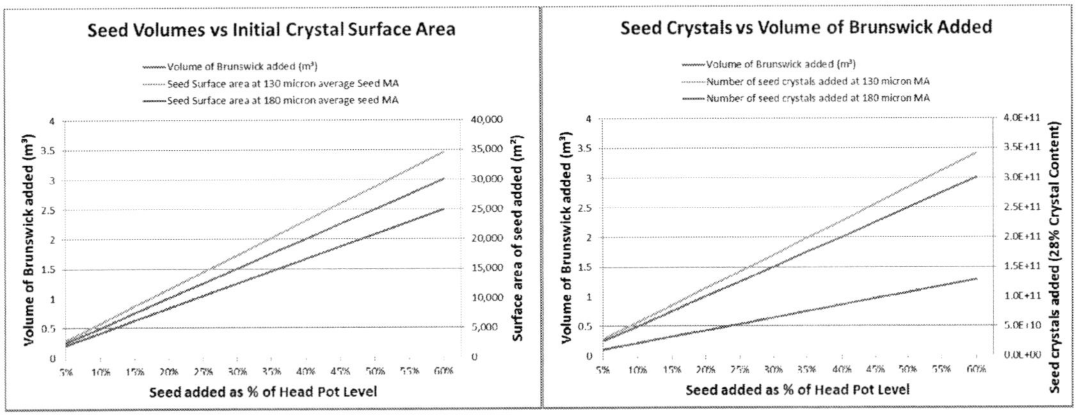

The theoretical impact of this increase in surface area, at the start of the growth phase, has the potential to range from 150kg crystal growth in the first 5 minutes of the pan with the 8% seed pot volume, to 1,024kg with the 56% seed pot volume. The model assumes a specified number of seed sites which equates to a seed pot volume of 26%. This provides the correct number of sites to match the predetermined evaporation rate, controlled from the steam and vacuum profiles. Therefore, any variation in the number of seed sites, and ultimately the crystal surface area, will introduce a variation in the crystallisation rate versus the evaporation rate. With higher seed volumes, the increased rate of crystal deposition post seeding could explain the decline in the mother liquor brix at the start of the growth phase (figure 3). Similarly, when the seed volumes were reduced towards the 8% level, the decreased rate of crystal deposition post seeding would result in an increase in the mother liquor brix. In this scenario, the rate of evaporation outstrips the rate of crystal growth and thus the mother liquor saturation increases, until false grain occurs and the pan corrects itself for the lack seed sites at the start (figure 4).

Unfortunately, at both extremes of the seed volume range, the final white sugar MA fell within the target limits of the site production requirements. This should not be the case unless the pan growth rates are inconsistent with the saturation theory, or there is a significant variation in the Brunswick seed production from the 130 μm MA, 28% crystal content scenario. To follow this uncertainty up, samples from batches of Brunswick seed were taken, separated, oven dried and then analysed for their MA and CV

The results from this analysis (figures 7 and 8) showed a significant variability in the MA and CV from batch to batch with the Brunswick seed production. With an average MA of 160 μm and a standard deviation of 32, it became apparent that the seed size was greater than the size modelled and could potentially lead to the variability in controlling the "delta brix" to the correct profile.

At this point in the trial the development of the linear growth program was suspended, whilst work on improving Brunswick Seed crystal size control and repeatability was undertaken. This work is currently ongoing and the results will be reported at a later date, along with the conclusion of the linear growth project.

Figure 7 - Brunswick batch particle size variability

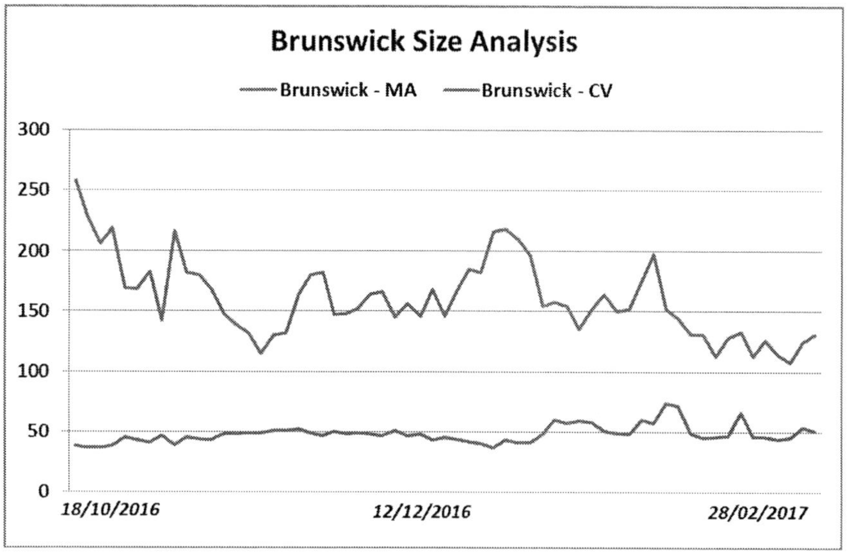

Figure 8 – Distribution of Brunswick seed size regularity

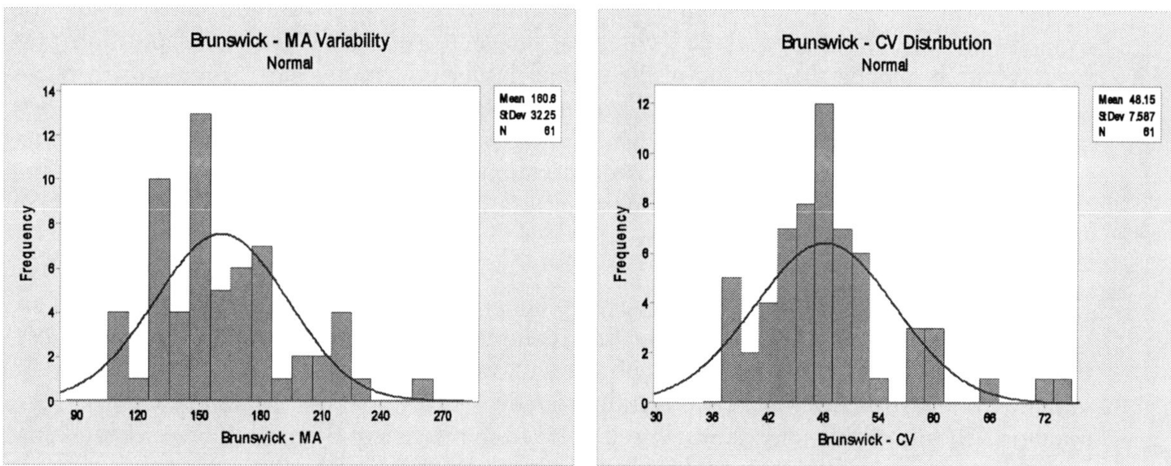

000

Observations and benefits of Linear Growth profiles

Whilst the development of the linear growth profiles are temporarily on hold, pending the stabilisation of the Brunswick seed system, there have been some observed improvements realised from this work.

The main discernible difference between the original pan profiles and the revised linear growth profiles can be seen in the massecuite brix and level profiles. In figures 9 and 10, the level profile (shown as the red line) shows a small rate of increase over the pan cycle. At the same time, the massecuite brix (blue line) tightens quicker in the early phase, showing the characteristics of a "tight" pan. This follows the principle that keeping the distance between crystals in the massecuite close, limits the space for false grain to form. However, by running at a higher brix the mobility of the crystal through the mother liquor is reduced, increasing the potential for localised oversaturation and the formation of false grain.

Figure 9 - Pan A1 - Original tight brix control profile

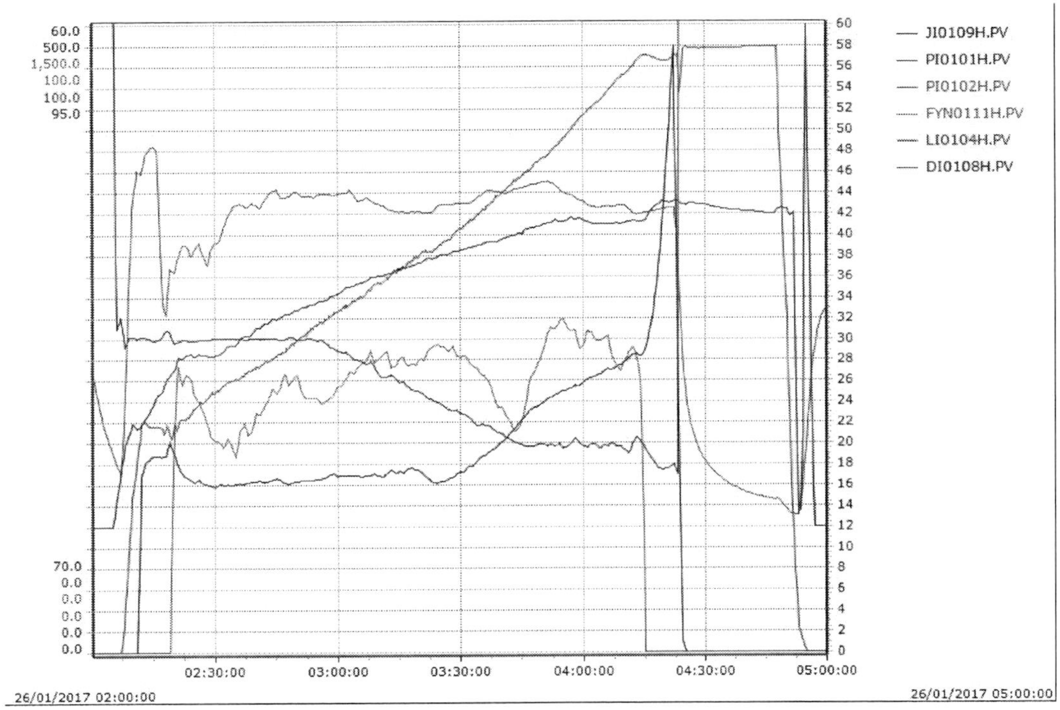

Figure 10 - Pan A6 - Linear Growth "open brix" profile

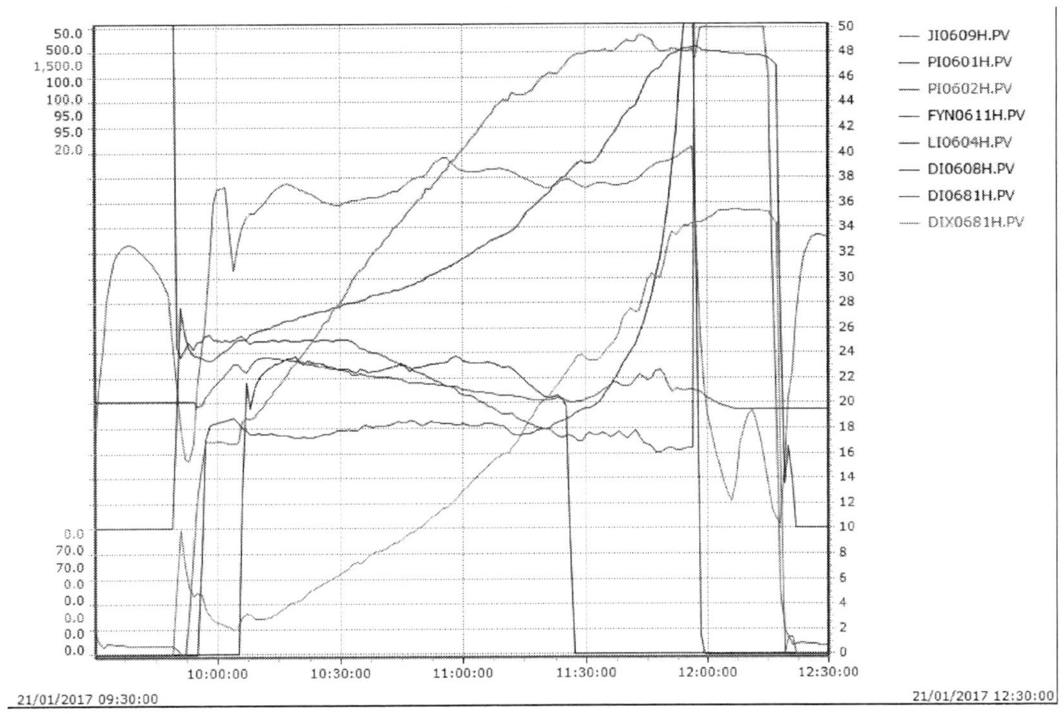

The linear growth profile (figure 10) operates with a more open massecuite in the early stages of the pan cycle, keeping the crystals more fluid within the mother liquor until they have grown sufficient surface area to control the mother liquor saturation and limit false grain formation.

With either an open or tight profile, getting the initial seed surface area correct is key to the final product quality of the pan. However, with the more open profile in the initial feed stage of the linear growth program, the risk of false grain forming early in the pan cycle is lower. Subsequently, with the linear growth profile, there is a point in the cycle at which the crystal surface area is sufficient such that the sucrose mass transfer from syrup to crystal exceeds the rate of concentration within the syrup, thus preventing false grain forming during the brixing up phase.

The evidence of this potential may be linked to the observed improvement of both the crystal CV (0.9 point) on the B-side pans when running the linear growth control profiles and the accompanying reduction in both high green and wash syrup purities (both 0.9 points lower) from the B-Side centrifuges and the 10% reduction in wash water to these B-side centrifuges.

Conclusions

Based on the observations and results reported so far, operating batch white pans following the linear growth profiles has the potential to deliver performance improvements in product quality, extraction and associated energy benefits. This would align with the earlier studies on the use of linear growth profiles.

The more open nature of the massecuite in the early feed phase under the linear growth profile appears to keep the massecuite clearer of secondary grain formation. The success in maintaining this improvement through to the end of each cycle depends on maintaining the ideal growth to the feed profile through to the transition point (the point at which the total surface area of the crystal exceeds the growth rate driving force in the massecuite).

The original linear growth concept was developed successfully to deliver an improved crystal quality based on pans seeded with a 10 μm seed slurry. The challenge for those operating an intermediate seeding system, such as the Brunswick system operated at Wissington, is to produce a consistent MA and CV seed for the white pans.

Acknowledgements

British Sugar wishes to thank Chris Rhoten for the development of the Linear Growth Profile model used in this study and the authors would like to thank the Wissington factory operational team for their support and resources used during the trial

References

1 Sucrose Crystallisation mechanism of Growth from aqueous solutions; Andrew Van Hook, J. Crystal Growth, 5 (1969), 305-311
2 Sugar Technology – Beet and Cane Manufacture, P.W van der Poel, H.Schiweck, T.Schwartz, Bartens, p655
3 Linear Growth Rate Based Adaptive Batch Crystallisation Control, Christopher D Rhoten, ASSBT Proceedings from the 38th Biennial Meeting, February 23 - February 27, 2015
4 Chernov A.A. The spiral growth of crystals, Soviet Phys - USPEKHI 4, 116-148 (1961)
5 G.H. Gilmer and P. Bennema, Simulation of crystal growth with surface diffusion, , J. Appl Phys, 1972, 43, 1347-1360

S.I.T. Paper #1154

Dynamic Imaging Particle Analysis Applications in Sugar Production

Alan O'Donoghue, JM Canty International Ltd.

Abstract

Dynamic Imaging operates on the basic principle of capturing a real time view of a process, to either support an operator in making process related decisions, or to feed image analysis software which performs specific process measurements. This technology can be used to great effect for particle analysis within the sugar industry. During crystallisation, the technology can provide a real time continuous view of the sugar crytals, from initial seeding right through to full crystal growth. Unique software allows for detailed crystal size and shape measurement during the early stages of crystallisation, while a size relatvie grid overlay on the real time image allows for size referencing throughout the process. Particle sizing can also be applied during the final QC stages of sugar production to ensure the size and shape distribtuion are within specification for the product begin produced. This paper outlines the mode of operation of each of the systems and include real industry case study information.

1 Background & Principle of Imaging Based Particle Analysis

Image based particle analysis has been in use for 20+ years, however it was not used online until relatively recently, due to the computing power requirements, which are now comfortably met by any modern day computer system. The vision based technique works on the basic principle of presenting the particles between a high intensity light source, and microscopic camera, and analysing the captured images using a specific image analysis software.

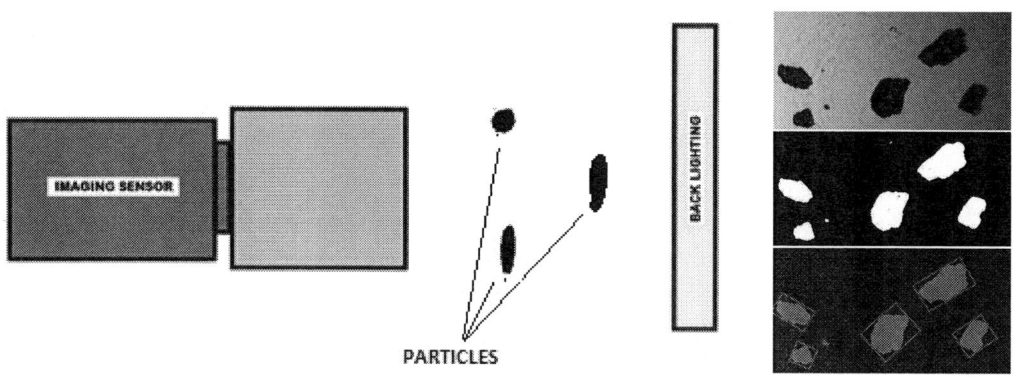

Fig.1 Image Based Particle Analysis - Principle

It is a direct measurment so is most representative of the physical size and shape of the shape of the particles. As it is a direct measurement, the data generated can be correlated directly to commonly used sieve analysis, the difference is that with the imaging based system, there is the capability to go online, with system options for operation on solids / liquid slurries (crystallisation process), and also on dry solids (final product QA/QC).

2 Specific Hardware Features

Every Canty dynamic imaging system is made up of 3 common hardware components, each with their own features and importance. The 3 components are a window / viewing medium, an imaging sensor, and process lighting, and careful consideration must be given to each of these to ensure the imaging system is suitable for the process conditions and application.

2.1 Window / Viewing Medium

All of Canty's range of dynamic imaging systems utilise a unique fused glass barrier through which the imaging sensors views, and the light system transmits. Fused glass is unique to Canty, and is produced by fusing glass to metal on a molecular level to form a one piece construction (no adhesives / seals / gaskets), capable of withstanding high process pressures and temperatures.

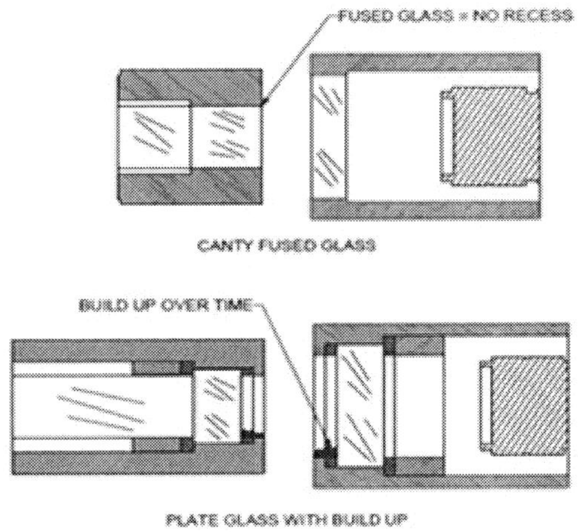

Besides allowing the system to toleratre process conditons, the fused glass also aids in keeping the view clear. Due to the fact that there are no crevices or spaces between the fused glass and metal, there is no where for product to begin to build up.
Fig.2 Fused vs Non-Fused

Non-fused glass and metal systems do not have a smooth transition from glass to metal, and it is in this step area that product would inevitably build up.

2.2 Imaging Sensor

The imaging sensor or camera is the simulation of the human eyes in the dynamic imaging system. All of Canty's systems used Gigabit Ethernet technology, incorporating high colour sensitivity, and designed specifically for process vision applications. The camera is an IP device with a simple RJ45 connection to allow for easy connection to the analyser network. The camera has the capability to take 30 frames per second, and can be utilised with a variety of lenses (micro, macro, standard) depending on the specific application.

2.3 Process Lighting

In any imaging system, the lighting is the key component, and without the correct amount or type of light, the imaging sensor cannot perform to it's highest specification ie. in order to magnify to a high level, or to capture dynamic particles in "freeze frame" by using an increased shutter speed, more light is required by any imaging sensor. Canty's background is in vessel lighting, with over 40 years of experience in the field. Each light system uses a high intensity LED source, and is specifically designed to suit the imaging sensor type it is combined with, and the application it is to be installed on.

3 Application of Technology for Sugar Crystallisation Process

Canty's dynamic imaging technology uses of a Sugarscope™ system to remote monitor the size and shape of sugar particles throughout the crystallisation process, and to provide specific size and shape measurement in the early stages of the process.

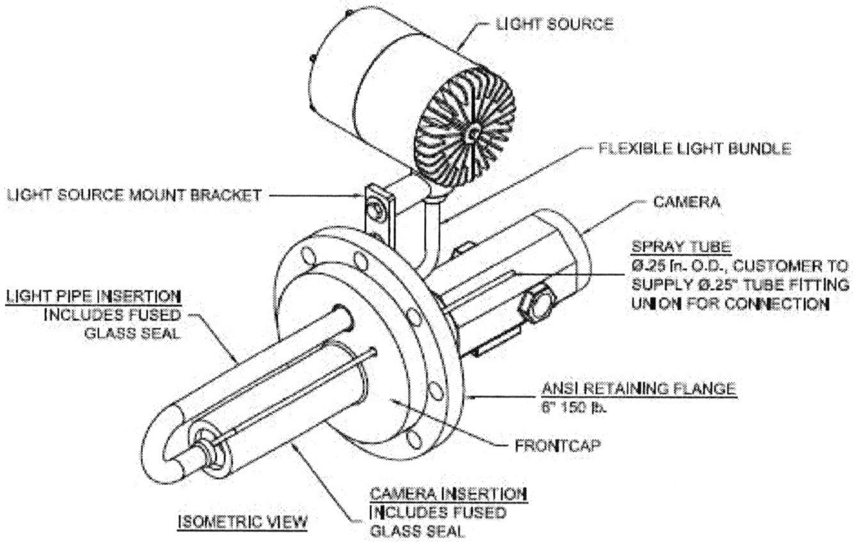

Fig.3 Sugarscope System

The imaging sensor mounted to a microscopic lens, combined with fused glass technology and the high intensity light source allows the operators to visualise sugar crystals from the early seeding stage, right through to full crystal growth. The system has the capability to visualise sugar crystals from 1μm upwards, with the overall range adjustable dependent on product type.

As seen in Fig.3 above the unit is mounted via flange to the crystallisation vessel, with the camera lens system and light pipe protruding into the process medium. This is an important feature as it ensure that the images captured, and any subsequent measurements performed, are representative of conditions in the vessel. It limits the effect of boundary layer conditions, as would occur if the images were captured through a sight glass mounted in the vessel shell.

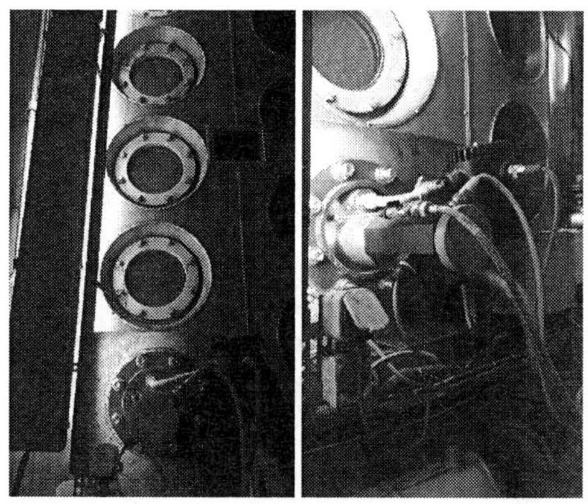

Fig.4 System Mounting

Cantyvision image analysis software allows for the real time measurement of the particle size distribution throughout the process, though even simply having a real time video stream available to the control room operators, can be hugely beneficial in controlling various process parameters, and improving overall efficiency.

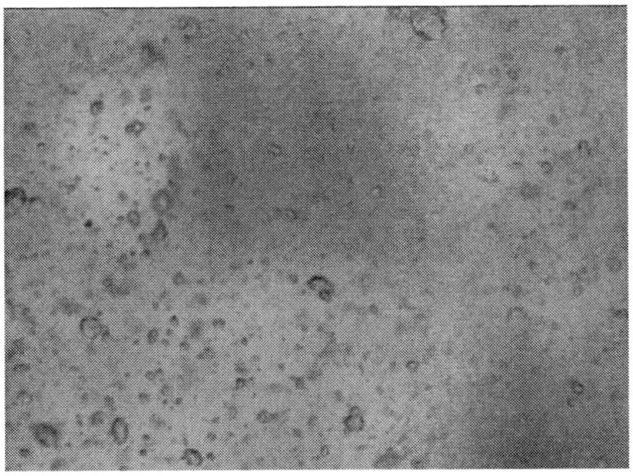

Fig.5 Seeding

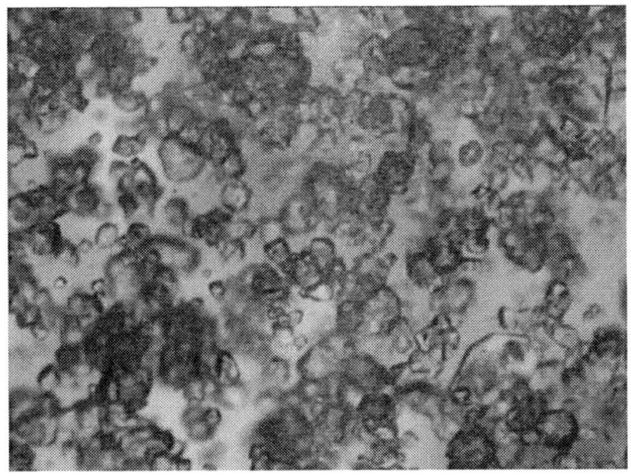

Fig.6 Seeding + 30 Minutes

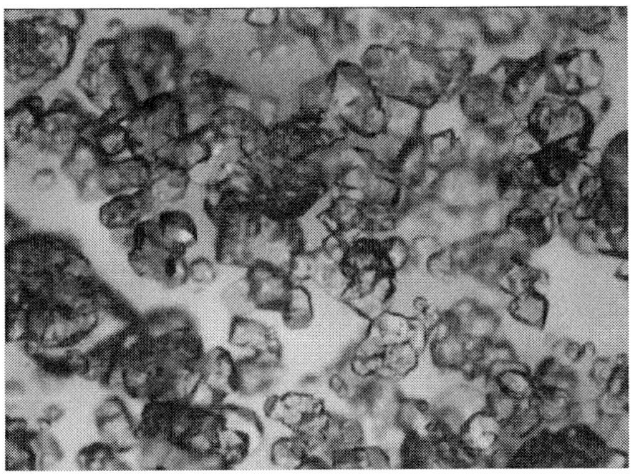

Fig.7 Seeding +60 Minutes

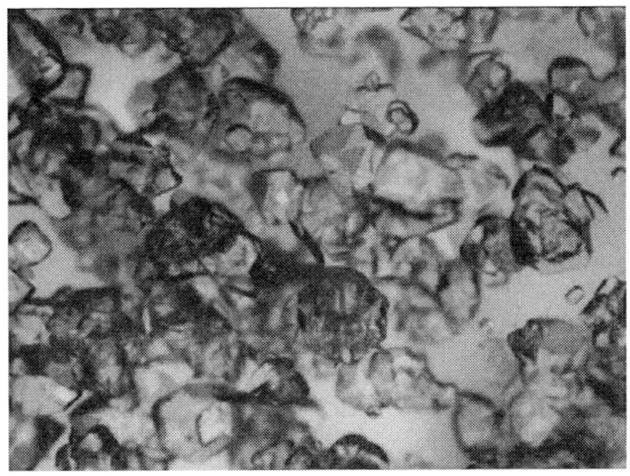

Fig.8 Seeding + 90 Minutes

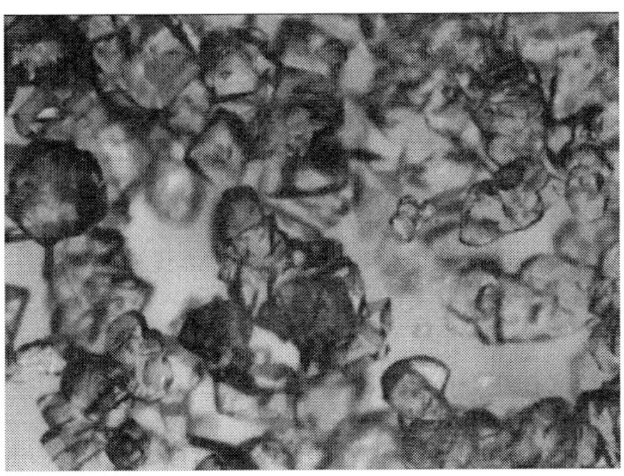

Fig.9 Seeding +120 Minutes

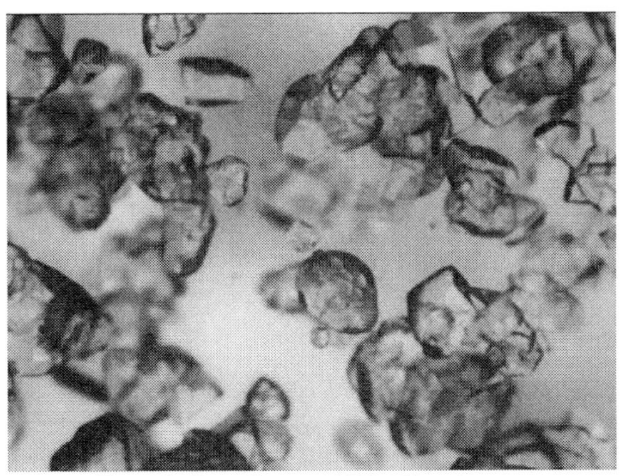

Fig.10 Seeding +150 Minutes

Sample Return on Investment - Sugarscope™

The below is a return on investment calculation performed by a sugar production facility, based on the installation of a Sugarscope™ on 2 of their vacuum pans;

With the annual increase in the production of large grain sugars, it is critical that we optimize our Pan Yields, Crystallization Times and Utilities in producing these products.By having operator's continuously view the crystallisation process, it is felt that the below can be achieved;

- *Reduce boiling times by 33% thus increasing number of batches*
- *Overall increase yield by 10% (250,000 lbs/day)*
- *Reduction in annual water / steam usage by 33% (3,200 gals to 2,110gals per strike, 545,000 gals/4,523,500 lbs. of steam per year*

Increasing by 250,000lbs/day, would save ½ a shift of processing labour per week, which is 9 less melt days required per year. This gives a saving on staff payments of ~$22k per year.

Reducing water / steam usage by 4,523,500 lbs would give a saving of ~$30k based on current enery costs.

Total savings for 1 year:	*~$52k*
Estimated total procurement / installation cost:	*~$20k per unit (x2)*
Payback duration:	*<10 months*

4 Application of Technology for Final Product QA / QC

Canty's dynamic imaging technology can be employed in 2 areas of final product QA/QC. A SolidSizer™ system can be used to measure the particle size distribution of the finished product, and / or the concentration of black speck / dis-coloured sugar crystals. The process works by loading a sample into the system hopper, where it lands onto a vibrating feed tray. This separates and transports the product in one even layer to the analysis zone.

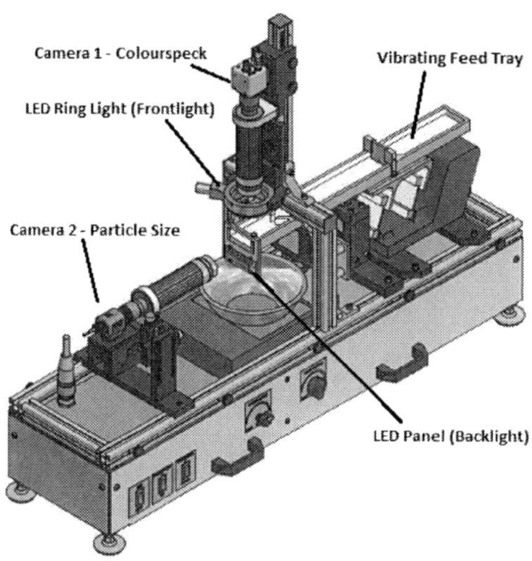

Fig.11 SolidSizer™ with Black Speck Detection – System Cover Removed

The first camera (see Fig. 11) uses a front lighting system and captures images which are sent to the analysis software for black speck detection. This provides a repeatable, subjective detection method, which requires limited operator / lab technician input.

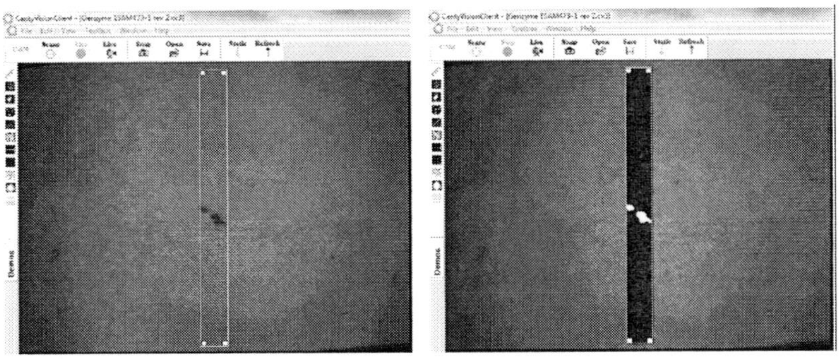

Fig.12 Black Speck Detection

After passing the first camera, the crystals reach the end of the vibrating feed tray where they free fall, or slide over and angled stage between the second camera and an LED panel. This results in sharp images of the crystal shapes being captured, which are then processed by the analysis software to provide size distribution and shape characterisation data

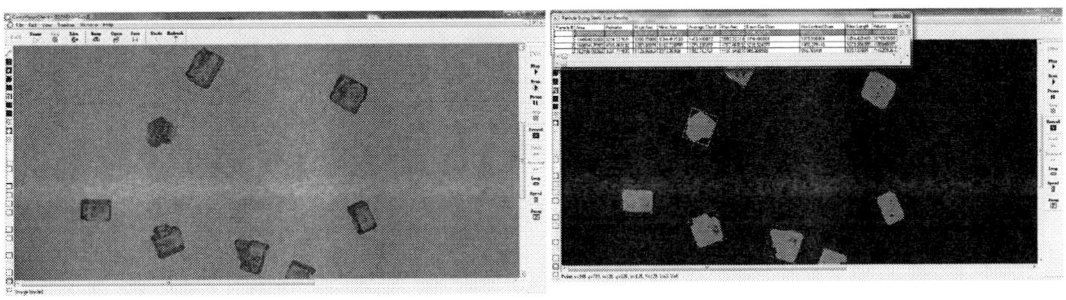

Fig.13 Analysis Screenshot – Sugar Crystals

As the system measures the crystal size under a large number of size and shape parameters, including minor axis (width), the data can be directly correlated to sieve analysis which is often used in sugar QA/QC labs. Using the SolidSizer™ system however, minimises the input required from the operator / lab technician and automatically generates the relevant data avoiding time consuming activities such as weighing sieve screens, and performing manual calculations.

5 CONCLUSIONS

While crystallised sugar has been around for close to 2000 years, and the basics of mass sugar refining have not changed significantly in the last 200 years, it is important to take advantage of relatively newer technology to allow the production processes to run more efficiently and produce a higher yield. Canty's range of dynamic imaging systems applied to particle analysisi can be used to do just that, by providing unparalleled views into the process, and generating useful data that can be used for process control and optimisation.

S.I.T. Paper #1155

Relationships Between the Physical Forms of Starch and Performance of Carbonatation Clarification and Press Filtration at the Refinery

Marsha Cole,[1,2] Gillian Eggleston,[1*] Eduardo Borges,[3] Jack Thompson,[4] Thomas Rathke,[5] June Naiki,[6] and Maria Diamond[7]

[1,*] USDA-ARS Southern Regional Research Center, 1100 Robert E. Lee Boulevard, New Orleans, LA, 70124, USA; gillian.eggleston@ars.usda.gov; +1 (504) 286-4446

[2] Fellow, Oak Ridge Institute for Science Education Research Participation Program at USDA-ARS

[3] Fermentec, Ltda,Av. Antônia Pazzinato Sturion 1155, Piracicaba, São Paulo, Brazil 13420

[4] Louisiana Sugar Refining, LLC, 1230 5th Ave, Gramercy, LA, 70052 USA

[5] Imperial Sugar Company, 201 Oxnard Drive, Port Wentworth, GA, 31407, USA

[6] American Sugar Refining, Inc., 1100 Key Hwy E, Baltimore, MD 21230

[7] American Sugar Refining, Inc., 7417 N Peters St., Arabi, LA, 70032, USA

ABSTRACT

Laboratory studies were conducted to underpin if and how the different physical forms of starch (insoluble or soluble) affected carbonatation clarification and press filtration, by using simulated melt liquors. Carbonatated liquors containing ≥250 ppm/Brix (USDA starch research method) of total starch and either insoluble or soluble physical starch forms (used separately or in a mixture at nearly equal amounts) were uniquely detrimental to the simulated carbonatation refining process and press filtration. High amounts of soluble starch accelerated carbonatation reactions by: (i) binding Ca^{2+} and removing it from solution, (ii) contributing to the production of $CaCO_3$ fines (<5 μm), (iii) causing negligible mud yields, and (iv) producing inadequately clarified liquors. On the other hand, soluble starch was not involved in the viscosity of liquors or problems with press filtration. In comparison, insoluble starch prolonged carbonatation clarification reactions, but had better $CaCO_3$ crystal sizes, mud yields, and impurity removal which influenced press filtration flow rates. Insoluble starch (precursor to swollen starch) also rapidly clogged filters and increased the viscosity of simulated melt liquors. Other studies using factory raw sugars confirmed that measuring filterability at the laboratory scale is complex and may not accurately reflect actual refinery performance without consideration of all impurities and carbonatation clarification reaction conditions since soluble starch, color, invert, dextran, and ash were indirectly involved. Results support the urgent need for new refinery methods that accurately measure both soluble and insoluble starch content in raw sugars and predict the filterability performance of raw sugars at the refinery with consideration of all impurities.

INTRODUCTION

Raw sugar filterability is considered one of the most important sugar quality parameters that directly affects refinery throughput. Many physico-chemical factors that are used to define raw sugar quality also affect raw sugar filterability, i.e., moisture, purity, the amounts of reducing sugars, ash, microbes and extraneous artifacts, type and amount of colorants, and, most recently reported, the physical form and concentration of starch (Cole et al., 2016a,b). There are two major reasons that determine the filterability properties of raw sugars and the overall efficiency of the refinery: 1) physico-chemical behavior of the melted raw sugar during carbonatation clarification reactions and 2) the mechanical removal of turbid particles with press filter units. With the knowledge that there are three physical forms of starch in raw sugars, i.e., (i) insoluble rigid granule (1 – 6 μm), (ii) swollen, amorphous granule (6 to 10 μm, the transformed product of the larger insoluble granules, Clarke & Godshall, 1988), and (iii) solubilized, (~17.5% amylose, Alves et al., 2014), newer perspectives on how starch affects raw sugar filterability were needed if solutions to problems were to be found.

Starch, a natural sugarcane impurity, exists as insoluble granules composed of tangled, soluble glucopolymer chains (amylose, linear and amylopectin, branched) that are extracted into factory juices during milling or diffusion. Most starch (>99%) in sugarcane crusher juices are insoluble with diameters of ~1 to 6 μm (Alves et al., 2014; Eggleston, 2017; Figueira et al., 2011; Park et al., 1985; Stevenson & Whayman, 1976), but range from 1 to 10 μm in heated juices (Eggleston, 2017). This is because, although some starch becomes solubilized during juice clarification at the factory, further solubilization is limited by the increasing syrup Brix in downstream evaporators and can lead to insoluble starch persisting into final raw sugars (Cole et al., 2015a). Despite some factories using commercial amylases to decrease starch levels in raw sugars, some raw sugars received at the refinery still contain total starch at problematic levels, e.g., >250 ppm/Brix at U.S. carbonatation refineries and >150 ppm/Brix at S. African refineries (Mickey Seither, personal communication, Domino Sugar). In a recent study, it was reported that the filterability of four raw sugars with known carbonatation and press filtration performance were strongly affected by both starch concentration and form when analyzed with the USDA Starch Research method, (Cole et al., 2016a). It was also reported that the current 250 ppm/Brix starch penalty limit set for the ICUMSA GS1-17 method may be too low. Using the Domino filterability method, Cole et al. (2015b, 2016a) determined that a filterability index (FI value) of 0.59 should distinguish poor filtering raw sugars from good filtering raw sugars based on their % insoluble starch content and the influence of other raw sugar impurities.

One major refining issue with starch is that, if present as an insoluble granule in the raw sugar, it can still transform into a swollen or soluble starch form during the refinery process, and all three physical forms affect carbonatation clarification and press filtration processes (Cole et al., 2016a,b). Although large particles and artifacts in raw sugars are mainly removed with screens prior to affination and carbonatation, smaller particles such as insoluble and swollen sugarcane starch can still increase the viscosity of the carbonatated liquors and adversely affect press filtration (Cain, 1984; Clarke & Godshall, 1988; Devereaux & Clarke, 1984; Hidi & McCowage, 1984; Nenadkevich, 1984). Cole et al. (2016a) reported that in simulated, affined sugars only the insoluble and swollen starch increased syrup viscosity and reduced filterability; soluble starch, however, decreased viscosity and facilitated filterability with increasing concentration. Therefore, insoluble (swollen) starch affects affined melt liquor filterability and viscosity the most.

Moreover, starch in melt liquors have also been reported to interfere with the milk of lime/CO_2 reactions that are needed to produce sufficient calcium carbonate ($CaCO_3$) for the carbonatation clarifying steps that precede press filtration (Murray et al., 1976a). Cole et al. (2016b) reported that soluble starch preferentially interfered with carbonatation clarification reactions and produced $CaCO_3$ fines (<5 µm) with consequences to downstream press filtration. On the other hand, insoluble starch only increased the viscosity of carbonatated liquors which contributed to reduced press filtration flow rates (Cole et al., 2016b). It was presumed, however, that starch generally did not affect phosphatation refining in the same way as carbonatation refining and new evidence is indicating that this may not be true. Since phosphatation is sensitive to overall impurity load, if large concentrations of starch enter the phosphatation refinery, the efficiency of phosphatation clarification can still be adversely affected by starch (Rein, 2007). High quantities of starch, particularly amylopectin, for example, have formed large, dense agglomerates that limits impurity removal and leaves unacceptable residual phosphate levels in the clarified liquors (Murray et al., 1976b). More recently, it is speculated that insoluble starch may be causing problems with ion exchange decolorizers as well (Jack Thompson, LSR, personal communication). Current efforts to minimize the impact starch and/or other soluble impurities have on the clarification reactions at all refineries and their subsequent downstream units have included predicting sugar filterability using analytical methods, blending multiple sugars to prevent major delays in sugar refining, and/or making impromptu modifications to the process to avert long delays in production (Maria Diamond, ASR, personal communication). Despite these processing strategies, both soluble and insoluble starch have still persisted through refining and into the refined sugar.

Since an in depth investigation had been previously conducted to identify the key contributors to poor sugar filterability with regards to the starch forms and concentrations (Cole et al., 2016a,b), an extension of studies was undertaken to identify how the different starch forms adversely and specifically affect carbonatation clarification reactions, which are reported in this paper.

EXPERIMENTAL

Chemicals and sugar samples

Corn starch, potato starch, sucrose (99.95%), and other chemicals were analytical grade and purchased from Sigma Aldrich Company (St. Louis, MO, USA).

Different Physical Forms of Starch

Simulated affined melt liquors with varying starch contents (0 to 1,000 ppm) and different physical starch forms, i.e., insoluble, swollen, and soluble, were prepared in 0.5 kg sucrose using the method of Cole et al. (2013) with slight modification. To simulate the effects of insoluble and swollen sugarcane starch, corn starch was used; whereas, simulated melt liquors with soluble starch were made using potato starch (Cole et al., 2015a).

Potentiometric Evaluation of Starch Complexation of Calcium Ions

An aqueous solution of calcium ions (Ca^{2+}) was prepared by dissolving 2 g $CaCl_2$ in 1L of distilled water as reported by Cole & Eggleston (2017). In 100 mL, 250 ppm starch suspension (insoluble or swollen starch) or solution (soluble starch) was prepared and allowed to stir for 30 min at room temperature to allow time for the starch to complex with the calcium ions. Then, starch-calcium complex underwent titration with 1M NaOH using a Dosino 803 and Titrando 836 autotitration system (Metrohm, Riverview, FL, USA), in which the uptake of 1M NaOH was recorded by the Tiamo 1.0 software (Metrohm, Riverview, FL, USA).

Statistical Analysis

All sugar data was analyzed using ANOVA in SAS 11.2 (SAS Institute, Cary, NC, USA). Means were separated using Duncan's New Multiple Range Test. Data for each analytical parameter was analyzed separately for each raw sugar. Means were ranked using Tukey's HSD test ($P>0.05$). Pearson correlation (R) coefficients were used to assess the relationship and significance among all sugar quality results.

Results and Discussion

Mechanism of How Starch Contributes to the Production of $CaCO_3$ fines

The underlying mechanisms as to why insoluble and soluble starch both uniquely interfered with carbonatation clarification and $CaCO_3$ formation needed be determined since there was irrefutable evidence of their involvement in carbonatation clarification and press filtration by Cole et al. (2016a, b). By using a titration (V_{NaOH}, pH) method, the interactions between physical starch form and MOL reactant, i.e., $Ca(OH)_2$ - Ca^{2+} ions in the presence of NaOH (Scheme 1), were studied. Several variables affect how MOL dissolves and reacts in affined melt liquors and these variables determine how much $CaCO_3$ mud is formed when MOL reacts with CO_2 during carbonatation clarification (Cole et al., 2016b). The reaction scheme to produce $CaCO_3$ is shown in Scheme 2. Therefore, the interactions between each starch form were evaluated in water and in 67 Brix sucrose solutions to identify how starch is directly and indirectly affecting carbonatation clarification reactions (Fig. 1). It should be mentioned that MOL is also used in phosphatation clarification and, although not the focus of this study, may undergo similar interactions as those presented herein with consequence to higher residual phosphate (Rundell et al., 1982) (Scheme 3).

$$Ca(OH)_2 \rightarrow Ca^{2+} + 2OH^- \quad (1)$$

$$Ca(OH)_2 \rightarrow Ca^{2+} + 2OH^- + CO_2 \rightarrow CaCO_3 + H_2O \quad (2)$$

$$3Ca(OH)_2 \rightarrow 3Ca^{2+} + 6OH^- + 2H_3PO_4 \; (Ca_3(PO_4)_2 + 6H_2O \quad (3)$$

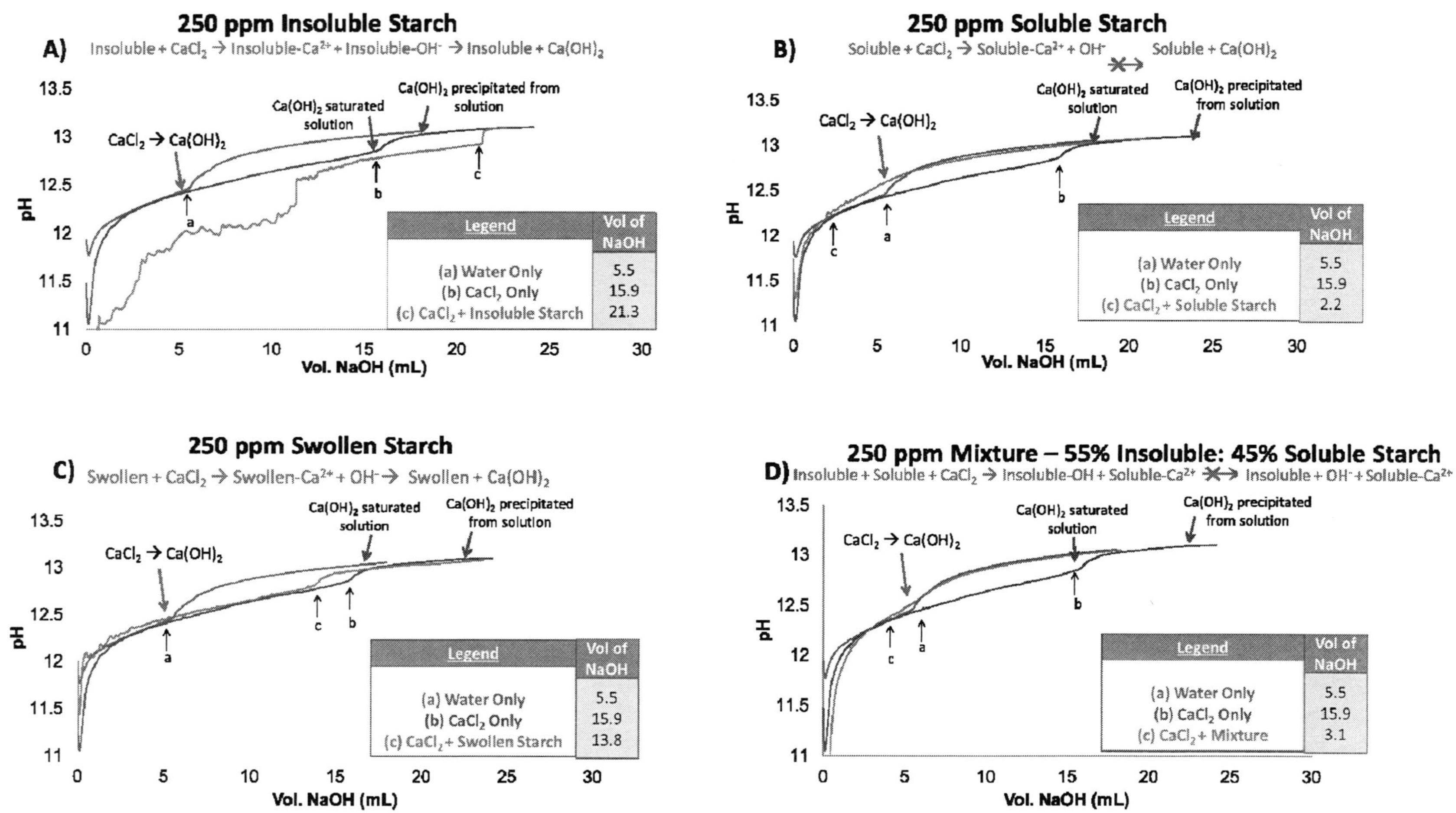

Fig. 1. Potentiometric evaluation of the interactions of 2,000 ppm Ca^{2+} 250 ppm and 250 ppm of different physical starch forms: A) insoluble starch, B) soluble starch, C) swollen starch, and D) mixture of 55% insoluble and 45% soluble starch in the presence of alkali (NaOH).

Insoluble starch

Insoluble starch interacted with both Ca^{2+} and OH^- ions (Fig. 1A), which agreed with the results of Baharuddin et al. (2014) who reported that insoluble starch forms strong complexes with Ca^{2+} because they fill the pores on the starch granule surface until the empty pores are filled. Fig. 1A shows that greater volumes of NaOH were needed to continue the simulated carbonatation clarification reactions when Ca^{2+} ions and insoluble starch were present, compared to when only Ca^{2+} ions and water were studied (Fig. 1A). Fig. 1A also shows that as the pH increased, the titration curve for insoluble starch approached the curve for Ca^{2+} but did not converge suggesting that the insoluble starch interacted weakly with the Ca^{2+} ions and did not inhibit the reaction needed to produce $Ca(OH)_2$. If higher dosages of alkali are added, however, this effect can be overcome (Fig. 1A).

In the studies conducted by Baharuddin (2014), the possible interactions that can also occur between the starch granule, its pores, and OH^- ions were not mentioned. It is well known that starch is net neutral and becomes more negatively charged as pH values approach 12.5 (Baharuddin et al., 2014), and therefore, it is unlikely that there would be a strong affinity between the positive calcium ion and the granule at pH 8.7 (pH of carbonatation clarification). It is more likely that, instead of only binding with Ca^{2+} ions, the insoluble starch would first absorb OH^- ions from the solution. Table 1 shows that similar to EDTA, a well-known strong Ca^{2+} ion complexing agent that removes Ca^{2+} from solution, insoluble starch absorbed both Ca^{2+} and OH^- from the titration solution. In turn, this would decrease the pH of the carbonatated liquor because the unreacted CO_2 will exist as the acidic ions HCO_3^- and CO_3^{2-}, require more MOL addition, and indirectly limit $CaCO_3$ formation or produce $CaCO_3$ fines, as reported by Cole et al. (2016b), and increase the internal pH within the starch granule (Fig. 2).

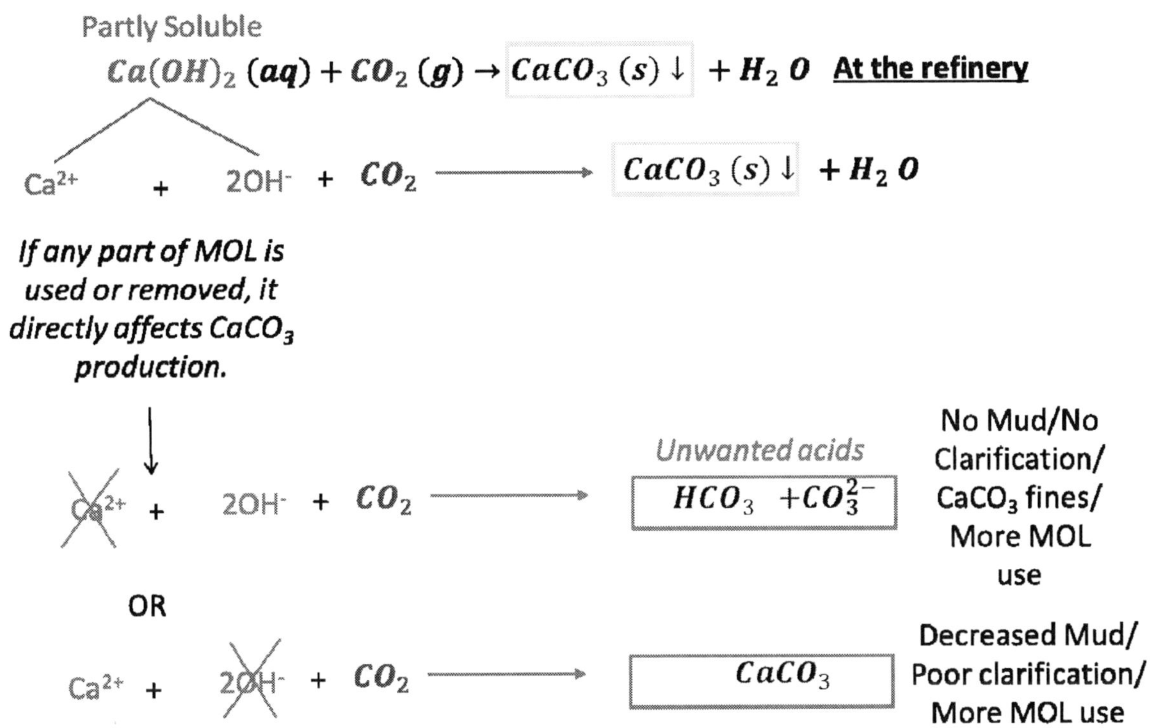

Fig. 2. Schematic illustrating the vulnerabilities of carbonatation clarification reactions.

Mangels and Bailey (1933) reported that insoluble starch granules and alkali interact very strongly and cause the starch granules to become swollen. Maher (1983) found that starch granules from twenty-one grass and tuberous roots could similarly bind OH^- ions, but each starch required a specific amount of OH^- to become swollen and increase the solution viscosity. This effect decreases by alkali strength with $Ca(OH)_2$> KOH> NaOH>>>Na_2CO_3 (Ragheb et al., 1995). Therefore, the results reported in this study using NaOH will be less intense than if MOL, i.e., $Ca(OH)_2$, had been used because the latter is a much stronger alkali than NaOH. Because of the strong attraction between the starch granule and both OH^- and Ca^{2+} ions in MOL and the intense reactivity of MOL, the Ca^{2+} and OH^- may be retained inside the starch granule limiting its reactions with CO_2. When this occurs, the starch granule will undergo physical and chemical changes, which in consequence, causes the starch granule to swell (due to charge-repulsion of the negatively charged amylose and amylopectin chains), increase the viscosity of the melt liquor (Larsson, 1980; Lindgvist, 1979; Tomasik & Schilling, 1998a,b), and create a chemical environment within the insoluble starch granule more favorable to Ca^{2+} ions that would enable stronger interactions. Cole et al. (2016b) also reported that carbonatated liquor viscosity increased when insoluble starch was present. Since large quantities of $CaCO_3$ were produced from simulated carbonatation containing insoluble starch, it is therefore, presumed that the interactions between Ca^{2+} and the negatively charged starch granule are reversible and weaker than the affinity for Ca^{2+} to react with CO_3^{2-}.

Soluble and swollen starches

Soluble (Fig. 1B) and swollen (Fig. 1C) starches interacted with Ca^{2+} ions much differently than insoluble starch. Table 1 shows that soluble starch did not interact with OH^- or require additional NaOH to produce $Ca(OH)_2$ when titrated to pH 13.65. This indicated that the OH^- were not physically absorbed from the solution which is logical sense soluble starch exists as free flowing chains of amylose and amylopectin rather than an intact granular structure. As seen in the titration curves in Fig. 1, very little difference in titration profiles occurred before pH 12.5 when in the presence or absence of soluble starch (Fig. 1B). Once the pH became >12.3, the pH where the soluble starch has a net negative charge and $Ca(OH)_2$ is formed and begins to precipitate, the strong interactions between the Ca^{2+} and soluble starch were apparent (Fig. 1B). After comparing the titration curves with soluble starch to the titration curve with only water, it was observed that the soluble starch bound all of the free Ca^{2+} ions in solution preventing it from forming $Ca(OH)_2$ (Fig. 1B). Table 1 shows that 11.3% less NaOH was needed to reach pH 13.25 than the control confirming that soluble starch bound and removed the Ca^{2+} from solution. It has been previously reported that soluble starch, mainly amylose (Tomasik & Schilling, 1998a,b), binds alkaline metals (e.g., $Ca^{2+}<Ni^{2+}< Zn^{2+}< Cu^{2+}$) and forms more soluble starch-Ca^{2+} complexes as pH increases (Trivunac & Stevanovic, 2006; Wettstein, Newkom, & Deuel, 1949). On the other hand, amylopectin inhibited $CaCO_3$ growth with ~2 orders of magnitude preferential affinity for calcium (Nielsen et al., 2012). Therefore, each soluble polysaccharides contributed to the production of $CaCO_3$ fines, where (i) amylose initially binds and removes Ca^{2+} from carbonatated liquors inhibiting it from reacting to produce $CaCO_3$ and (ii) amylopectin blocking crystal growth patterns by coating and become entwined in the crystal.

According to Nielsen et al. (2012), however, neither soluble polysaccharide fully inhibits the growth of $CaCO_3$ or produces crystal sizes that interfere with press filtration, instead a different $CaCO_3$ crystal shape, i.e. cube, needle, spherical, may be produced and the source of poor filterability (Dickinson & McGrath, 2004). This is because Ca^{2+} and CO_3^{-2} ions are only consumed by the growing crystals, i.e., not by the nucleation of new particles, and by the temperature and dosing rate of MOL/CO_2 (Beck & Andreassen, 2012). Thus, any binding between soluble starch and Ca^{2+} will only limit crystal size, but not necessarily crystal quantity, an artefact of nucleation. Cole et al. (2016b) found that residual Ca^{2+} concentrations in carbonatated filtrates containing soluble starch did not differ from the non-starch control filtrates and the nucleation phase was -55% shorter than controls. It was found that carbonatated liquors and $CaCO_3$ filter cakes with soluble starch were nonexistent or possessed very small crystals with needle- and flake-like shapes (data not shown); whereas, insoluble starch possessed larger rounded-cubes. Beck & Andreassen (2012) found that plate- and needle-like $CaCO_3$ crystals have worse filterability than spherical and cube-like crystal shapes of $CaCO_3$ because overall smaller crystal size and the crystal fragility of the needle-like $CaCO_3$ crystals that are easily fractured during filtration. Considering the findings by Dickinson & McGrath (2004) and Beck & Andreassen (2012), the mechanism at which soluble starch affects press filtration is now believed to be due to $CaCO_3$ crystal distortion and size in addition to the lack of $CaCO_3$ mud produced. All of which are because of initial soluble starch-Ca^{2+} binding that occurs during the nucleation and induction phases of the carbonatation clarification.

The titration curve for swollen starch in Fig. 1C showed that $Ca(OH)_2$ was still formed but with a small adverse effect. Table 1 shows that titrations for swollen starch without Ca^{2+} similarly required ~11% more NaOH like insoluble starch. When Ca^{2+} was present, however, the swollen starch used less NaOH than insoluble starch, indicating that less Ca^{2+} was present. This result was attributable to decreased porosity of the swollen starch granule that limits the ability of the ions to diffuse in and out of the granule. Fig. 1C also shows that swollen starch was minimally affected by OH^-. This is because the granule was already gelatinized by heat, and the additional exposure to OH^- only changed the starch's internal structure and charge, thus requiring either i) very little OH^- to become negatively charged or ii) the swollen starch granule does not easily diffuse OH^- ions into its internal surface leaving only the surface glucose molecules to become negatively charged. The difference in titration curves at pH >12.5 shows that trace amounts of Ca^{2+} ions were actually bound to the swollen starch, but the majority was converted to $Ca(OH)_2$. Therefore, swollen starch minimally interfered with $Ca(OH)_2$ formation by reversibly binding Ca^{2+} and exhibited behavior similarly to insoluble starch.

Other polysaccharides have been generally reported to also strongly bind Ca^{2+} ions and inhibit $CaCO_3$ formation and/or precipitation reactions (Borman et al., 1982; Butler et al., 2006; Giuffre et al., 2013). Therefore, all polysaccharide impurities present in carbonatated liquors may yield similar results if not removed during affination. In full view of these soluble starch results, it has now unequivocally been proven that soluble starch can remove Ca^{2+} ions from carbonatated liquors, catalyze the $CaCO_3$ precipitation reactions by limiting the free and available reactants, distort $CaCO_3$ crystal shapes, and contribute to the formation of unwanted $CaCO_3$ fines.

Table 1. Percent change in 1M NaOH volume needed to reach pH 13.25 in the presence of 250 ppm starch and EDTA with or without 2000 ppm Ca^{2+}. EDTA was used to compare Ca^{2+} binding effects and is fully deprotonated with a -4 charge at pH 13.2; glucose is net negative at pH>12.2 at 25 °C. Published with permission from Cole et al. (2017).

Complexing Agent	Concentration of Ca^{2+} (ppm)	Vol of 1M NaOH needed to reach pH 13.25 (mL)	% Change *(Water – 0 ppm Ca^{2+})* Calcium Buffering Effect	% Change from Water + Ca^{2+} *Ca^{2+} in water vs Ca^{2+} with EDTA/Starch*
Water (no sequestrant)	0	38.23	+0.0	-
	2000	43.02	+12.5	+0.0
% Difference			Δ +12.5	
250 ppm EDTA	0	42.19	+10.4	-
	2000	43.70	+14.3	+1.6
% Difference			Δ +3.9	
250 ppm Soluble starch	0	38.27	+0.1	-
	2000	38.16	-0.2	-11.3
% Difference			Δ -0.1	
250 ppm Insoluble starch	0	42.21	+10.4	-
	2000	47.20	+23.5	+9.7
% Difference			Δ + 13.1	
250 ppm Swollen starch	0	42.41	+10.9	-
	2000	46.71	+22.2	+8.6
% Difference			Δ +11.3%	

Mixture of 55% Insoluble and 45% Soluble Starch

Cole et al. (2014) reported that many raw sugars contained ~35-65% insoluble starch. In the filterability and simulated carbonatation clarification and press filtration studies, Cole et al. (2016a,b) also found that when insoluble and soluble starches were together in raw sugars at such nearly equal ratios, their press filtration flow rates were reduced ($P<0.05$) and had marked amounts of $CaCO_3$ mud formed. In this study, Ca^{2+} solutions containing a mixture of different starch forms (55% insoluble: 45% soluble starch, 250 ppm) exhibited similar behavior as the soluble starch and insoluble starch (Fig. 1D). At low doses of NaOH, the pH of the starch solution was more acidic than when only Ca^{2+} was studied, indicating that the insoluble starch fraction was absorbing OH^- ions from solution. As the volume of NaOH was increased, at 3.1mL the solution containing the starch mixtures increased above the curves with water or Ca^{2+} only showing that the Ca^{2+} in this sample was removed by the soluble starch. Although both starch forms were present, results indicate that the soluble starch component dominated the effects of insoluble starch since very little OH^- was needed to raise the pH to 13.65 while also removing the Ca^{2+} ions from solution. A detailed mechanism of how starch affects carbonatation clarification and press filtration can be found in Cole et al. (2017).

In conclusion, the soluble starch prevents Ca^{2+} ions from reacting and inhibits the formation of $CaCO_3$ mud and causes unwanted $CaCO_3$ fines to be produced. The soluble amylose polysaccharide interacts and strongly binds with Ca^{2+} from the MOL while the soluble amylopectin polysaccharide interferes with crystal growth becoming embedded in growing $CaCO_3$ crystals. As a result, different and unfavorable $CaCO_3$ crystal shapes and sizes are produced that have poor filterability characteristics. The insoluble starch preferentially consumes OH^- ions from the MOL (further decreases pH), and as a swollen granule, increases liquor viscosity (decreases press filtration rates) and refining costs. The presence of both starch forms also requires more MOL use, further increasing costs to the sugar refinery. Together or separately, both starch forms adversely affect carbonatation clarification and press filtration, but soluble starch is the worse component because it disrupts the refining efficiency of the carbonatation clarification reactions. This does not exclude the impact large quantities of insoluble starch may have on the viscosity of carbonatated liquors and its adverse effect in press filtration. Likewise, these results indicate that the presence of other soluble sugar impurities with similar chemical functionality as starch, i.e., dextran and polyphenolic colorants, may behave similarly as the soluble starch, and also exacerbate clarification reactions if present. This is expected especially in filtrate flotation clarification and syrup clarification processes used for Blanco Directo sugar or liquid sugar, among other membrane separation processes (Jensen, 2007; Lancrenon & Paillat, 2005; Rein, 2009).

CONCLUSIONS

- Predicting raw sugar filterability and performance at the refinery is complex.
- Soluble starch and other soluble impurities preferentially affect carbonatation clarification reactions by removing free Ca^{2+} ions from MOL solution and contributes to the production unwanted $CaCO_3$ fines, whereas insoluble (swollen) starch absorbs OH^- ions from MOL and causes high liquor viscosities which mainly inhibits press filtration.
- New methods to measure raw sugar filterability with strong emphasis on total starch concentration and form are now warranted, as other methods do not consider the contribution of insoluble (swollen) starch to carbonatation filtration and all the soluble impurities on preceding clarification reactions.

ACKNOWLEDGEMENTS

This research was supported in part by an appointment to the Agricultural Research Service (ARS) Research Participation Program administered by the Oak Ridge Institute for Science and Education (ORISE) through an interagency agreement between the U.S. Department of Energy (DOE) and the U.S. Department of Agriculture (USDA). ORISE is managed by ORAU under DOE contract number DE-SC0014664. All opinions expressed in this paper are the author's and do not necessarily reflect the policies and views of USDA, ARS, DOE, or ORAU/ORISE. Mention of trade names, commercial products, or methods in this article is solely for the purpose of providing specific information and does not imply recommendation or endorsement by the U.S. Department of Agriculture (USDA). USDA is an equal opportunity provider and employer.

REFERENCES

Alves, F. V., Polesi, L. F., Aguiar, C. L., & Sarmento, S. B. S. (2014). Structural and physicochemical characteristics of starch from sugar cane and sweet sorghum stalks. *Carbohydr. Polym.*, 111(10), 592–597.

Baharuddin, N. H., Nik Sulaiman, N. M., & Aroua, M. K. (2014). Unmodified starch as water-soluble binding polymer for chromium ions removal via polymer enhanced ultrafiltration system. *J. Environ. Health Sci. Eng.* 12, 61–61.

Beck, R., & Andreassen, J.-P. (2012). Influence of crystallization conditions on crystal morphology and size of CaCO3 and their effect on pressure filtration. *AIChE Journal*, *58*(1), 107–121.

Borman, A. H., de Jong, E. W., Huizinga, M., Kok, D. J., Westbroek, P., & Bosch, L. (1982). The role in $CaCO_3$ crystallization of an acid Ca^{2+}-binding polysaccharide associated with coccoliths of *Emiliania huxleyi*. *Eur. J. Biochem.*. 129(1), 179–183.

Butler, M. F., Glaser, N., Weaver, A. C., Kirkland, M., & Heppenstall-Butler, M. (2006). Calcium carbonate crystallization in the presence of biopolymers. *Crys. Growth Des.* 6(3), 781–794.

Cain, C. W. (1984). Sugar liquor clarification using diatomite filter aid. *Proc. 1984 Sugar Proc. Res. Conf.*, *ARS-49*, 176–185.

Clarke, M. A., & Godshall, M. A. (1988). *Chemistry and Processing of Sugarbeet and Sugarcane*. New York: Elsevier.

Cole, M., & Eggleston, G. (2017). How the forms of starch affect filterability at a carbonatation refinery. Part III: Mechanism on Ca^{2+} and OH^- scavenging in carbonatated liquors. *Internat. Sugar J.*, in review.

Cole, M., Eggleston, G., Borges, E., Thompson, J., Rathke, T., Naiki, J., & Triplett, A. (2016a). How the physical forms of starch affect filterability at a carbonatation refinery. Part I: Filterability of industrial sugars. *Internat. Sugar J.*, 118(1407), 204–213.

Cole, M., Eggleston, G., Rathke, T., Naiki, J., Triplett, A., & St. Cyr, E. (2016b). How the physical forms of Starch affect filterability at a carbonatation refinery. Part II: Simulated carbonatation filtration. *Internat. Sugar J.*, 118(1413), 662–673.

Cole, M., Rose, I., Chung, Y. J., & Eggleston, G. (2015a). A structured approach to target starch solubilization and hydrolysis for the sugarcane industry. *Food Chem.*,166, 165–172.

Cole, M., Eggleston, G., Thompson, J., Rathke, T., Naiki, J., Triplett, A., & Zatlokovicz, J. (2015b). Battle of the starches: Insoluble versus soluble at the refinery. *Proc. Sugar Ind. Technol.*, *74*(1116), 184–208.

Cole, M., Eggleston, G., Gilbert, A., & Chung, Y. J. (2014). Development of a research method to measure insoluble and soluble starch in factory and refinery products. *Proc. Sugar Ind. Technol.*, *73*, 212–229.

Cole, M., Eggleston, G., Gilbert, A., Rose, I., Andrzejewski, B., St. Cyr, E., & Stewart, D. (2013). The presence and implication of soluble, swollen, and insoluble starch at the sugarcane factory and refinery. *Internat. Sugar J.*, *115*(1380), 844–851.

Devereaux, J. A., & Clarke, M. A. (1984). Observations on filtration impedance in raw sugar. *Proc. 1984 Sugar Proc. Res. Conf.*, *ARS-49*, 209–230.

Dickinson, S. R., & McGrath, K. M. (2004). Aqueous Precipitation of Calcium Carbonate Modified by Hydroxyl-Containing Compounds. *Crystal Growth & Design*, *4*(6), 1411–1418.

Eggleston, G. (2017). *Cane Sugar Processing* (Referee's Report No. General Subject 7). Warsaw Poland: ICUMSA. 1–28.

Figueira, J. de A., Carvalho, P. H., & Sato, H. H. (2011). Sugarcane starch: Quantitative determination and characterization. *Food Sci. Technol. (Campinas)*, 31(3).

Giuffre, A. J., Hamm, L. M., Han, N., De Yoreo, J. J., & Dove, P. M. (2013). Polysaccharide chemistry regulates kinetics of calcite nucleation through competition of interfacial energies. *PNAS*, 110(23), 9261–9266.

Hidi, P., & McCowage, R. J. (1984). Quantification of the effects of different raw sugar impurities on filtration rates in carbonatation refineries. *Proc. 1984 Sugar Proc. Res. Conf.*, *ARS-49*, 186–208.

Jensen, C. C. (2007). Direct white sugar manufacture in the cane sugar industry via membrane filtration and continuous ion-exchange demineralization. *German Ass. Sugar Technol.*, *132*(6), 446–452.

Lancrenon, X., & Paillat, D. (2005). Production of liquid sugars from fine liquors in the refinery. *Proc. Sugar Ind. Technol.*, *64*(870), 1–18.

Larsson, K. (1980). Inhibition of starch gelatinization by amylose-lipid complex formation. *Starch - Stärke*, 32(4), 125–126.

Lindgvist, I. (1979). Cold gelatinization of starch. *Starch - Stärke*, *31*(6), 195–200.

Maher, G. G. (1983). Alkali gelatinization of starches. *Starch - Stärke*, 35(7), 226–234.

Mangels, C. E., & Bailey, C. H. (1933). *J. Am. Chem. Soc.*, *55*, 1981.

Murray, J. P., Runggas, F. M., & Shepherd, G. S. (1976b). Influence of raw sugar quality on phosphatation process. *Proc. S. Afr. Sugar Technol. Assoc.*, *50*, 179–183.

Murray, J. P., Runggas, F. M., & Vanis, M. (1976a). Filtering quality of raw sugar: mechanism of starch influence in carbonatation. *Proc. S. Afr. Sugar Technol. Assoc.*, *June*(50), 1296–1306.

Nenadkevich, N. (1984). Laboratory filterability test methods. *Proc. 1984 Sugar Proc. Res. Conf.,* ARS-49, 162–175.

Nielsen, J. W., Sand, K. K., Pedersen, C. S., Lakshtanov, L. Z., Winther, J. R., Willemoës, M., & Stipp, S. L. S. (2012). Polysaccharide effects on calcite growth: The influence of composition and branching. *Crystal Growth & Design*, 12(10), 4906–4910.

Park, Y. K., Martens, I. S. H., & Sato, H. H. (1985). Enzymatic removal of starch from sugarcane during sugarcane processing. *Process Biochem.*, *20*, 57–59.

Ragheb, A. A., Abdel-Thalouth, I., & Tawfik, S. (1995). Gelatinization of starch in aqueous alkaline solutions. *Starch - Stärke*, 47(9), 338–345.

Rein, P. (2007). Cane Sugar Engineering. Berlin, Germany: Bartens.

Rein, P. (2009). White cane sugar production. *Internat. Sugar J.*, *111*(1321), 1–15.

Rundell, J. T., Pottage, P. R., & Harradine, R. J. (1982, August 24). Clarification of glucose syrups. United States. Retrieved from https://www.google.com/patents/US4345947

Stevenson, J., & Whayman, E. (1976). Cane starch -1. Isolation and iodine affinity. *Internat. Sugar J.*, *78*(926), 44–46.

Tomasik, P., & Schilling, C. H. (1998a). Complexes of starch with inorganic guests, *Adv. Carbohydr. Chem. Biochem.*, *53*, 263–343.

Tomasik, P., & Schilling, C. H. (1998b). Complexes of starch with organic guests, *Adv. Carbohydr. Chem. Biochem.*, *53*, 345–426.

Trivunac, K., & Stevanovic, S. (2006). Removal of heavy metal ions from water by complexation-assisted ultrafiltration. *Chemosphere*, 64(3), 486–491.

Wettstein, F., Newkom, H., & Deuel, H. (1949). Cationic-starch and strong phosphate. *Helv. Chim. Act.*, *1961*, 44.

S.I.T. Paper #1156

CONVERSION OF LOUISIANA SUGAR FACTORIES EXCESS BAGASSE INTO SYNTHESIS GAS—A VISIONARY RESEARCH PROJECT

Jeffery Leblanc, Ph. D. Researcher, City College of New York.
Dr. William Chirdon, Associate Professor, University of Louisiana at Lafayette.
Dr. Chung Chi Chou, President of Dr. Chou Technologies, Inc. Texas.

As we entered the 21st century, the US sugar industry found itself at a cross road facing many difficult challenges: diminishing resources (fuel, water, etc.), environmental pressure, health and safety issues, regulatory compliance, nutritional values of refined sugar under attack, and global competition (WTO). To survive and prosper in this new millennium, the industry must make significant changes to remain profitable including a) reevaluating and improving conventional practices including organization/management b) simplifying manufacturing processes, c) integrating cane mill and refinery operation to produce refined sugar directly from sugar mills, d) capitalizing on the best available technologies, and e) developing value-added products to maximize profits. The work presents a novel concept for converting carbon dioxide and bagasse into hydrogen-rich synthesis gas at a sugar mill. No commercial technology currently offers the utilization of both carbon dioxide and biomass waste to produce synthesis gas.

In the United States, CO_2 accounts for 82% of the total greenhouse gases (GHG) emission.[1] To improve the rate of progress on reducing GHG emissions, additional CO_2 utilization technologies must be developed and commercialized. Currently, most CO_2 storage solutions are geographically specific (i.e., near shales, aquifers or saline formations). Therefore, a need for novel methods of producing value-added products from CO_2 will be addressed by the proposed work. Waste biomass, such as bagasse, may provide a solution to this problem. The process consists of three main steps in order to convert the two waste products into synthesis gas; (1) bagasse dryer, (2) microwave gasification unit and (3) water-gas shift reactor. In biofuel production processes, synthesis gas production is estimated to account for 50-75% of the production cost.[2] The production of synthesis gas also preserves the opportunity to generate other bio-products or direct-use as a clean gaseous fuel.

It is well known that carbonaceous materials can be reacted with CO_2 to produce carbon monoxide via the Boudouard reaction ($C + CO_2 \rightarrow 2CO$) at temperatures of 700 °C leaving an ash residue.[3] Experiments have demonstrated that the reaction between solid carbonaceous materials and carbon dioxide can occur at 213 °C when heated by microwave irradiation.[4] This finding provides significant energy savings in the thermal conversion of the two undesired reactants. In addition, microwave heating is volumetric and is tolerable of moisture (in comparison to conventional thermal heating).

As stated in the process concept, the carbon monoxide effluent from the Boudouard reaction can be co-reacted with water, recovered from drying the bagasse, through the water-gas shift reaction ($CO + H_2O \leftrightarrow H_2 + CO_2$) over a Fe-Cr based catalyst at 500 °C.

The Fe-Cr based WGSR system heated by microwaves has already been experimentally proven.[5] The synthesis gas can be concentrated by removing water from the product stream. The hydrogen-rich synthesis gas can be used as fuel or to make value-added bio-products or liquid biofuel.

Phenomenological experimental results have been obtained on microwave pyrolysis and oxidation of fresh (2016) and biodegraded Louisiana bagasse. Fresh bagasse oxidizes more readily due to tar production during microwave heating. The addition of activated carbon to biodegraded bagasse increase the maximum temperature reached during microwave heating. Future research objectives will include;

1. Determining the gas species production profile from the microwave-induced Boudouard reaction on bagasse,
2. Determine the effect of moisture in the bagasse on conversion by the microwave-induced Boudouard reaction;
3. Investigating the microwave-induced catalytic WGSR and catalyst performance;
4. determining the optimal hydrogen and carbon monoxide mole fractions achieved from the combined microwave-induced Boudouard reaction and WGSR steps.

The microwave-assisted pyrolysis and catalytic water-gas shift process will be unified in a single reactor design in order to operate at low temperature for higher energy efficiency. The proposed process is an on-site or transportable modular process to be used at sugar mills. Pilot scale tests on a 3 TPD bubbling fluidized bed reactor will be carried out at the University of Louisiana Energy Institute as a proof of concept. The investigation will provide a guide for the process design, carbon dioxide life-cycle analysis and economic evaluation of the technology.

[1] U.S. Greenhouse Gas Inventory Report: 1990-2014. EPA. Environmental Protection Agency, 19 Dec. 2016. Web. 31 Dec. 2016.

[2] P.L. Spath, D.C. Dayton, Preliminary Screening - Technical and Economic Assessment of Synthesis Gas to Fuels and Chemicals with Emphasis on the Potential for Biomass-Derived Syngas, NREL. (2003).

[3] S. Ergun, Kinetics of the reaction of carbon with carbon dioxide. The Journal of Physical Chemistry, 1956, 60 (4), 480-485.

[4] J. Hunt, A. Ferrari, A. Lita, M. Crosswhite, B. Ashley, A.E. Stiegman Microwave-specific enhancement of the carbon-carbon dioxide (Boudouard) reaction. The Journal of Physical Chemistry C, 2013, 117 (51), 26871-26880.

[5] W.-H. Chen, J.-G. Jheng, A.B. Yu, Hydrogen generation from a catalytic water gas shift reaction under microwave irradiation. International Journal of Hydrogen Energy, 2008, 33, 4789-4797.

Feasibility Studies of 3D Sugar Object by Additive Manufacturing

Yu-Hao Huang [1], Kai-Cheng Zhang[1], Jin-Xiang Huang[1], Wei-Hsiang Lai[2*]
Tso-Lin Cheng[2], Tsung-Han Lin[4]
[1]Department of Aeronautics and Astronautics, National Cheng Kung University
[2]Taiwan Sugar Research Institute, Taiwan Sugar Company, Taiwan

Abstract

The aim of this research is to develop and evaluate its feasibility to make a 3D object by 3D printing technology for sugar powder. Because modern people pay more and more attention to the foods preparation, a 3D object made of sugar powder is full of potential to the market. This study is to redesign a machine to achieve the needs of modern foods. In this research, the concept of the commercial printer and 3D printing are used in this 3D printer. A commercial available printer to print the pattern on the sugar, and then it used the method of the binder jetting which is one of additive manufacturing to layer of the stack to establish the solid model. This study has successfully completed a desktop of the 3D printer, the printer used the method of the binder jetting and with disPIC4011 to control the machine, the physical dimension of this printer is 600 mm×470mm×420 mm, and its build volume is 200mm×200mm×100mm. Result of this study is fruitful, and 3D sugar model has been successfully printed by sugar powder. In the powder material selection, the white powder sugar with the size range of 200μm -300μm can be formed after the sieve is screened by 28 Mesh sieve. In the three-dimensional sugar experiment, the superfine white powder sugar is mixed with the corn starch powder with a ratio of 3:1 is prepared.

Keywords3D Printer, sugar, additive manufacturing, powder technology

1 Postgraduate, 2* Corresponding author and Professor, 3 Researcher, 4 Director

1. Introduction

The modern are more and more attention to the foods and it is not long for applying three-dimensional printing technique to food industry, so the product of three-dimensional food is not mature, so this study will use the binder jetting as the foundation to develop a desktop 3D printer for printing sugar.

2. Material and Methods

This research includes two parts. In the first part, use the binder jetting as the foundation to develop a desktop 3D printer with volume of 600mm(X) x 470mm(Y) x 420mm(Z) and its build chamber volume of 200mm(X) x 200mm(Y) x 100mm(Z). The second part is to develop food powder under platform of 3D printer which is just mentioned in the first part. The sugar powder is selected to be main developing material and analysis its physical property.

In order to reduce the cost, its printing module is from ink-jet printer CANON IP110. The signal of printer to control the process of 3D printer is also simulated through microchip dsPIC30F4011.

3. Research on powdered sugar

3.1 Powder particle size measurement

The powder particle size was measured by using an optical microscope to observe the shape of the powder, Insitec company produced by Ensemble particle concentration and size (Fig.3-1) to measure the particle size distribution range of the powder.

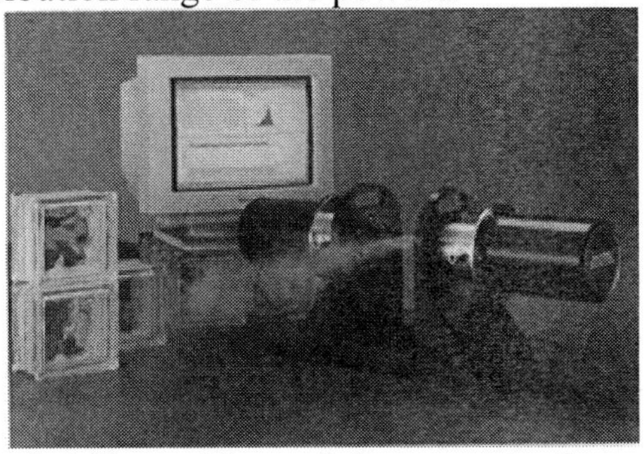

Fig.3-1 Ensemble particle concentration and size

The results show that the average particle size of the intermediate particles is not large, and the powder has no agglomeration distribution.

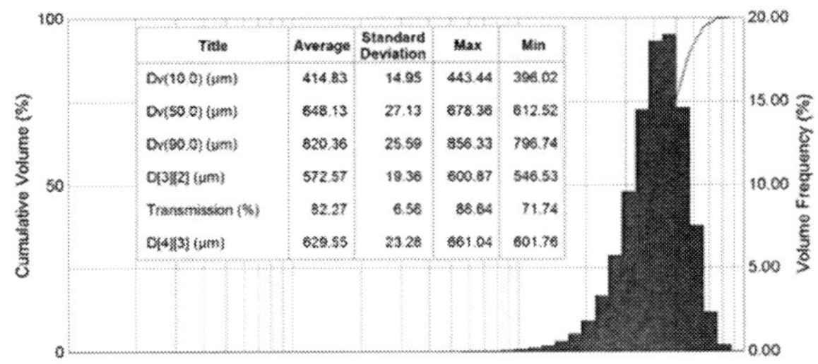

Fig.3-2 Refined Golden Sugar

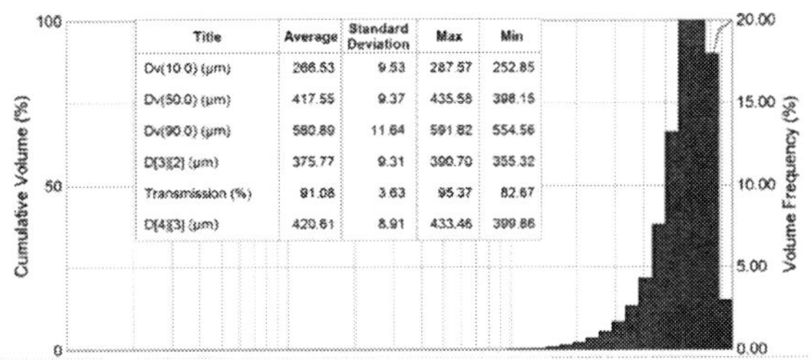

Fig.3-3 Dark Brown Sugar

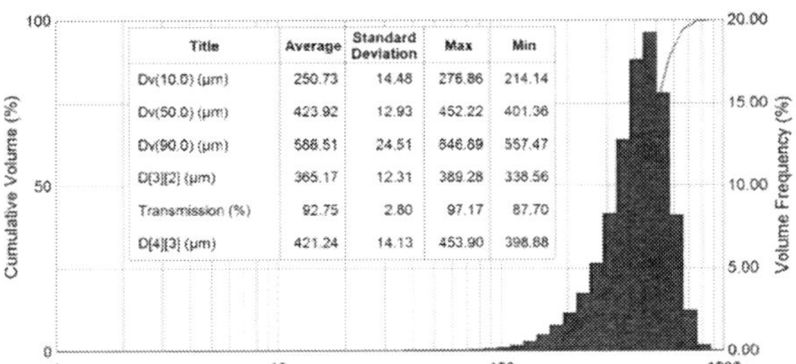

Fig.3-4 Refined Fine White Sugar

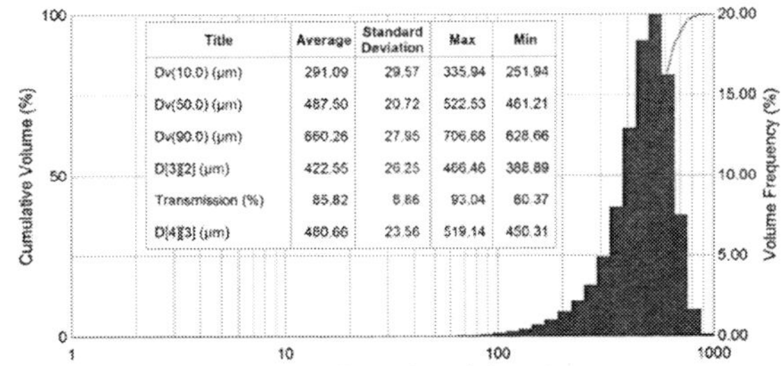

Fig.3-5 Refined White Sugar

3.2 Powder flow measurement

The powder flowability (Angle of repose) was measured using an instrument for Hall Flowmeter (Fig.3-6), the basic composition of the standard funnel, fixed bracket and measuring cup. The measurement step to take 50 grams of powder into the standard funnel, so that the natural outflow, the measured drop out of the powder due to natural stacked with the horizontal plane produced by the angle, called Angle of repose.

Fig.3-6 Hall Flowmeter

Angle of repose:

The free surface of the powder buildup layer in a static equilibrium state, the maximum angle formed with the horizontal plane is called the angle of repose. It is by a specific way to make the powder naturally fall to a specific platform formed.

The angle of repose has the greatest influence on the fluidity of the powder, the smaller the angle of repose, the better the powder's fluidity.

The measurement results are as follows:

Table 3-1. The powder angle of repose

The powder	Angle of repose
Refined Golden Sugar	40°
Dark Brown Sugar	31°
Refined Fine White Sugar	29°
Refined White Sugar	31°

3.3 Powder apparent density measurement

Powder as the density measurement instrument is also used Hall Flowmeter, the basic composition of the standard funnel, fixed bracket and measuring cup, The measurement step is to place the powder into the standard funnel, so that it naturally falls into the bottom of the measuring cup, the volume of the cup was 150 cm ^ 3. When the powder began to drop out of the cup, the powder was stopped and the powder was scraped along the cup. The weight of the cup was deducted by the precision balance and the density was calculated. Because the measured powder type is not the same, so stacked together after the presence of pores, so the measured weight of the powder volume is **Powder volume V_p plus pore volume V_h** ,t he calculation of the density is not True Density ρ_t of the powder but rather as the apparent density ρ_a, the calculation and results are as follows:

The density measurement results are as follows:

$$\rho_a = \frac{W_p}{V_p + V_h} \qquad (1)$$

Table 3-2. The powder apparent density

The powder	Apparent density
Refined Golden Sugar	0.76
Dark Brown Sugar	0.36
Refined Fine White Sugar	0.84
Refined White Sugar	0.88

3.4 Powder printing test

The above sugar powder was used for the printing test.(Fig.3-6)

Refined Golden Sugar | Dark Brown Sugar | Refined Fine White Sugar | Refined White Sugar

Fig.3-6 Powder printing test

Experimental results (considering pave, humidity, particle size and other factors) Refined Golden Sugar particles are too large, too small for the graphics can not be completely printed, and powder paving is not good, not the whole close to the plane,and Dark Brown Sugar printing situation is good, compared to other powder, a higher chance of a trip into the complex graphics, but because of high moisture absorption, there will be caking when the powder situation.So consider the humidity and the size of the powder, and finally to TSC powder Sugar provided as the main experimental powder. (particle size between 200-300μm).

4. Sugar edible grade binder development

4.1 Material selection and application

The objective of this study was to develop an edible grade of binders in the Binder Jetting technology. In order to make the thermal bubble type printhead used in this laboratory can be sprayed with binders, will be on the market to eat sticky binder viscosity test.

1. Corn Glue
2. Guanhua bean gum
3. Taiwan sugar chemical starch

4. Deionized water

4.2 Viscosity testing

After the selection of this study, we selected the following four food grade binders using RV viscosity tester (Fig.4-1) for viscosity analysis, Since the viscosity of the binder from the Sprinkler head when the eruption, there is a certain speed, so also to change the Shear Rate, to observe the different eruption speed, viscosity changes.

Fig.4-1 RV Viscosity Tester

4.2 Corn Glue Adjusting the proportion of test

Table 4-1. Corn Glue & Pure Water

Proportional unit / RPM	1:400	1:600	1:1000
30	50	26	16
60	31	16	10
100	23	12	8

4.3 Guanhua beans Adjusting the proportion of test

Table 4-2. Guanhua beans and pure water

Proportional unit / RPM	1:400	1:600	1:1000
30	18	10	7
60	15	8	5
100	12	5	3

4.4 Taiwan sugar chemical starch Adjusting the proportion of test

Table 4-3. Taiwan sugar chemical starch and pure water

Proportional unit / RPM	1:100
30	6
60	4
100	3

4.5 Deionized water test

Table 4-4. Deionized water

Proportional unit / RPM	1000
30	1
60	1
100	1

In order to meet the requirements of 1 ~ 3 (mPa · s) viscosity, the four viscosities of different binders at different shear rates were tested,can be found in the dissolved organic matter powder into a liquid are too viscous, and there will be sediment, considering the original of the thermal bubble type printhead can accept a viscosity of 1 to 2, if the binder ratio of less than 1% of the viscosity is higher than 3 not used, the Taiwan sugar chemical starch is still in the experimental stage.Finally, the deionized water as the main binder.

5. Design of Food Grade 3D Printer

5.1 Laboratory equipment

This section is based on binder spray printing method based on the initial design of the food three-dimensional printing machine appearance specifications of the length and width of each 60cm X 47cm X 42cm; color inkjet printer body selection CANON IP110 inkjet printer(Fig.5-1), designed to contain:System X-axis platform, the machine body structure, for the powder platform, printing and powder institutions, Z-axis movements, recycling tank institutions, mechanical and electrical systems and architecture, machine control required for the software interface eight categories , The development process shall be subject to food specifications.

Fig.5-1 CANON IP110

5.2 Machine construction and optimization

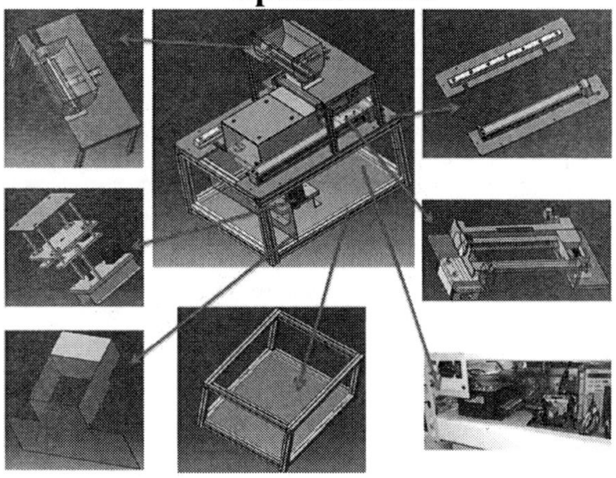

Fig.5-2 3D Printer Development for sugar

Optimization section:
1. Contact surface design
Because food-grade three-dimensional printer must consider food safety issues, reference to "Food Hygiene Management Law" which,food contact directly or indirectly with the surface of the machine.Under normal operating conditions, the liquid or vapor coming out of it will be in direct contact with the surface of the machine to be constructed of prescribed materials such as fresh cover, lid, etc.
So in the design of Z-axis contact surface and the construction of trough and other major food contact surface for the stainless steel-based materials.

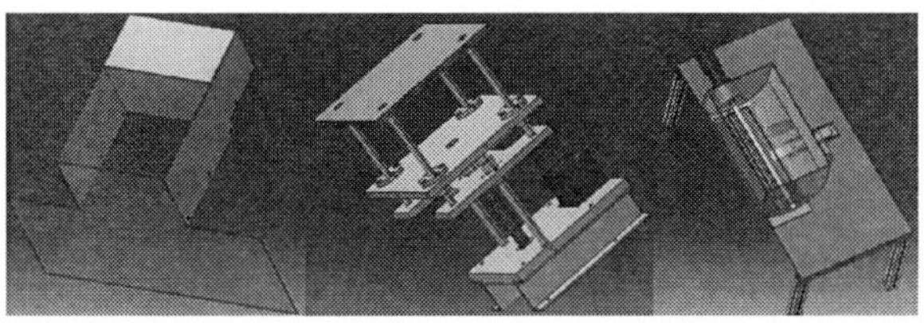

Fig.5-3 Construction groove, Z-axis contact surface, for the powder tank (from left to right)

2. Servo motor and Clean screw slide

In the past, the three-dimensional printer in this experiment was driven by the stepping motor in the X-axis direction. However, the biggest disadvantage of the stepping motor is that if the printing is interfered by external factors such as dust, the machine is out of step, It is not possible to have the feedback function as a servomotor. Therefore, during the printing process, the state of the machine must be ensured, and the servo motor and cleaning screw rails should be selected to improve the effect of out-of-step and food safety.

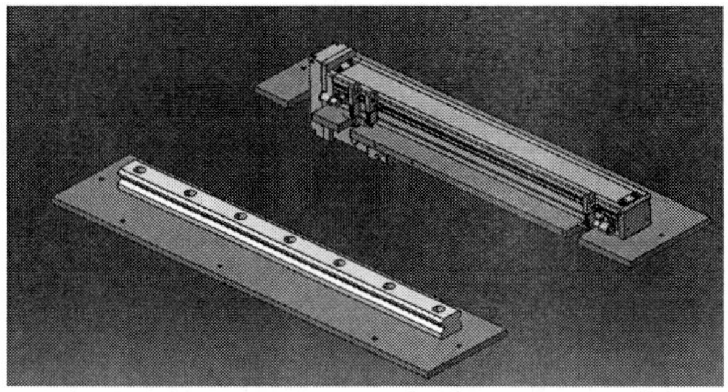

Fig.5-4 Servo motor and Clean screw slide

6. Results and discussion

The following experiment is based on food-grade three-dimensional food printer, powdered sugar and other materials, as well as the use of deionized water as binder, through different molding experiments, the best molding results and best post-treatment methods.

6.1 Whole sugar molding experiment

Thin plate specimen: length and width of the 20mm, high 1mm, 2mm, 3mm, 4mm, 5mm.

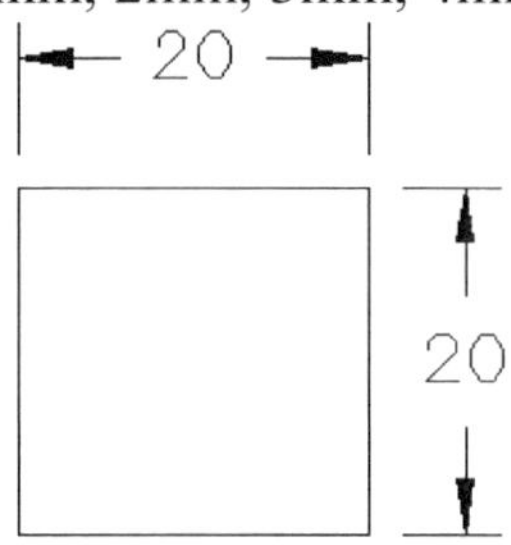

Fig.6-1 Schematic sheet

Experimental results and analysis:
The experiments were carried out by printing each ink layer at a thickness of 0.1 mm,

After the completion of the printing sample to be placed for 30 minutes to do the drying process, and then remove the waiting for hardening, remove the sample after the need to blow off the excess powder and check without breaking, and can be easily removed from the construction tank, and can be used. The surface quality and the dimensional deformation amount thereof will be observed.

According to the following figure, it can be observed that the deformations of the white powder sugar in the X direction and the Y direction are smaller than those in the Z direction, that the deformation in the X and Y directions is more uniform and that the surface quality is good and that the edges do not come into contact with the generated angle, And the surface of the object will easily remove excess sugar powder.

Table 6-1. Test sheet of white granulated sugar

CAD (mm)	Actual size	Deviation(%) x:y:z
20x20 x 1	20.4x20.51x1.82	0.02:0.0255:0.82
20 x 20 x 2	20.4x20.4x3.2	0.02:0.02:0.6
20 x 20 x 3	20.52x20.6x4.4	0.026:0.03:0.466
20 x 20 x 4	20.5x20.52x5.8	0.025:0.026:0.45
20 x 20 x 5	20.52x20.5x6.84	0.026:0.025:0.368

Fig.6-2 White sugar samples from left to right thickness of 1mm-5mm

Sugar is a single molecule composed of carbon, hydrogen and oxygen. Hydrogen and oxygen atoms in water and sugar are adsorbed in an electrostatic manner because of the attraction of hydrogen and oxygen atoms to each other. Water and sugar mixture, the two will be formed between the hydrogen bonds linked to each other, these links need high-temperature heating in order to destroy, so sugar will firmly grasp moisture in food, so the molding condition is quite good, but the powder itself is easy to absorb moisture in the atmosphere, resulting in agglomeration, and the powder size is not the same, in the process of printing the paving effect is not good, the finished surface is too rough and fragile. So decided to join other food powder, increase its mobility, paving, hardness and fine degree.

6.2 Mixed with other food powder molding experiments

Molding test pieces:length and width of the high 50mm (TSC logo round hollow sugar)

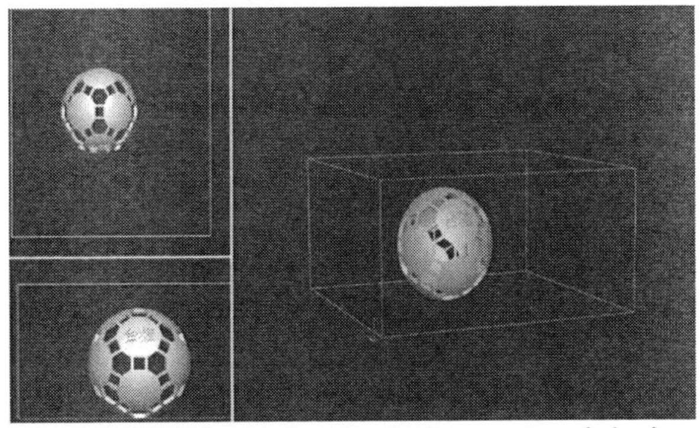

Fig.6-3 TSC logo circular hollow carved design drawing sugar

Selection of printed food powder:
1. powder tea
2. corn starch

Three-dimensional candy using different proportions of powdered sugar, tea powder, corn starch to print out to observe the various molding conditions. The following are the results of individual experiments:

Table 6-2.The proportion of powder sugar and tea powder

proportion	powder Sugar	powder tea	photo
Powder ratio1	10	3	Fig.6-4
Powder ratio2	10	5	Fig.6-5

Fig.6-4 Powder ratio 1

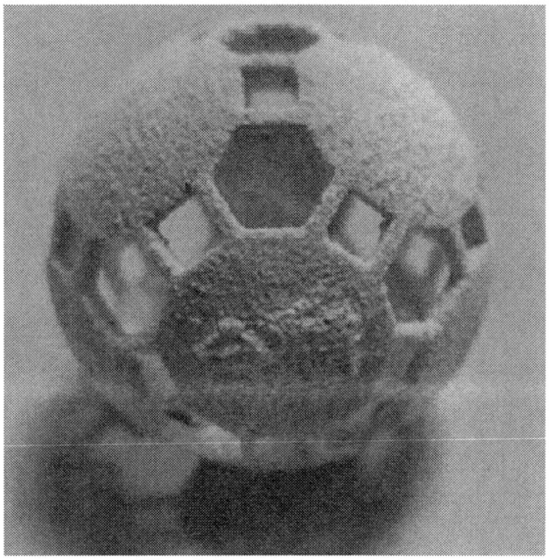

Fig.6-5 Powder ratio 2

Table 6-3.The proportion of powder sugar and corn starch

proportion	powder Sugar	corn starch	photo
Powder ratio3	9	5	Fig.6-6
Powder ratio4	3	1	Fig.6-7

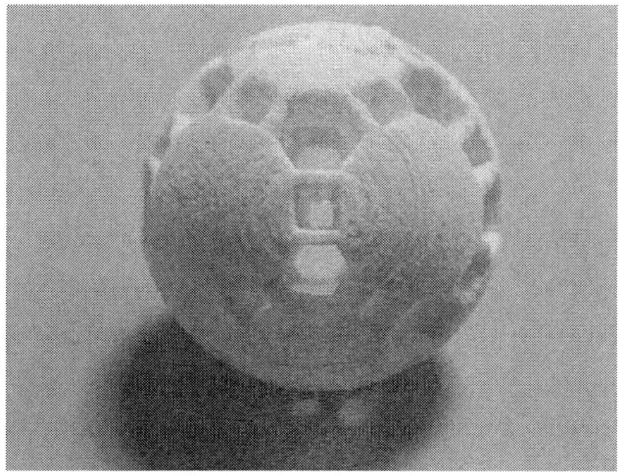

Fig.6-6 Powder ratio 3

Fig.6-7 Powder ratio 4

The figure shows that the proportion of tea powder can fill the gap between the sugar powder, increase mobility, but severe oxidation, the color turns yellow, difficult to save. The use of corn starch, the finished product is clearly visible surface detail, and finished the strength, integrity and color is quite good, print the place is obvious, although compared with the tea powder, powder condition slightly worse, the characteristics of more Brittle, but still meet our requirements, so the powder ratio of 4 the best ratio.

7. Conclusions

After a year of study, the conclusions are as follows:

1. To determine the shape of the powdered sugar can be formed specifications (provided by the Taiwan sugar), particle size between 200-300μm.
2. The use of deionized water can be used as a preliminary edible binder formular, remains to be further research and development.
3. 3D three-dimensional sugar technology parameters of the conditions:
   The sugar powder, corn starch to 3:1 ratio with deionized water as binder successfully printed three-dimensional sugar .
4. 3D stereoscopic molding machine and equipment for the body structure Size: 60 (X) x47 (Y) x42 (Z); Construction slot size: 20 (X) x20 (Y) x10 (Z) Preliminary completion of the production and evaluation of the case.
   Designed to include food specifications groove construction, for the powder tank to stainless steel production, and servo motor drive clean screw slide (to prevent food powder erosion institutions)

Acknowledgements

The authors would like to pay appreciation to TSC for providing the funding for this project and thank to the director Cheng and Mr. Tsung-Han Lin for their assistance in making this project success.

References

[1]Lin Ding-sheng, "three D print the status of the development of" *scientific development of* 503, 2014.

[2]Mina Aliakbari, Additive Manufacturing: State-of-the-Art, Capabilities, and Sample Applications with Cost Analysis, Master of Science Thesis, Production Engineering and Management, Department of Industrial Production, KTH.

[3] L. Serenó and J. Delgado J. Ciurana, "Customizing Food with an Additive Manufacturing Technology," *Materials Science Forum,* Vol. 713 pp.43-48, 2012.

[4]Available:http://chocedge.com/shop/index.php?id_category=12&controller=category

[5]Available :http://www.cnet.com/products/xyzprinting-food-printer/

[6]Available :http://www.kwsmfg.com/services/engineering-guides.htm

[7]Available: http://www.me.vt.edu/dreams/

[8]You-Lin Huang, "3D Printer Development for Starchy Foods Powder", thesis, *Department of Aeronautics and Astronautics, National Cheng Kung University,*2016.

[9]Cheng-wei Hsia, "Development of Three-Dimensional Printer for Stero Formation of Food Powder", thesis, *Department of Aeronautics and Astronautics, National Cheng Kung University,*2015.

[10] Yi-Chen Tseng, "A Study on Material Property and Process Parameter Optimization For Power Type Rapid Prototyper", thesis, *Mechanical and Automation Engineering, National Kaohsiung First University of Science and Technology*, 2001.

[11] Hsin-Hsiung Wang, "Analysis and implementation on the 3D Printer System by Using Inkjet Printing Module", thesis, *Department of Electrical Engineering, Southern Taiwan University of Science and Technology*, 2008.

[12] Delta Motor, ASDA-AB Series, Advanced Universal Servo Driver Application Manual.

[13] C.X.F. Lam , X.M. Mo, S.H. Teoh , D.W. Hutmacher , "Scaffold development using 3D printing with a starch-based polymer, " *Materials Science and Engineering* C 20 (2002) 49–56.

S.I.T. Paper #1158

Modification of the Recovery House Boiling Scheme To improve Refinery Productivity

A. S. Vawda and A. E. Muneef

United Sugar Company, amuneef@savola.com

Abstract

The United Sugar Company operates a 3500MT per day destination refinery employing conventional carbonatation and granular activated carbon decolourization system. Like many other sugar refineries built in the nineties, it has been designed to use a traditional sequence of strikes or boilings to produce white sugar and recycle all recovery house sugars to the melter. It is well known that efficient sugar end operation is achieved by avoiding unnecessary recirculation of colour, non-sugars, and sugar from the recovery house. A new system was proposed and implemented where none of these sugars are recycled to the melter, however, the first crop sugar is recycled to R1 continuous crystallization. This new boiling scheme was implemented with the objective to improve overall productivity of the refinery without negatively affecting the final quality of product sugar which is EEC2 specification. This paper describes the operating experiences and advantages and disadvantages with respect to:

- Sugar quality
- Mass balance
- Energy efficiency
- Process control

Keywords: Recovery house, Recycle, Energy efficiency, Quality, Sugar colour.

Introduction

The United Sugar refinery is a standalone factory located in Jeddah Saudi Arabia. Its process 1.2 mMt of VHP raw sugar mostly from South America. The processes employed are carbonatation and adsorption using activated carbon. There are four white boilings and three recovery boilings. The white boiling is characterised by continuous boiling for R1 massecuite and batch for the other three boilings[1].

A new boiling system in the recovery house was established to eliminate any sucrose recycle to the melter. This modified scheme required only minor equipment addition and changes to pipe-work, while it has delivered lower processing costs without any serious detriment to sugar quality.

Discussion

Most sugar refineries dissolve the first stage recovery sugar and recycle this to the melter. This is the function of any recovery house in a sugar refinery. The recovery house sugar, known as First Crop or A-sugar is sent to the main melter.

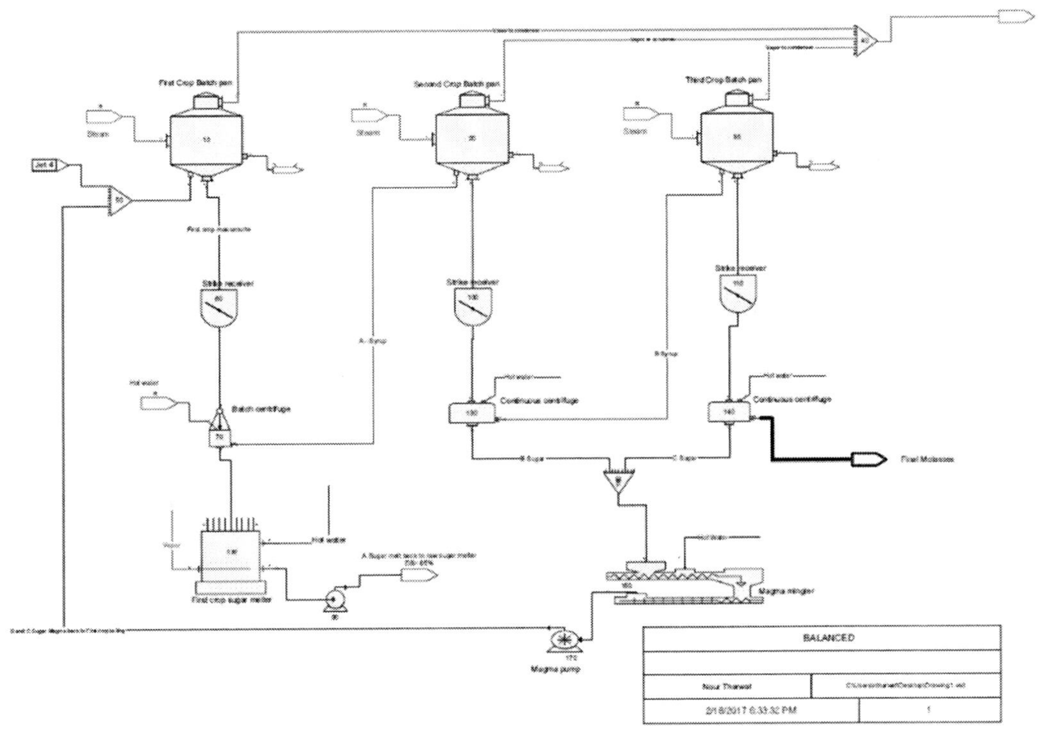

Figure 1 The old recovery house boiling scheme.

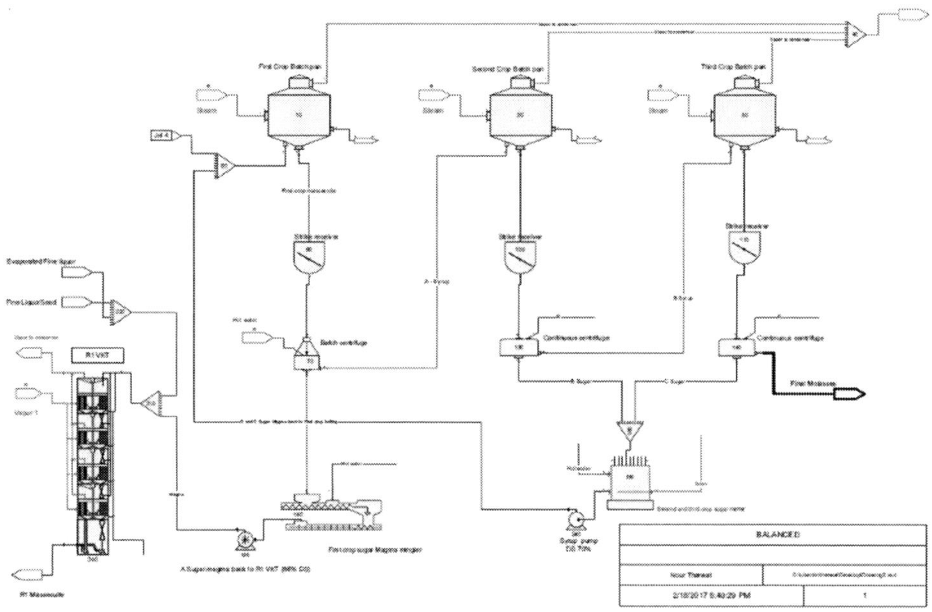

Figure 2 The new recovery house boiling scheme

The new system avoids this recycling step and produces an A-sugar product of a quality deemed sufficiently good enough to be used as seed to the R1 continuous crystallization stage.

Process Modification

In order to achieve the desired quality, several changes were made in the recovery house.

1. A massecuite cured in a batch centrifugal rather than a continuous centrifugal
2. The A-sugar is magmatized and pumped to the pan floor.
3. The magma is used as a footing in a batch pan to create a seed magma to be pumped into the continuous pan.

In order to send a recovery sugar product to the refinery without jeopardizing the final sugar quality, it is imperative that the separation process of the crystal from the mother liquor is very effective. In a somewhat similar exercise conducted at the Noodsburg sugar refinery in South Africa, B-sugar was affinated and centrifuged. Both purgings employed continuous centrifugals.[2] Most sugar refineries employ continuous centrifugals which can handle viscous massecuites and smaller sugar crystals. These continuous machines produce broken crystals which does not help in the production of a uniform crystal sugar size.

There are several well-known disadvantages of recycling sugar:

1. Increased energy costs: Recycled and fresh sugar require the same energy for processing.
2. Sugar losses: The more sugar is re-processed, the greater the losses.
3. Capacity reduction: Recycled and fresh sugar occupy the same refinery capacity.
4. Process control: decolourization of recycled products is less effective than for fresh sugar.

The Magma System

The key to solve the problem of achieving a good A-sugar quality, lies in good separation without excessive washing. This was achieved by switching from continuous to batch centrifugals. A simple test to see the potential of ideal separation was conducted in the laboratory by affination and separation as shown in Table 1. Clearly, a well purged A massecuite can deliver a reasonable A-sugar quality to the R1 boiling.

Table 1 Colour removal by laboratory affination

	A Sugar from Batch Centrifugal	Lab affined A Sugar
Purity %	99.8	99.9
Colour IU	302	220
Ash %	0.010	0.008
Invert %	0.072	0.048

The quality of the A-sugar is slightly inferior to the R1 seed it is incorporated into, however it is relatively small in percentage of the total sugar, about 3 %. It was anticipated and eventually seen that no processing or quality problems occurred. Two initial trials were conducted manipulating the existing equipment, using continuous centrifugals. The average results in Table 2 looked promising enough and based on these numbers, process modifications were approved.

Table 2 Initial trials on the A sugar recycle system using continuous centrifugals and batch pans

A Sugar		Seed Magma Without A-Sugar		Seed Magma With A-Sugar		Fine Sugar		
Colour	Ash	Colour	Ash	Colour	Ash	Colour	MA	Ash
413	0.119	475	0.165	868	0.251	33	0.63	0.014

These initial trials were done without any plant modifications using continuous centrifugals. The preliminary results were promising enough to select a batch centrifugal and install a magma mingler for the next trial and to send the A-magma to the R1 continuous pan. This magma is used to prepare the footing for the continuous pan.

Sugar Quality

This A magma recycle system contributes material to the R1 continuous boiling system. In this system, two batch pans produce seed magma to supply seed magma to the R1 continuous pans[1]. The seed magma contributes 33% of the total R1 massecuite. The solids emanating from A-sugar contributes 2.9% of the R1 massecuite output.

Table 3 Typical effect of A Sugar recycle on massecuite and sugar colour

A-Sugar	Seed Magma Without A Sugar	Seed Magma With A Sugar	R1 Sugar Colour IU
424	237	513	26

One month average data.

As the process was optimised, and lower fine liquor colours were achieved, the situation improved as shown in Table 3.

Table 4 Data showing the colour of continuous pan output colour

Seed Magma With A Sugar	R1 Continuous Pan Output	R1 Sugar Colour IU
522 IU	426 IU	17 IU

In the first trial of A-magma recycle, a quality problem emerged which was over looked. The recovery section of the refinery is made of carbon steel and the refinery sugar magnets were inundated with rust particles.

After a thorough investigation the source was found to be the A strike receiver. The roof and upper section of the receiver had a significant amount of loose rust. This loose rust was physically removed using needle descalers.

A particle size analysis of the rust captured on magnets showed that 99% of rust particles could be filtered using a 25 micron filter. This was implemented by fitting stainless steel filters at the A and B pan syrup feed inlet. Magnets have been ordered for the pan feed liquid, to be positioned after the particle filter.

Table 5 Comparison between sugars produced by both systems.

Month	Colour IU	MA mm	CV	Ash %
Sep 2016*	34	0.62	37	0.012
Mar 2017**	36	0.64	38	0.012

*September 2016 no magma recycle system in place

**March 2017 magma recycle system in place

Mass Balance and Energy Balance

The effect of A-sugar recycle removed 4.9 Mt of sugar recycled to the melter. The effect of this was a reduction of recycle to the main melter, reduction in chemical and energy cost. Steam savings were smaller due to the 25% increase in wash water in the A Sugar curing.

Process Control

The system to recycle A-sugar involved some modification involving the use of a batch centrifugal, a magma receiver and a pair of massecuite pumps. Appropriate instrumentation was quickly installed to control the new operation. The laboratory reported an increase in carbonatation colour removal by 10 percentage points. This was significant and a welcome development because USC generally suffered lower colour removal numbers compared to similar design and age refineries.

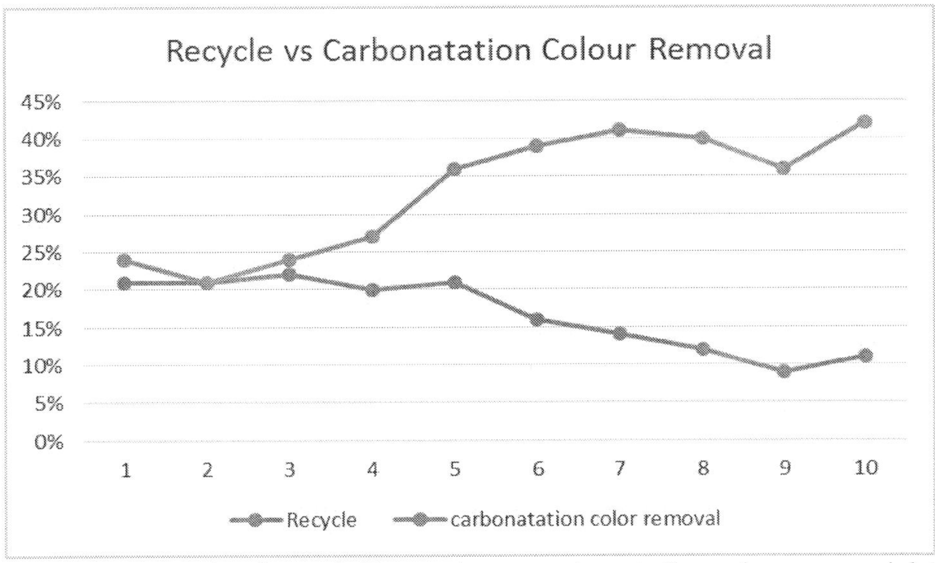

Figure 3 Ten weeks of recycle% to melter vs carbonatation colour removal data

On the operational side, the batch pan that prepared the seed magma had a quicker cycle time compared to slurry seeded pans because it was speedier to develop a full crop of crystals. On average the cycle time for a batch pan with magma was 50% compared to slurry seeded pans.

There was a 25% increase in the wash water requirement from 3 to 4 seconds, however, the R1 sugar colour average rose slightly from 16 IU to 21 IU. Overall product sugar colour averages 36 IU which is the same as in the past. There were no customer complaints as a direct result of moving to this system.

Conclusion

The use of batch centrifugals for curing A massecuite in the recovery house allowed for better utilization of this sugar than simply diverting it to the main melter. The use of this A-sugar as a footing for the seed pan reduced considerably the boiling time of the seed pans. The contamination by iron due to recovery house carbon steel equipment was a serious challenge that had to be solved by filtration of the pan feed and rigorous application of the in line magnets. The elimination of recycle to the melter also improved the carbonatation process.

Recommendations

For any refinery considering such a modification, the main concerns to be aware of are rust contamination and limitation of the recovery house pan capacity. The benefits though small, are significant. It unlocks extra melt capacity at the recovery front end.

Acknowledgements

Permission to publish from the management of United Sugar and the valuable contribution of refinery production, engineering, laboratory and technical team towards the preparation of this paper is thankfully acknowledged.

References

1. Vawda, A. S, Muneef, A. E, and Voelling, M. (2011) Experience with Continuous

Crystallization of Refined Sugar at United Sugar Jeddah. Proceedings of the Sugar Industry Technologist Conference.
2. Julienne, L. M. S. A (1990) Towards a More Productive Use of B-Sugar. Proceedings of the South African Sugar Technologists Association.

S.I.T. Paper #1159

Slow Calorie for health consciousness

Daisuke Tashiro, Kiyoaki Miyasaka, Daichi Miyazaki, Yukie Nagai and Masahiro Okuno

Mitsui Sugar Co., Ltd., Japan

1. Introduction

In response to rising health consciousness, some general consumers think that carbohydrates including sugar are a cause of "hyperglycemia" and "obesity." Especially, since Japanese people ingest a lot of carbohydrates as a staple food, the idea that weight can be lost by reducing or excluding the carbohydrates is easy to spread. Thus, "carbohydrate restriction" which restricts ingestion of sugars and carbohydrates becomes popular in Japan. It is said that the market size associated with carbohydrate restriction comes up to 300 billion yen, and carbohydrate restriction is practiced by some people due to its easiness of the way that anything can be eaten except excluding carbohydrates. However, since it is reported that about half of the people who experienced carbohydrate restriction experienced bad physical conditions and that most of the people further underwent rebound (1), many nutritionists express negative opinions on "carbohydrate restriction." In addition, there is a data that, although the consumption of sugar per person in Japan decreased by about 30 percent in the last 30 years in association with aging and reduction in population, the number of diabetic patients and prediabetic individuals is increasing (2). Therefore, there is some question to think that a reduction in the amount of sugar-ingestion simply makes Japanese healthy.

In response to this problem, we searched for a probability of a product that can be ingested by health-conscious people without worrying about their health and which can put the brake on a decrease in demand for sugar. Furthermore, we paid attention to isomaltulose and devised a concept of Slow Digestion (we call it Slow Calorie).

2. Slow Calorie

2-1. Isomaltulose (Palatinose)

Isomaltulose, which is a structural isomer of sucrose (Figure 1) and naturally contained in honey, is industrially produced from sucrose using enzymatic reaction. Isomaltulose is degraded by isomaltase localized in the small intestine, but the activity of this enzyme is low. Isomaltulose is digested and absorbed in the small intestine at a speed of about 1/5 as compared with that of sucrose, and thereafter completely absorbed (3). In addition, isomaltulose also has an effect of inhibiting activities of other α-glucosidases and delaying digestion and absorption of other saccharides ingested at the same time (4). That is to say, a mixture of isomaltulose and, for example, sucrose is slowly absorbed.

This effect is very important when isomaltulose is actually used. It is because isomaltulose has about half sweetness as compared with that of sugar and isomaltulose alone has processing characteristics different from those of sucrose. Thus, by mixing isomaltulose with sucrose, healthy functions of isomaltulose and a sweet taste close to that of sugar (sucrose), and ease of use similar to that of sugar (sucrose) can be combined.

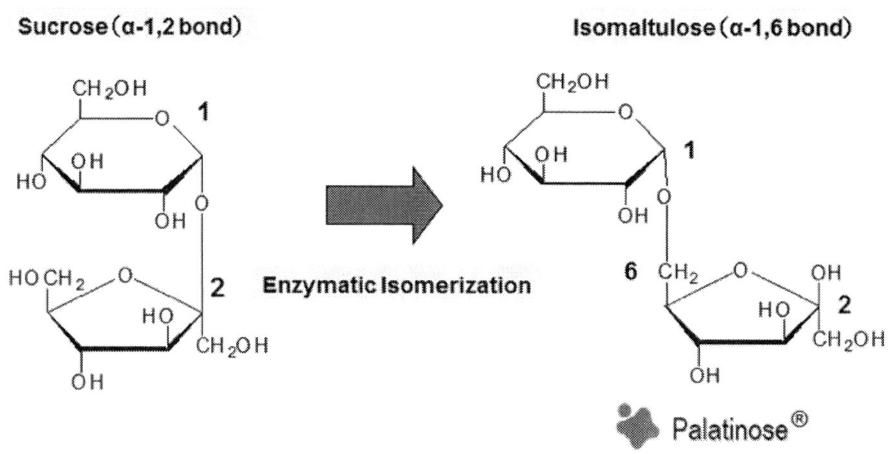

Figure.1 Structural formula of isomaltulose

2-2. Benefits of Slow Calorie

Among a lot of benefits that we can obtain from slow absorption (slow calorie), we will introduce the following three benefits this time.

The first is duration of a sense of satisfaction. According to the study that we carried out, healthy adult subjects were asked to eat 50 g each of chocolate made by substituting about half of sucrose with isomaltulose or chocolate made by using only sucrose, and blood GLP-1 concentration and duration of a sense of satisfaction were compared. As a result, when the subjects ate the chocolate in which isomaltulose was blended, GLP-1, which is a hormone which prolongs feeling of a sense of satisfaction, was secreted more and the actual feeling of the sense of satisfaction lasted longer (Figure 2)(5).

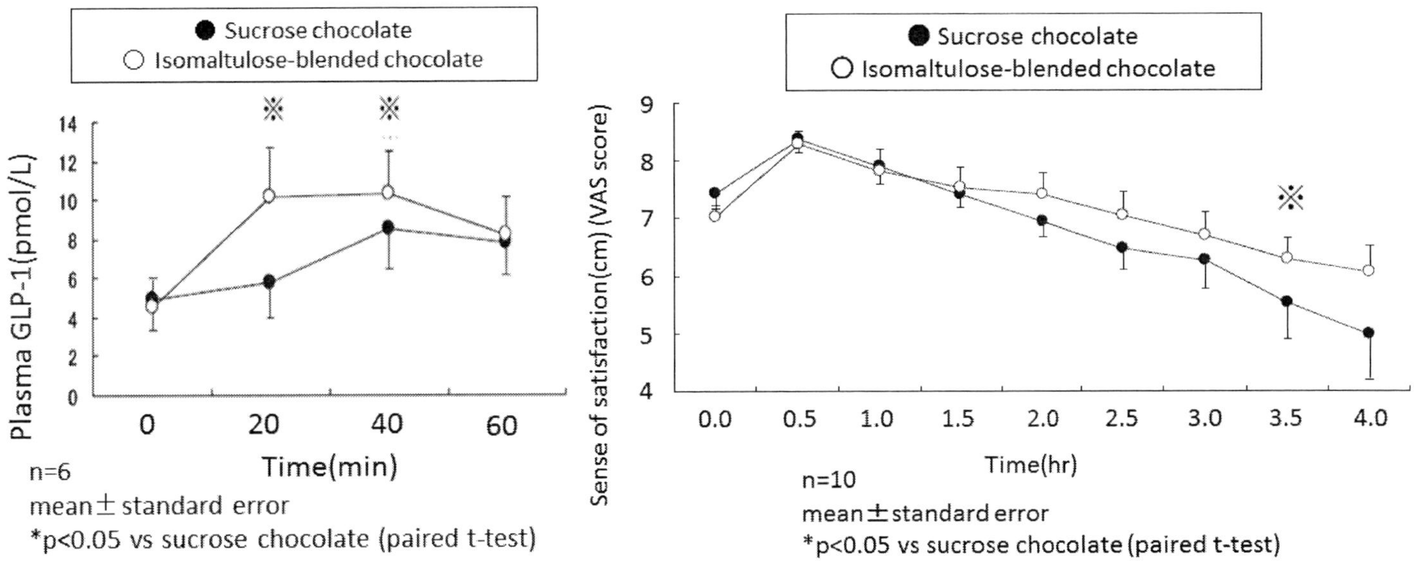

Figure 2: Change in blood GLP-1 concentration and sense of satisfaction when isomaltulose-blended chocolate was ingested (revised)

The second effect is continuous supply of saccharide on exercise. According to the study by Daniel König et al., athletes were asked to ingest 750mL of a beverage containing either 75 g isomaltulose or maltodextrin 45 min before the test, and performed 90 min cycling followed by a time trial test. As a result, it was confirmed that, when isomaltulose is ingested, the amount of lipid burning and blood glucose level during exercise tend to be maintained at a significantly high level and the time trial test results tends to be good (6).

The third is an effect of suppressing a rapid increase in postprandial blood glucose level. According to the our study, Japanese healthy adult subjects were asked to put 15 g of a mixture prepared by mixing isomaltulose and sucrose in the same amount (PS group) or 15 g of sucrose (S group) in coffee for breakfast, respectively, on a separate day and to ingest the coffee together with breads and margarine. As a result, postprandial hyperglycemia was significantly suppressed in the PS group (Figure 3) (7).

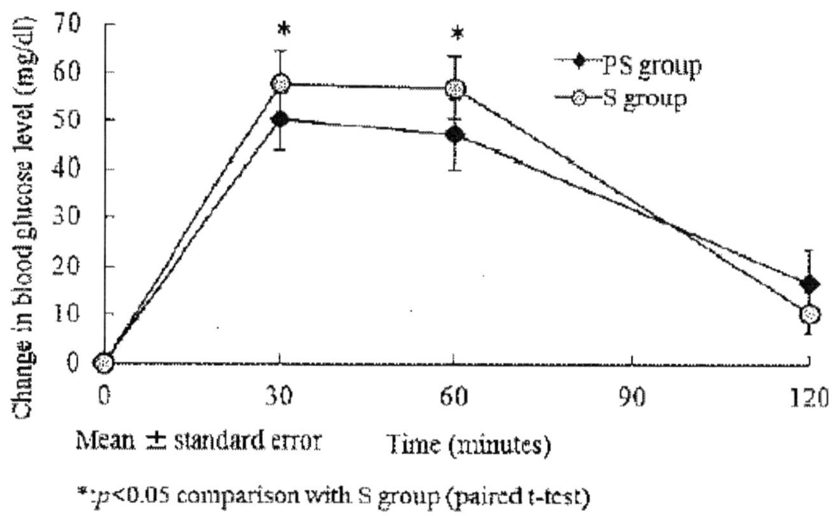

Figure 3: Change in postprandial blood glucose level when isomaltulose was ingested at breakfast

3. Attention to Slow Calorie in Japan

3-1. Attention from companies

As we have made efforts to disseminate the Slow Calorie concept since 2007, the following three industries took an interest in the above-mentioned benefits of Slow Calorie.

The first is the food industry. More than 20 stores of confectionary came to have an interest in the point that confectionary which is delicious and gives a more sense of satisfaction can be made, and started selling Slow Calorie products. As a result, Slow Calorie confectionary began to be sold as souvenirs in Tokyo station. In addition, events using slow calorie confectionary are held in various regions, and the situations are broadcasted via the mass media.

The second is the sports industry. Since the Slow Calorie concept using isomaltulose enables continuous supply of saccharides on exercise, there are an increasing number of cases where isomaltulose and other saccharides are combined for use in confectionary and protein products directed at endurance sports such as marathon. Sports instructors also showed a good response, and the idea of Slow Calorie is regularly used by world tournament-level players.

The third is the health care industry. Slow Calorie has been evaluated as countermeasures for hyperglycemia, and began to be used in hospital diet and food for patients. For example, Prof. Nishimura, a Director of The Japan Dietetic Association, has started supplying feeding containing isomaltulose in the hospital to which he belongs, and this idea started to extend to other hospitals.

3-2. Attention from general consumers

We developed a product in a small package for home use prepared by mixing sucrose and isomaltulose in the same amount, "Slow Calorie Sugar," for general consumers. Although we started sales of the product through limited distribution channels, the product is now sold nationwide in the supermarket that has the number-two domestic sales in Japan, due to recent increase in the interest in Slow Calorie.

4. Conclusion

While negative view for saccharides is spreading in Japan, we devised an idea called Slow Calorie, and have made efforts to disseminate the idea. We have sold isomaltulose for companies, and Slow Calorie Sugar, which is a mixture of isomaltulose and sucrose, for general consumers. As a result, Slow Calorie draws attention from food, sports and health care industries, and the market started to be vitalized.

Isomaltulose is a sugar product of which raw material is sugar. Thus, we think that an increase in consumption of isomaltulose due to such dissemination of Slow Calorie concept leads not only to health promotion of people but also to an increase in consumption of sugar.

5. Reference

(1) NEO MARKETING INC. □Syokuji seigen diet ni kansuru tyousa□ https://www.i-research.jp/report/?p=4375 (2017/Mar/8th)

(2) Takada A et al.□Satou sessyu to tounyoubyou no kankei□ https://www.alic.go.jp/joho-s/joho07_001285.html (2017/Mar/8th)

(3) Tsuji Y et al. Digestion and Absorption of Sugars and Sugar Substitutes in Rat Small Intestine. *J Nutr Sci Vitaminol* (1986) **32** 93-100

(4) Kashimura J et al. Inhibitory Action of Palatinose and its Hydrogenated Delivatives on the Hydrolysis of α-Glucosylsaccharides in the Small Intestine. *Journal of Agricaltural and Food chemistry* (2008) **56** 5892-5895

(5) Maeda A et al. palatinose® haigou chocolate no kettouchi jousyou yokusei kouka oyobi manpukukan jizoku kouka ni kansuru kentou *Nihon byoutai eiyougaku kaishi* **16** 77-84

(6) König D et al. Substrate Utilization and Cycling Performance Following Palatinose™ Ingestion : A Randomized, Double-Blind, Controlled Trial. *nutrients* (2016) **8** 390

(7) Miyasaka K et al. Effect of Isomaltulose-Blended sugar on Postprandial Blood Glucose Levels (Randomized Double Blind Cross Over Study) *Pharmacometrics* (2016) **91** 55-60

S.I.T. 2017
TAINAN
Energy Symposium

Improving Energy Economy in a sugar refinery – "From fundamentals to innovation"

S.I.T. Paper #1160

David Love *Tongaat Hulett Sugar*

INTRODUCTION (what I intend to cover)

- Energy or Monergy
  - Some energy is cheaper than other energy
  - Some money is "cheaper" than other money
- Back to basics
  - Thanks to Oliver Lyle in particular
- Back to the future!
  - Basics brought up to date
  - Recent innovations
- To Infinity and beyond!
  - Looking into the future
- What can I do in my refinery ?

Energy or Monergy ?
(how do I compare performance?)

- Energy (kW vs kW, kJ vs kJ))
  - Some energy is cheaper than other energy
  - Process steam or electricity
  - Conversion of fuel to steam
  - Use HP steam to generate electricity and still have exhaust steam available as process steam.
- Money ($ vs $)
  - Operating costs versus Capital Costs
  - Tax incentives (capital vs operating costs)
  - Company finances (debt to earnings)
  - Carbon taxes or carbon credits
- Comparison with myself and an objective standard for myself.

Back to Basics
(thanks mainly to Oliver Lyle)

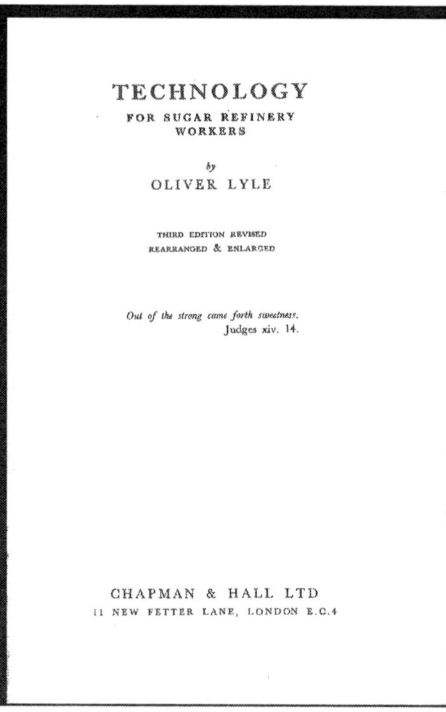
TECHNOLOGY
FOR SUGAR REFINERY WORKERS

by
OLIVER LYLE

THIRD EDITION REVISED
REARRANGED & ENLARGED

Out of the strong came forth sweetness.
Judges xiv. 14.

CHAPMAN & HALL LTD
11 NEW FETTER LANE, LONDON E.C.4

Back to Basics (thanks mainly to Oliver Lyle)

THE EFFICIENT USE OF STEAM

BY

OLIVER LYLE

"In short, I expect almost totally to prevent waste of steam"

JAMES WATT. Letter to Dr. Lind, 1765

LONDON

HER MAJESTY'S STATIONERY OFFICE

1947

Back to Basics (Oliver Lyle's Recommendations)

- Reduce both primary and secondary heat losses
- Primary heat losses
  - Steam leaks
  - Safety valve emissions
  - Faulty steam traps
  - Radiative and convective heat losses (poor or no lagging).
- Secondary heat losses
  - Not re-using waste heat e.g.
    - Recovery of heat from boiler blow down
    - Use of low grade vapours and flash vapour
    - Heat recovery from condensate streams

Back to Basics
(Oliver Lyle's Techniques for energy saving)

- Reducing processing temperatures where possible
- Minimise the recycling /reprocessing of material
- Reduce residence time of material in the factory (for example, by minimising the size of tanks so as to minimise heat losses)
- Increasing the brix of processed material as high as possible.
- Minimising water addition, to reduce evaporation requirements.
- Increasing the number of evaporator effects, and making extensive use of low grade evaporator vapours for heating purposes. *****

Back to the Future

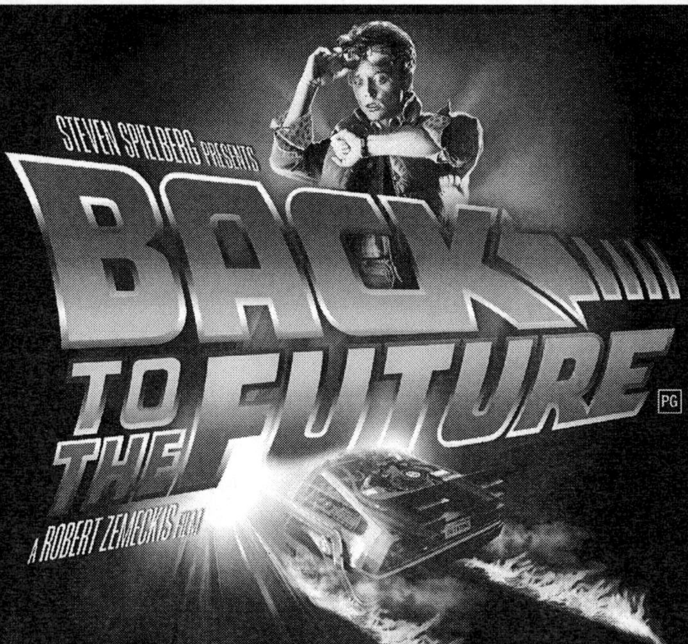

Back to the Future (Basics brought up to date)

- Reducing drive speeds to save electrical energy
  - Electronic VSDs
- The heat balance as a standard for evaluation (Lyle's concept of a Bogey)
  - Spreadsheet heat balances
  - Process modeling software (e.g. Sugars)
  - Modern instrumentation and DCSs for reporting
- Modern instrumentation and automatic control
  - Pan control
  - Liquor brix control

Back to the Future (Basics brought up to date)

- New Liquid sugar products
  - Turning return syrup into liquid sugar products and eliminating the recovery house.
- Pinch Technology - Formalising the principles of Lyle (?!)
  - Pinch Technology as a rigorous thermodynamic technique (stories of amazing insights).
  - Pinch Technology and multiple effect evaporation
  - Application to sugar refining (SIT papers)
  - Gives insights relative to specifics of each refinery
  - Limitations of Pinch Technology (does not question or optimise underlying manufacturing process)

To Infinity and Beyond!

To Infinity and Beyond!
(Radical new approaches for the future)

- The future is now – Look to pioneers of new approaches

To Infinity and Beyond!
(Radical new approaches for the future)

- The future is now - Pioneers of new approaches.
- Continuous pans and double effect pan boiling.
- Mechanical Vapour Recompression and thermocompression (combined with continuous pan boiling)
  - Al Khaleej as pioneers and leaders.
- GREEN Refining Technology (THS)
  - Decoupling of Crystallisation and Evaporation
  - Crystallisation by cooling with colour transfer benefits
  - Multiple effect evaporation in falling film evaporators
  - Intermediate decolourisation of jet

What can I do in my refinery?
(Suggestions for up to date basics and beyond)

- Conduct a pinch technology study to identify energy saving oportunities.
  - Savings achievable with existing equipment
  - Strategy for replacing equipment as appropriate
    - Short payback time
    - As equipment ages – performance drops, maintenance increases
- Use modern DCS for on line performance monitoring and comparison with an objective standard (Bogey).
- Investigate new technology options
  - Studies to look at "Brown Fields" opportunities.

S.I.T. PAPER #1160

ENERGY SYMPOSIUM

AN INTRODUCTION TO REFINERY ENERGY MANAGEMENT

- Indrajit Singh

PERFORMANCE INDICATORS FOR ENERGY CONSUMPTION

1. To measure is to know. "If you can't measure it, you can't improve it."
   LORD KELVIN 1883
2. Incoming Energy to refinery: Fuel and Electricity
3. In a refinery, Energy is used in 2 basic forms: Steam and electricity. Typical ratio* of steam : electricity energy input is 90:10 range.
4. Steam per ton rso
5. Electricity (kwh) per ton rso
6. Water consumption per ton rso
7. Measurement of steam at boiler and power at generator are essential. Downstream measurement at selected points is desirable for better control.
8. Measurement of energy consumption of back end refinery and its comparison with stand alone is more complicated.

* contribution of electricity can be higher when use ACCs and MVR , RSO= Refined sugar output

COMPARISON /BENCHMARKING OF ENERGY CONSUMPTION ACROSS REFINERIES:

Following factors contribution need to be accounted while making comparisons :

1. Refineries operating with or without affination section
2. Raw sugar quality
3. Refined Sugar quality
4. Process technologies
5. Maturation facility
6. Packaging
7. Waste water treatment plant
8. Permissible discharge limits for pollutants
9. Refineries design for future expansion might underperformed at phase I capacities
10. Capacity ? one of the reason total hot surfaces / unit RSO is much higher in small refineries (≈500tpd)

TECHNOLOGIES CONTRIBUTING TO ENERGY IMPROVEMENTS IN SUGAR REFINERIES:

1. Double effect evaporation in crystallization realized by the application of CVPs / VKTs.
2. Vapor compression:
   a. Thermal : Motive steam
   b. Mechanical : Electricity
3. Multiple effect Falling Film evaporators
4. Plate heat exchangers with wide gap technology
5. Richter tubes
6. Pinch technology*
7. Flash steam usage
8. Steam transformers /water evaporators
9. Heat recovery from spent brine
10. Automation

* It's a methodology for minimizing energy consumption of chemical processes by calculating thermodynamically feasible energy targets which leads to optimum utilization of heat.

OLIVER LYLES PRINCIPLES ON ENERGY CONSERVATION:

1. Preventing the escape of heat
   - a. Radiation losses
   - b. Leakages (steam, air)
   - c. Steam traps
   - d. Atmosphere vent
   - e. Flue gas losses
   - f. Condensate loss
2. Reducing the work to be done by heat
   - a. Lower the process temperature
   - b. Minimize water addition
   - c. Optimize sugar recirculation
   - d. Keep brix as high as possible
   - e. Reduce the electrical load
   - f. Productivity
3. Using the heat over again
   - a. Multiple effect evaporators
   - b. Flash steam

ENERGY MANAGEMENT PROGRAM:

1. Good energy program is more then simplistic house keeping approach
2. Focus should be on total energy scenario including objectives[2]:
   - To establish clear energy goals
   - To structure the program to achieve the goals
   - To give mgr authority to implement the program.
3. Energy audits program should be establish to keep the program ongoing:
   - Walk through audit: Monthly, look at leaks (steam, water, air) insulation, steam traps etc.
   - Mini audit: Quarterly, Integrity of measuring/recording devices, performance in energy targets.
   - Maxi Audit (Energy balance): Annually, Energy saving projects identification.
4. Outcome of energy audit would be an action log with clear responsibilities and target dates.

ENERGY MANAGER :

1. Technical *know how* to identify the energy intensive process streams
2. Technical *know how* to select the correct measurement points
3. Selection of the correct measurement points for measuring instruments/hardware's:
   - Permanently installed
   - Portable
   - Subcontracted
4. Technical support
   - Internal and External
   - Models for Mass & Energy balance like Sugars / In house
5. Energy report: Clear and concise evaluation of recorded data in daily, weekly and monthly format. Energy cost for refinery start up and shutdown.

EVALUATION OF ENERGY SAVING PROJECT'S:

1. Several existing and old refineries modernize by adopting the latest technologies evolved or est. at later stages.
2. Energy saving projects are often validated based on ROI. Fuel Price constitute a critical input data for energy saving project realizations.
3. Recent past has witnessed the large swings in energy prices, multiple drops in crude oil prices creates a seismic effect on natural gas, coal etc.
4. In many (developing) countries, uncertainty over long term energy prices affects funding of energy saving projects and hence modernization.
5. Policy decisions consistency like reduction in overall green house emissions/ Carbon footprint and related incentives.

ROI=Return on Investments

CONCLUSION:

1. Total energy input to refinery calculated in kcal /ton rso is good performance indicator of energy usage.
2. Comparison of energy consumption should account for factors which are not universally present but have substantial effect on energy indices.
3. Regular evaluation of suitability of emerging technologies.
4. Energy management program in place with clear objectives and a leader.
5. Careful evaluation of investment return of energy saving projects.
6. Possible adoption of best practices from European Beet Industry.

REFERENCES:

1. **Technology for Sugar refinery workers** by Oliver Lyle , 3rd edition
2. Energy Conservation for sugar refining, in **Handbook of sugar refining** by Dr. Chou
3. **SASTA Energy workshop 2008**, Challenges in Energy reporting in a Sugar Mill by I. Singh
4. **Sugar Technology** by P.W. Poel, H Schiweck, T. Schwartz

Benchmarking and Modelling to drive energy improvements

8 May 2017

Peter Simpson
Wilmar Group Technical Manager, Sugar refining

S.I.T. Paper #1160

wilmar

Abstract

How benchmarking and detailed SUGARS ™ modelling was used across 3 refineries to create sound justification for projects, by quantifying the size of the improvement possible for each potential improvement, but, more importantly, create competitive motivation to improve.

This is about incremental improvement with minimal capital investment. Its about finding out how to squeeze more performance out of your existing factory.

Where capital investment is required, modelling helps provide the justification.

wilmar

CONTENTS

- Introduction
- SUGARS process modelling
- Benchmarking with models
- Improvement Projects
- Lessons Learned

wilmar

How to make energy reductions happen more quickly?

Every refinery has improvements that can be made to reduce energy use. There are many reasons why projects have not been done yet.

•It can be hard to put numbers on the financial benefit. Modelling can help quantify the impact of a change.

•People may be doubtful something is possible. Benchmarking can help by demonstrating the potential performance is achieved at another refinery.

•Sometimes (always) people are busy. Teamwork with multiple refineries can accelerate projects.

wilmar

Make use of multiple refineries

- The Wilmar group now has 8 refineries which creates a large pool of skilled people and also the ability to benchmark and share knowledge.
- The exercise I will talk about today involves three refineries in Australia and New Zealand.
- Comparing performance we could see that all sites could benefit from sharing best practice.
- We set up a team involving people from each refinery, and used SUGARS ™ software to model the refinery processes and help understand improvements.

wilmar

Models – can we trust them?

"All Models are wrong, some are useful"

George Box – Industrial Statistician

wilmar

Models make it easier to make mistakes!

- We are suspicious of complicated spreadsheets.
- With SUGARS software it is easy to create an impressive looking model.
- Because the flowcharts look like refining equipment, it is easier to believe the results. But they may be wrong!
- Even a model that has passed careful testing and validation can suddenly become seriously in error after a minor change to one variable.
- Insist on key results being estimated independently of the model – nothing wrong with pencil and paper!

wilmar

SUGARS models are static, real processes are variable

- A model may show a batch pan steam flow to be 1.2 t/h.
- In reality the flow may never be 1.2 t/h. It will be zero between batches, maybe 4 t/h during seeding and 10 t/h during boiling.
- We need to understand the limitation of static models.
- But a static model is still a very powerful tool to quantify impact of changes to a process.

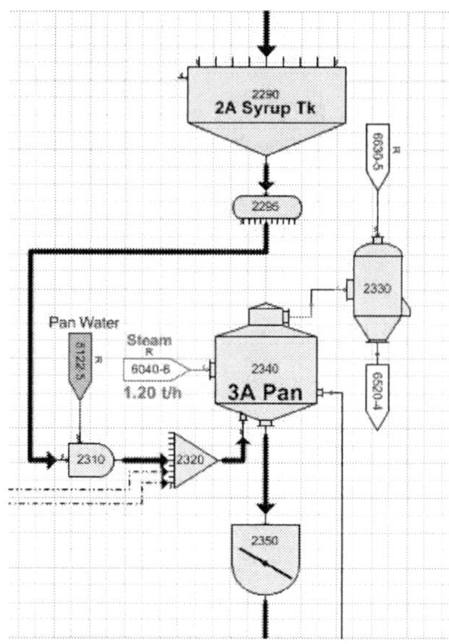

wilmar

One model - three different refineries

- For consistency we used the same model for three different refineries. For example two used affination, one did not.
- Built a model that included all processes and products.
- Provided "switches" that could turn processes on or off, such as an affination bypass.
- Using these switches the model could be made to represent each refinery.

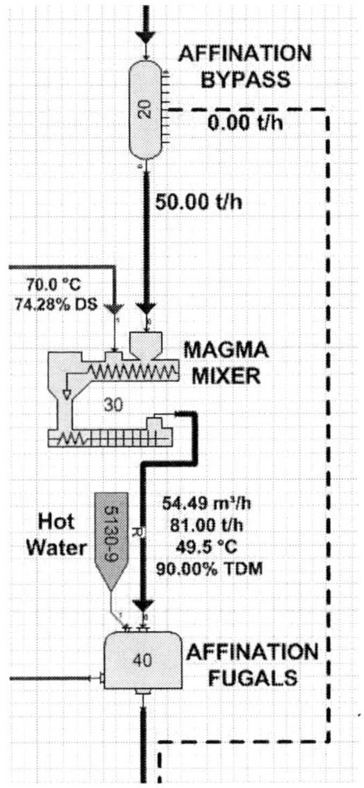

Teamwork and involvement

- Set up a team of 6 people, 2 from each refinery
- Worked together to improve and validate the model
- Made 3 copies of the model, one for each refinery
- 3 teams of 2 to work on these models, each team had one person not from the refinery being modelled.
- Entered current actual process data into the model and tuned the model until the model output matched current actual results to within 2%.
- Team accepted the three refinery base case models reasonably represented current performance, suitable for "what-if" analysis

wilmar

Model Validation

- A "perfect" model typically understates steam use by about 20%. Don't trust a model with 20% error!
- Force the model to include inefficiencies.
- First we have to be able to estimate them.
  - For example check rate of temperature drop in a melter when there is a refinery stop, to quantify rate of heat loss.
  - By measurement or careful observation, make estimates of activities such as washing fugal baskets.
  - Measured heat gain by air leaving refinery building. Equivalent to 140MJ/t and approximately matched the "unexplained" heat loss from that refinery.
- Use an in-model calculation to compare model results with known factory results and **generate automated flags on excessive deviation**

wilmar

Use Model to understand differences

- We now had models which estimated steam use of each refinery to within 2% of real results.
- Next step – find out what to improve.
- Made a table of the process parameters which will have the greatest impact on steam use. Such as Brix
- Collected the data from each refinery to find the best result for each parameter
- These were accepted as achievable because they were real results, already achieved somewhere
- Put these best results into each refinery model, noting the steam reduction for each item.

wilmar

Process parameters

- A sub-section of the table

VARIABLE	A	B	C	"BEST"	COMMENTS
FL Bx	63.2	64.7	65.3	65.3	now achieving 67+
HFL Bx	74.2	69.9	74.7	74.7	now achieving 77-78
1A Syrup Bx	72.1	70.7	72.8	72.8	now achieving 74.8
2A Syrup Bx	72.7	71.4	72.9	72.9	
3A Syrup Bx	72.1	71.3	73.2	73.2	

- Question and answer session to understand why results were different between refineries.
- For example centrifugal runoff syrup Brix.
  - One refinery had no heating, used dilution to control saturation
  - Another had heating, used live steam injection
  - Non-contact heating was identified as best practice.

wilmar

Improvement Projects

- Each refinery had a list of parameters to improve
- Steam saving estimated by the model for each parameter improvement
- Projects developed, prioritised by the value of the improvement
- Plans presented to managers, using the models to explain and justify projects.
- Running the model to demonstrate how changing one parameter impacted on total refinery steam and water use was a very powerful communications tool.

wilmar

Projects for Refinery A.

Note two different projects to increase syrup Bx in stages.

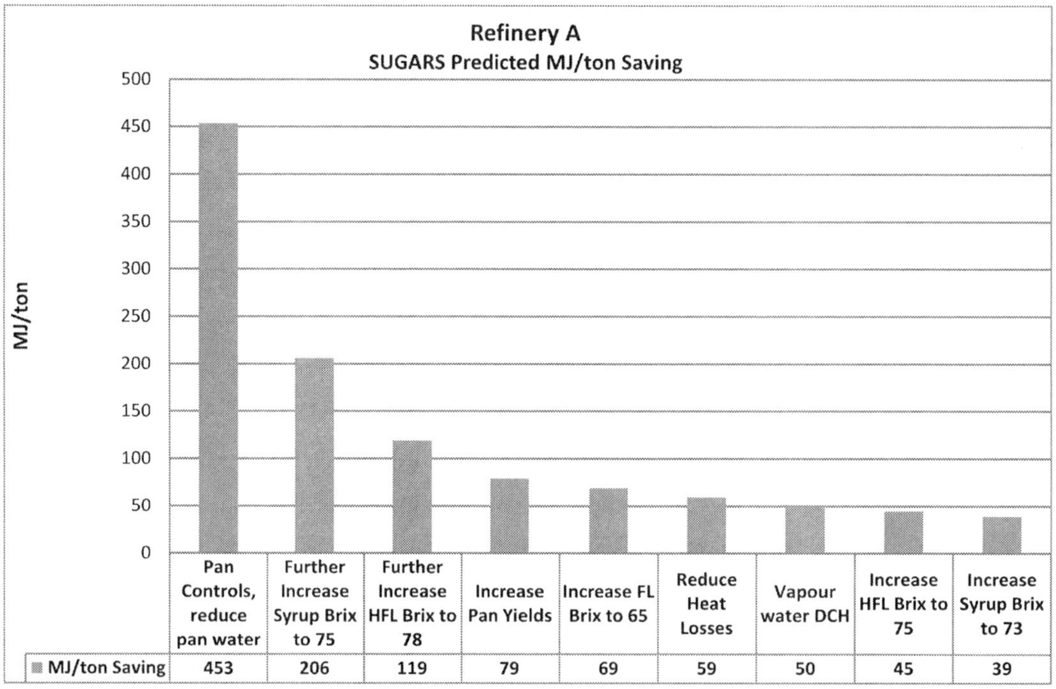

Refinery B

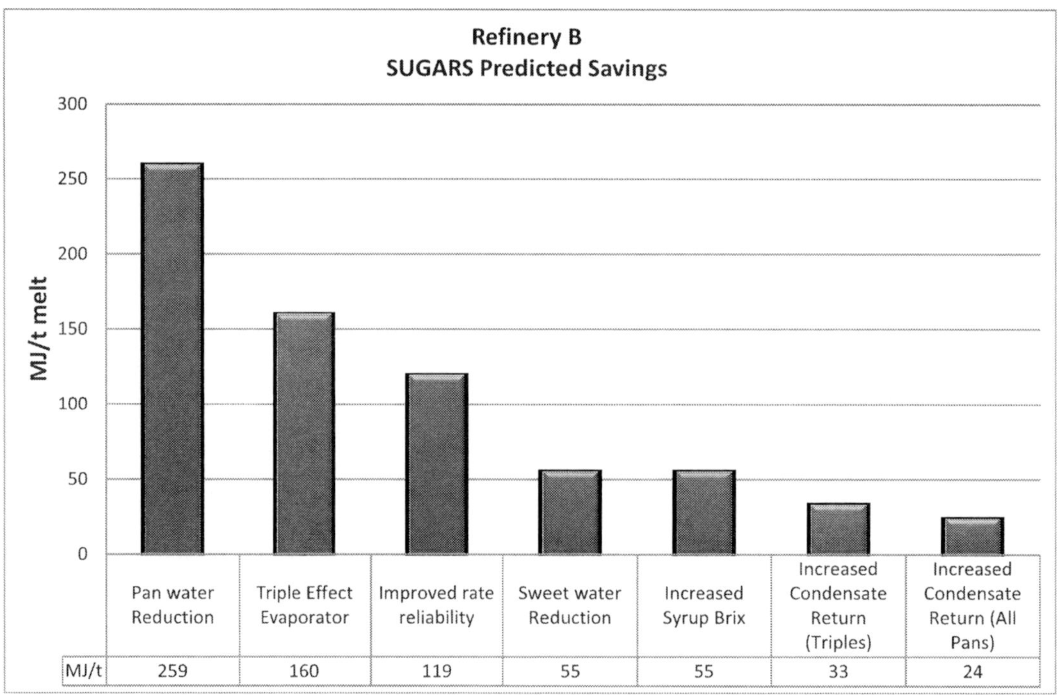

Refinery C

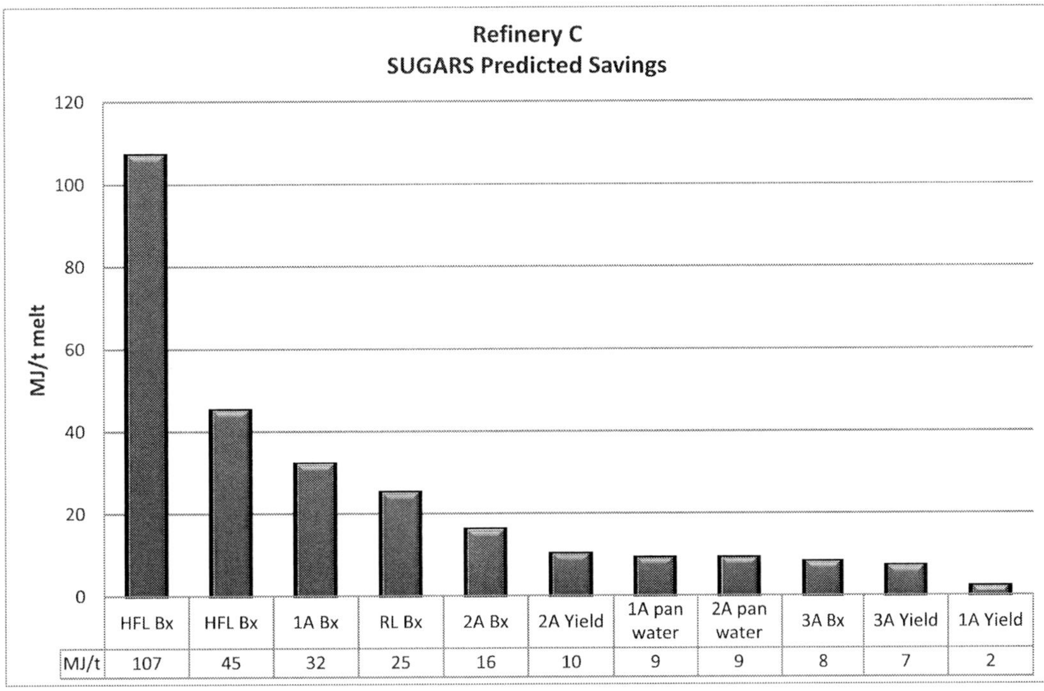

A plan must have dates!

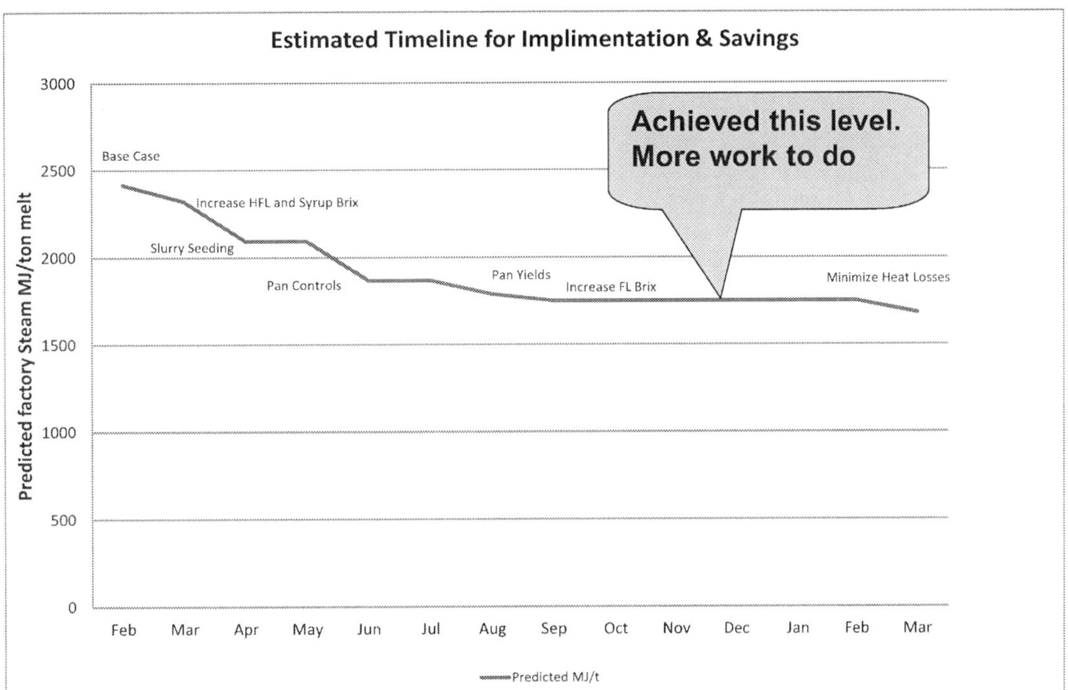

Lessons Learned

- Every refineries was best at something
- Even the refinery with the lowest steam improved
- Benchmarking provided the motivation
- Modelling helped understanding and justified projects
- Teamwork shared knowledge – fun working together
- Using actual results as targets helped people believe it was possible to achieve or do better than target
- Not every project worked as expected, as situations changed – new products introduced or different quality raws being melted. But overall targets achieved as other projects were developed

wilmar

This exercise proved the benefit of benchmarking and learning from the refineries with the best results.

We see the best refineries have also improved, so we need to keep learning and improving.

Thank You

wilmar

PASSION FOR PROGRESS

© BMA

BMA

S.I.T. Paper #1160

S.I.T. Conference 2017

Panel discussion "Energy Economy"

Use of energy-friendly technology and process equipment

Uwe Schwanke – BMA

PASSION FOR PROGRESS

© BMA

Panel discussion
Energy-friendly technology and process equipment

BMA

- Steam demand
- Automation
- Continuous crystallisation
- Boiler house capacity
- Design of new refineries

PASSION FOR PROGRESS

Steam demand for pans and heaters

BMA

- **Basic refinery concept**

  More than 1.06 t_{steam} / $t_{raw\ sugar}$

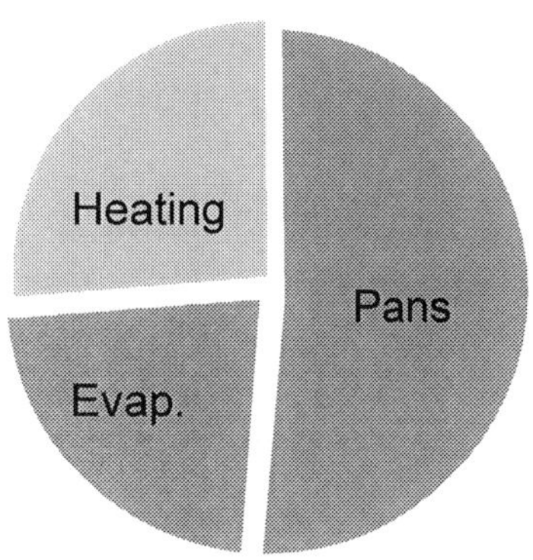

- **Advanced refinery concept with simple multiple heat use**

  Approx. 0.86 t_{steam} / $t_{raw\ sugar}$

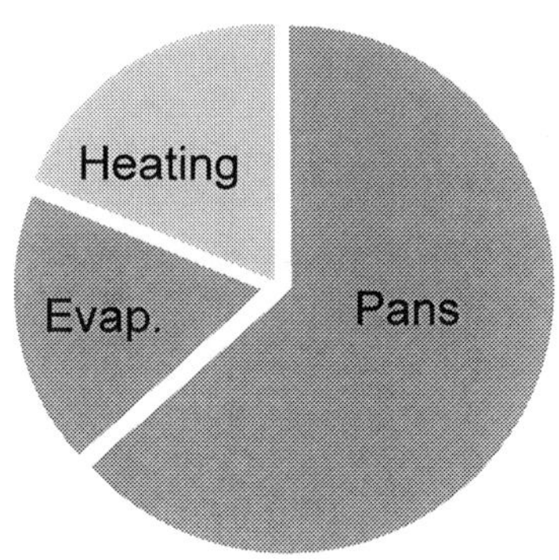

PASSION FOR PROGRESS

Advanced energy saving installation with continuous crystallisation

BMA

- Approx. 0.66 ton steam per ton raw sugar

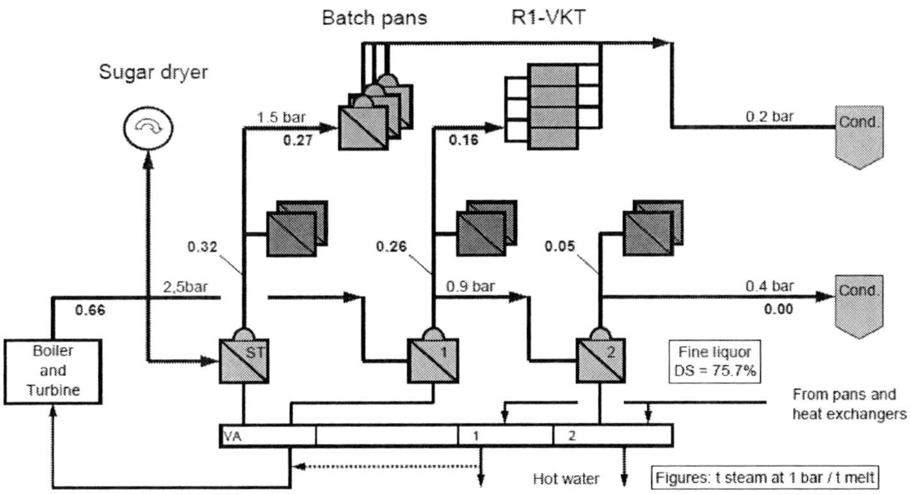

PASSION FOR PROGRESS © BMA

Controlled crystallisation with automation

BMA

PASSION FOR PROGRESS © BMA

Example improved automation with batch pans

- Development of batch pan crystal content over 3 months after installation of the new Control System and average vs. previous year

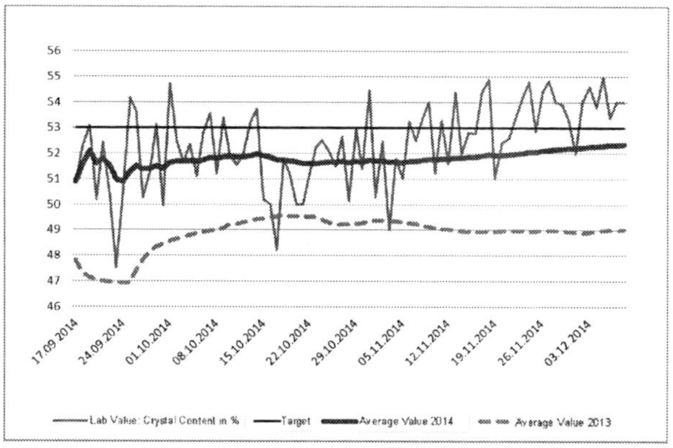

Crystal Content was increased by approx. 4%

-resulting in increased energy efficiency,
-higher capacity utilization
-and improved sugar yields

Boiler house capacity

- … a restricting factor

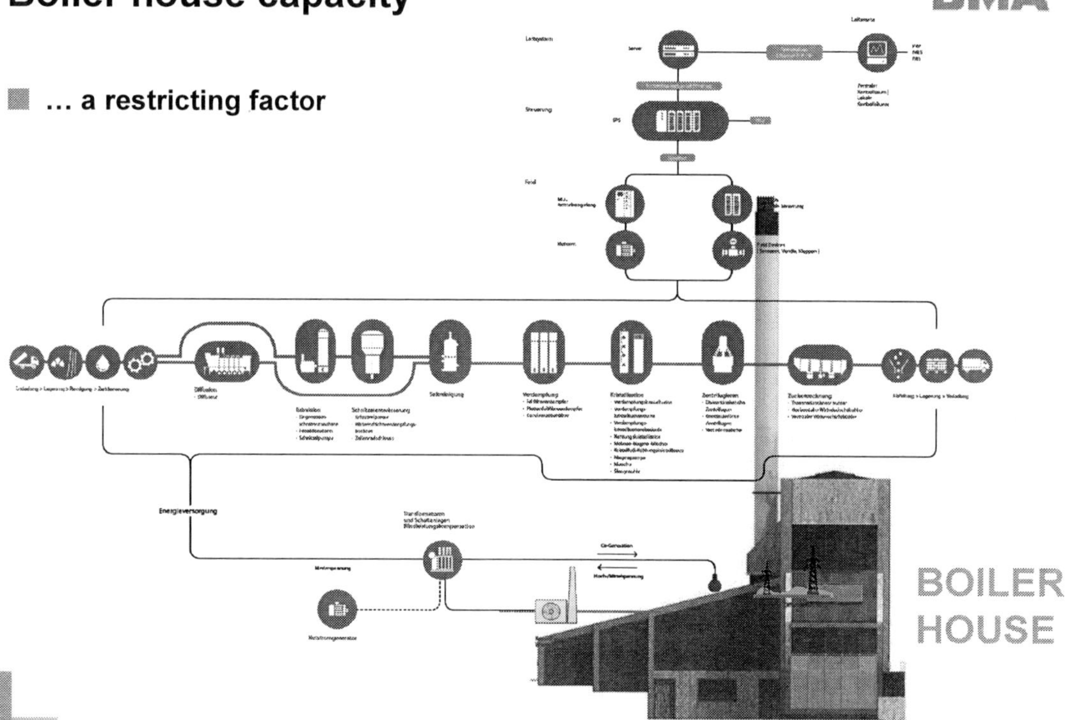

PASSION FOR PROGRESS

© BMA

Source: e-architect

Design of new refineries: Reliability and use of energy-friendly technology

BMA

PASSION FOR PROGRESS

© BMA

Summary Energy-friendly technology and process equipment

BMA

- Even simple multiple heat use reduces steam demand
- Improved crystallisation and reduced energy consumption by state-of-the-art automation
- Continuous crystallisation enables more steam savings
- The restriction of boiler house capacity can be the key for upgrading refinery capacity along with improving energy efficiency
- The design of new refineries must incorporate reliable processes and energy-friendly technology

Ref.: Hempelmann, R.; Lehnberger, A.: Design of sugar refineries - practical experience from executed projects. SIT Conference 2009, New Orleans, #970; Al Ghurair, J.; Singh, G. C.: Al Khaleej Sugar – An Energy Efficient Refinery. S.I.T. Paper 751, 1999, #751; Hempelmann, R.: Continuous crystallization of refined sugar – Present situation and concepts for the future. SIT Meeting 2005, Dubai / VAE, #877

PASSION FOR PROGRESS

© BMA

Thank you for your attention!

PASSION FOR PROGRESS

S.I.T. Paper #1161

CARBO SOLUTIONS INTERNATIONAL

HIGH TECH SUGAR SOLUTIONS

CARBONATATION ENHANCEMENT WITH HIGH PERFORMANCE DECOLORANT AND ADSORBANTS.

Sarir, E.M., Pabon, B.R.; Omar, G.
Carbo Solutions International
sarire@carbo-solutions.com

ABSTRACT

Carbonatation is a primary decolorization process used in both cane and beet sugar refineries and has been used in cane sugar for well over a hundred years. Carbonatation is a simple, low cost and robust process for color and impurities removal. The process requires the adding of Ca (OH) 2 to melt sugar liquor, and then CO_2 is bubbled into solution under controlled pH, temperature and reaction time. Impurities and color are absorbed and enmeshed in formed calcium carbonate that is filter removed. Color removal between 50 and 60% can be expected after filtration.

Despite its robustness, the carbonation process can be affected by very high color sugars and / or the presence of polysaccharides as dextran or starch on melt. This affects the efficiency of subsequent stages in the refinery and increases sucrose losses and production costs.

CSI has designed technology solutions using process aids to mitigate the effect of these undesirable conditions

The use of patent pending high performance adsorbents FRS series from CarboUA in many sugar refineries achieved enhanced daily production rates, higher process efficiency and filterability improvement of syrups. Enhancements in carbonatation color removal up 30% is also possible by use of MAJ series high performance decolorants, these are proprietary products which provide additional color removal to carbonation

Use of this process aids enhance performance of secondary color removal process and even with use of HPA absorbents to by-pass unit operations such as granular activated carbon or ion-exchange resin are possible. The benefits are based on improved final liquor colors, improved daily throughput of sugar solids, improved sugar recovery (t raw sugar/t refined sugar), reduction for syrups recycled in the white sugarhouse and reduction in energy consumption

Keywords: Carbonatation enhancement, adsorbants, decolorants, color removal technologies, sugar refining

INTRODUCTION

Carbonatation is one of the oldest processes in sugar technology for the refining of raw sugar because it is an inexpensive and reliable process that can be adapted to many variable conditions. The process consists of adding a slurry of calcium hydroxide in water to the raw melt solution. Then carbon dioxide gas is bubbled into the sugar solution in saturators, under controlled conditions of pH and temperature. The process is highly competitive when compared to other discoloration technologies such as phosphatation and sulphitation.

Generally, CO_2 is added to the saturators in two stages, with the major part of the gassing carried out in the first saturator. The impurities are both absorbed by, and enmeshed in, the conglomerated particles of the calcium carbonate precipitated in the reaction.

The carbonatation has proven to be a very effective process for the removal of color, silica, turbidity, polysaccharides and many types of inorganic ashes. The results of laboratory tests done by Moodley et al. (2002) have shown that carbonatation is very effective in the removal of the following impurities:

- About 40 to 50 % of the color was removed.
- Most of the turbidity (95%), starch (93%), sulphates (86%) and phosphates (100%) were removed.
- A small amount (29%) of the gums was removed.
- A high proportion (67%) of magnesium was removed.

Color removal performance using the carbonatation process can depends on initial melt color, lime dosage, reaction time and final exit pH prior to precipitate separation. The mechanism for color removal is via adsorption and absorption into the crystal (Bennett, 1974)

In sugar refining, the two key objectives of carbonatation are :

(i) To clarify raw sugar melt liquor and
(ii) To produce calcium-carbonated liquor that has good filterability characteristics (Murray and Runggas, 1975).

PROBLEMS IN CARBONATATION

Despite its robustness, carbonation does not always achieve the same expected results on impurities removal. Some of the factors affecting carbonatation performance are pH control, temperature control, and quality of lime, CO_2 supply, Impurities load and type on melt. For this reason a wide variation can be observed in the removal of color throughout the harvest season or when comparing the results in different countries

Raw sugars with a high content of polysaccharides (dextran, starch) affect the performance of carbonation, resulting in short filtration cycles and a higher operating cost. As a solution, refineries decrease the brix of melt but this decreases the production capacity and increases the consumption of steam.

The effects of different impurities on precipitation has been widely studied. (F. H. C. Kelly, Mak Fong 1972). Ionic impurities such as sodium sulphate, sodium hexa metaphosphate, sodium chloride and ammonium ions were found to inhibit to a varying extent the precipitation of calcium carbonate in an aqueous resulting in fewer but larger calcium carbonate crystals with distorted shapes.

The ease of separation of impurity from carbonated liquor in refinery practice is determinated by the filterability of the precipitated $CaCO_3$.

CSI PROCESS AIDS FOR CARBONATATION

CSI has designed technology solutions using process aids to solve or mitigate the effect of these undesirable conditions.

The use of patented high performance adsorbents and color precipitants from CarboUA in sugar refineries achieved enhanced daily production rates, higher process efficiency and filterability improvement of syrups. Improvements in carbonation can be achieved by using one of the following alternatives:

- Addition of high performance decolorant MAJ, ST series in carbonatation stage
- Addition of High performance adsorbant (FRM, RFB series) after last saturator
- Addition of High performance adsorbant (FRM, RFB series) after primary filtration
- The 2-stage addition of the adsorbants and colorants also allow to enhance or even to by-pass further color removal unit operations such as granular activated carbon or ion exchange resins at very low cost.

The benefits are based on improved final liquor colors, improved daily throughput of sugar solids, improved sugar recovery (t raw sugar/t refined sugar), reduction for syrups recycled in the white sugarhouse and reduction in energy consumption. HPA FRS- series is extremely efficient in removal of impurities as starch, dextran, gums, and polysaccharides, among others that causes poor filterability. The removal of these impurities allows an increase in operation Brix and/or flowrate through filtration/I.E.R./G.A.C achieving improved total process efficiency

HIGH PERFORMANCE ADSORBANTS (HPA) FOR CARBONATATION

High Performance adsorbents (HPA) have evolved from powdered activated carbons (PAC). Although HPAs are similar to PAC, the major difference between them is the ability of HPA to remove color with a greater degree of potency thereby significantly reducing the quantity of HPA required. HPA also requires less body feed due to its superior filterability. Furthermore, HPA offers great flexibility as its addition can be turned up or down based on incoming color intensity.

HPAs are similar to PAC; adsorption is the main mechanism of color removal for both PAC and HPA. Adsorption is a surface phenomenon in which an adsorbate is held onto the surface of the activated carbon by Van der Waal's forces and an equilibrium point (Lima et al. 2016) represents saturation. HPA, however, is an improvement on PACs, from both structure of raw material and its processing effectiveness. While regular carbons have surface area of 800 – 1,200 m2/g, a HPA is typically 1,200 – 2,000 m2/g (Lima et al. 2016). Additionally, the pore structure is especially designed to have a better configuration (ratio between "micro" and "meso" pores) to allow more efficient adsorption in sugar solutions. HPA balances very high surface area (1,300-2,000 m2/g) with coarse particle (average sizes can be up to 275 micron) to promote filtration and high adsorption rates (Lima et al., 2016).

Results of HPA properties measured at USDA are summarized in Table 1.

TABLE 1. HPA PHYSICAL PROPERTIES

HPA	Surface area m2/gr	Micropore volume cc/g	Micropore area m2/g	Total negative surface charge meq H+/g	Medium particle size um	Mean particle size um
RFB-EC	1294 +/- 34	0.38+/- 0.01	852 +/- 5	0.93+/-0.06	28.2 +/- 1.7	54.4 +/- 16.3
RFB-SE	1190 +/- 17	0.36+/- 0.02	792 +/- 46	0.68+/-0.31	60.0 +/- 13.0	97.0 +/- 12
RFM-JH	1764 +/- 80	0.52+/- 0.03	1111 +/- 56	0.94+/-0.07	78.3 +/- 8.6	275.1 +/- 24
FRS-3	666 +/- 59	0.18+/- 0.02	413 +/- 43	1.29+/-0.04	59.8 +/- 7.9	209.7 +/- 71
FRS-3A	857 +/- 11	0.25+/- 0.01	529 +/- 20	1.05+/-0.07	37.6 +/- 2.8	127.6 +/- 18.8

Average of duplicates with standard deviation shown

The treatment with high performance adsorbants improves the filterability due to the difference of particle size between the carbon and the calcium carbonate, which produces a filter cake with greater porosity. The adsorbant has the ability to remove polysaccharides, especially insoluble starch

Test performed at USDA (Egglestone 2016) reported 20-70% removal on insoluble starch and 10-20% on total starch content for sugar liquor treated with 292 ppm RFM series adsorbants. Results are summarized on table 2

TABLE 2 STARCH REMOVAL ON REFINERY STREAMS USING CARBON ADSORBANTS

Parameter	Before CarboUA HPA	After CarboUA HPA	% Change
Raw Sugar syrup 1			
Color (ICU)	1218	985	19.1
Insoluble Starch (ppm/brix)	121.7	98.2	19.3
Total Starch (ppm/brix)	130.4	117	10.3
Raw sugar syrup 2			
Color (ICU)	1163	1152	0.9
Insoluble Starch (ppm/brix)	1015.6	232.3	77.1
Total Starch (ppm/brix)	1431.4	1200.5	16.1

HPA dosage: 292 ppm, T= 85C, Time 60 min.
Average of duplicate results

Basic principle for HPA use as carbonatation improvement is related to enhancing the filterability of raw liquors in sugar refineries. Studies carried out in refineries in the Middle East showed that the use of HPA improves the filtration factors and flow rate in carbonation process compared to the traditional process (Sarir ISJ 2008). Figure 1

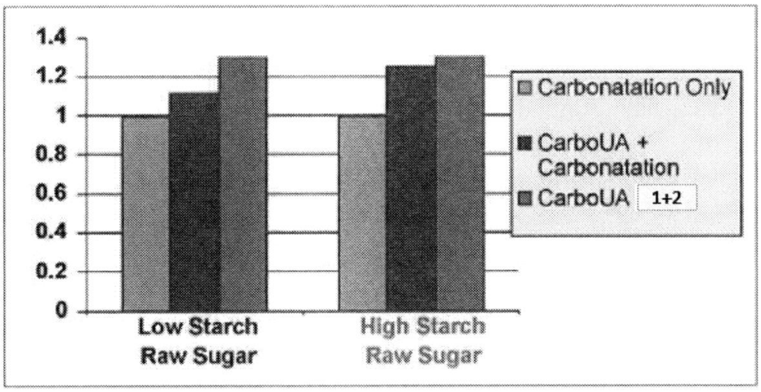

Comparison of filtration factors and flow rates in lab scale

Process Method	Estimated Filtration Factor (F)	Observed flow rate (relative to carbonatation only)
Carbonatation Only	~7	1.0
Process Aid #1 only	~5,000	3.6
Combination of Process Aid #1 and #2	~1,250	5.7

FIGURE 1. INCREASING AVERAGE FLOW RATE THROUGH CARBONATATION FILTERS WITH TWO CARBO UA PROCESS AIDS

Below are case studies showing the improvements obtained in refineries of carbonation with process aids and laboratory / factory tests that show the high potential of the HPA technology to solve the problems of the carbonation process when operating with low quality sugar and / or when high degree of impurities removal is required.

The presented data correspond to the average duplicate samples obtained during testing period using refineries control & sampling plan. The analytical and sampling frequencies were the same before and during the test. Color, pH, brix analyzes were carried out following the official method ICUMSAGS 1/3-7, GS 2/3-10 and GS 2/3-10

Case Study1: Refinery 900 MT/day Carbonatation + IEX Philippinnes

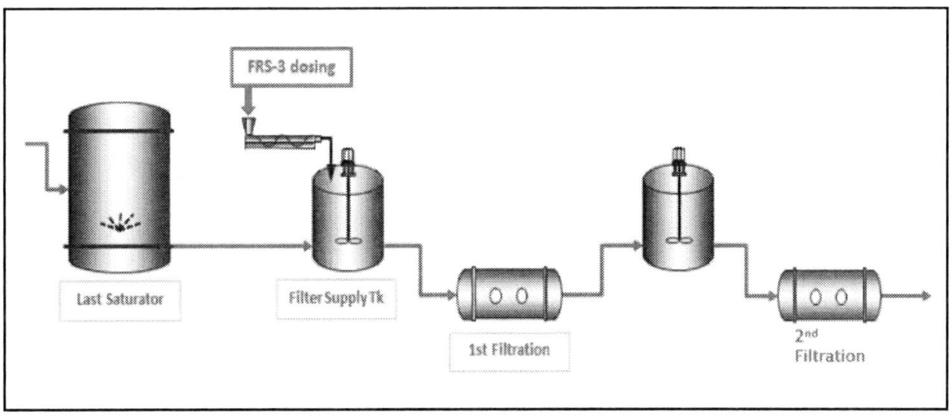

FIG 2. HPA FOR CARBONATATION ENHANCEMENT PROCESS DIAGRAM

The purpose of the test was to increase the brix of the melt liquor without affecting the filtering operation and to achieve an additional removal of color

Carbo UA FRS series is added in a special designed reaction tank just after last saturator where it mops up impurities that were not occluded in the calcium carbonate crystal. Dosage is in range 500 – 1000 ppm basis melted sugar. This improvement in color removal and filterability has allowed the refinery to operate at higher capacity due to higher brix profile, with a higher volumetric flow rate, lower steam consumption and reducing recycle. Results are summarized on table 3

Table 3 HPA FRS-Series factory results in Philippines

Process Parameter	Traditional Process	FRS3	% Variation
Melt sugar (MT/day)	**779**	**930.9**	**19.5**
Total sugar produced (MT/day)	**755.8**	**903**	**19.47**
Carbonated liquor IU	**945**	**780**	**17.5**
Carbonated liquor brix	**57.67**	**58.95**	**1.28 units**
Fine liquor color	**228**	**185**	**- 19.0**
Fine liquor brix	**57.87**	**59.11**	**1.24 units**
Total massecuite strikes	**27**	**38**	**26.3**
Bunker fuel (Liter/LKg refined sugar)	**4.07**	**1.43**	**- 64.9**
R1 sugar color IU	**37.81**	**36.8**	**-2.7**
Sweet water recycled (IU)	**2186**	**1649**	**24.5**
Capacity utilization %	**84.22**	**99**	**17.5%**
HPA dosage	**-**	**750**	

Case Study 2: 1500 MT/day sugar refinery Carbonatation + IEX Thailand

The purpose of the test was to evaluate the color removal potential of HPA technology in carbonation refineries.

Carbonated liquor was treated with 1000-ppm RFS series HPA before primary filtration, product was added in stirred reaction tank, and 80C, contact time 20 minutes, and after that filtered. All the HPAs tested reduced the color of the clear liquor (blank), particularly HPAs RFB-SE_{TM} and RFB-EC_{TM} (Fig. 1). More results are listed in Table 4. On each day of the refinery trial, color was markedly reduced (64.3-75.9%). There was also a marked reduction in turbidity (55.6-89.6%) when measured with either a spectrophotometer (I.U. values) or nephelometer (NTU values) (Table 4) Unwanted insoluble matter was also strongly removed by the HPA treatments.

Table 4. The effect of using HPAs at 1500 MT/day carbonatation refinery a sugar refinery in Thailand.

Parameters	Test 1		Test 2		Test 3	
	Carbonated Liquor	Treated with 500 PPM RFB-SE	Carbonated Liquor	Treated with 750 ppm RFB-EC	Carbonated Liquor	Treated with 750 ppm RFB-SE
Color (I.U.)	743	253	621	222	697	168
Turbidity (I.U.)	979	469	604	255	741	104
Turbidity (N.T.U.)	54	24	54	19	48	5
Insoluble matter (ppm)	155	43	86	55	106	2
Brix	63.7	60.3	63.4	61.6	63.3	62.4
pH	7.3	7.6	7.3	7.5	7.3	7.7
Color Removal	65.9%		64.3%		75.9%	
Turbidity Removal (N.T.U.)	55.6%		64.8%		89.6%	
pH variation	-4.1%		-2.7%		-5.5%	

FIGURE 3 HPA TEST AT THAILAND REFINERY

HIGH PERFORMANCE LIQUID DECOLORANTS (HPD) IN CARBONATATION

Carbo Solutions International Company develops High Performance Decolorants (HPDs). They are engineered process aids that can be added to carbonatation / phosphatation processes to remove color. HPD are typically added to carbonated liquor tank for additional color removal in filtration. High active chemical radicals on HPD trap colorants and these are then eliminated during filtration together with calcium carbonate. HPD can enhance up 30% color removal than traditional carbonatation. HPD are food grade liquid products with high advantages in safety handling.

Case study 3: Liquid decolorant test at 1000 MT/day refinery in Thailand (Carbonatation + IEX)

The liquid decolorant MAJ - Series was evaluated during 2 days in 1000 MT carbonatation + IEX refinery. Product was added in doses of 300-400 ppm after the last saturator, test objective was to evaluate the additional color removal reached in the carbonation process before treating the Liquor with ion exchange resins. Results obtained are summarized in table 5.

Table 5. The effect of using HPDs in carbonatation refinery in Thailand.

DATE & TIME	CARBONATED PH	CARBONATED COLOR	FILTERED PH (WITH CUA)	FILTERED COLOR WITH CUA	% COLOR REMOVAL	REMARKS
4 MARCH						**TEST START MAJ-1 300 PPM**
9 AM	8.31	1354	8.29	1182	13	
10 AM	8.92	1320	9.49	1218	8	
11 AM	9.57	1447	9.05	1126	22	
12 PM	8.28	1315	7.94	878	33	**MAJ-1 DOSE INCREASE TO 400 PPM**
2 PM	7.7	1134	7.77	836	26	
4 PM	7.8	1052	7.89	813	23	
5 PM	7.88	1195	7.87	908	24	
6 PM	7.78	1212	7.8	1007	17	
7 PM	7.56	1256	7.69	856	32	
8 PM	7.68	1330	7.81	914	31	
10 PM	8.5	1131	8.14	812	28	
11 PM	8.47	1063	8.96	863	19	
5TH MARCH						
8 AM			8.57	630		
9 AM	8.18	833	8.17	586	30	
10 AM	7.63	758	7.76	495	35	
11 PM	7.61	806	7.7	435	46	**TEST END**

The liquid decolorant MAJ- series achieved good results with color removal mostly more than 30%, which assisted in reducing the color of carbonated liquor input to IER resin and consequently the color of R1 massecuite and R1/R2 sugar as well.

It was clear during the plant test that MAJ- series has proven its efficiency in reducing the color of fine liquor when the color of raw liquor is too much high and secured the process

Case study 4: Liquid decolorant test at 800 MT/day refinery in Lebanon (Carbonatation + PAC)

The liquid decolorant ST- series was evaluated in doses of 250-300 ppm basis sugar after the last saturator in a Lebanon refinery with a capacity of 850 tons per day, in order to evaluate the additional color removal reached in the carbonation process before treating with powder activated carbon. Results obtained are summarized in table 6.

Table 6. The effect of using HPDs at a sugar refinery in Lebanon

Color of raw liquor	Color of limed filtrated liquor	% Color removal	Color after filters using HPD	Turbidity (IU)	Remarks
1125	372	66	183	11	Start of test
899	412	54	188	12	
1003	325	67.5	193	9	
1003	372	63	195	16	
968	360	63	141	11	85% total color removal
1020	371	63.5	148	10	85% total color removal
1085	393	64	183	13	Test Finish
1104	418	62	208	20	Traditional Process

The liquid decolorant ST- series achieved very good results with color removal mostly more than 50% and some peak values up 85%, which assisted in reducing the color of carbonated liquor input to PAC process.

CSI TWO STAGES PROCESS

CSI has successfully developed a two-stage application that is very simple with low Capex that aims to:

Improve the color removal capacity at very low Operational cost for those refineries handling very high color sugar

Replace IER or GAC process for those refineries having clarified liquor color typically below 700 IU

Alternative process with low to no capital investment for new refineries being designed.

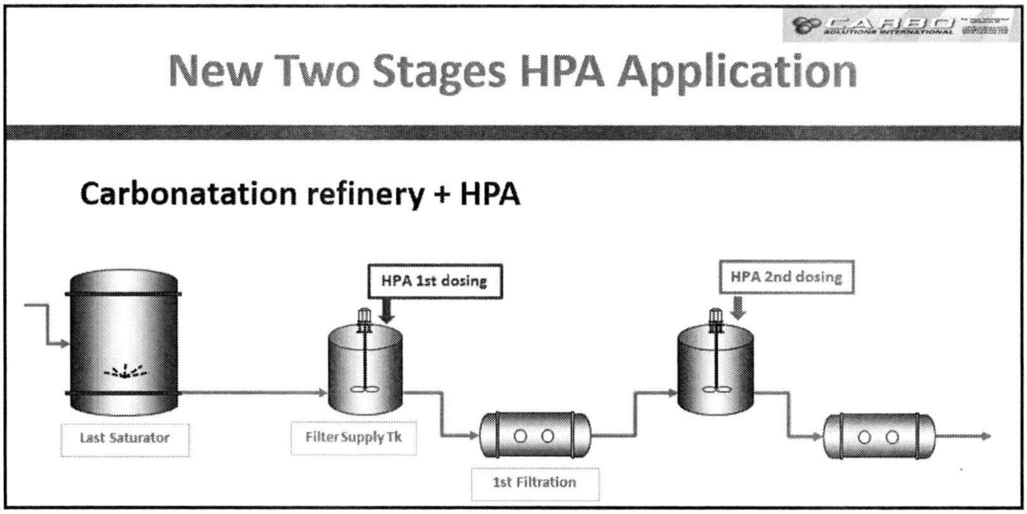

FIGURE 4 HPA TWO STAGE SYSTEM

First dosing is a special decolorants at 250-300 ppm sugar basis followed by a carbon based HPA at 750 - 1500 ppm batch and precoat decolorization.
In some applications where a high removal of color or ash is required, precoat adsorption is often used.
This involves placing a second adsorbent precoat layer built up over the first filter aid precoat.

Adsorbant Precoating is accomplished by circulating a slurry of HPA adsorbant and filtered or clear liquid between the filter and the precoat tank immediately after having made the filter aid precoat.

Pilot plant and factory test performed in Thailand has demonstrated superior capacity of two stages process to fir color removal up 80% and open opportunities for a novel and environmental friendly process to complement or replace traditional color removal technologies.

Results for factory test using two stages process in Thailand 1500 MT per day carbonatation refinery are summarized in Table 7:

Table 7: Two stages process factory test at Thailand refinery 2016

TREATMENT	BRIX	PH	COLOR	% COLOR REMOVAL
Carbonated liquor (blank)	61.2	8.43	850	
MAJ-1 350 ppm	60.9	8.16	543	36.11
RFB-SE 1000 ppm	60.7	7.7	186	78.11
RFB-SE 1250 ppm	60.43	7.55	170	80.00
RFB-SE 1500 ppm	60.7	7.22	146	82.8
RFB-SE 2000 ppm	60.22	7.13	109	87.17

Average of duplicate results

Case study 5: Liquid decolorant & HPA adsorption at 1000 MT/day carbonatation refinery in Indonesia

Sugar refineries in Indonesia face serious problems because of the high quality variation of raw sugar. The results obtained are presented for tests of different doses of adsorbent and decolorizing products in a refinery at Indonesia in 2016.

Different treatment for 2 stages process were performed under lab conditions, results are summarized on Table 8.

TABLE 8 HPA TWO STAGES PROCESS LAB TESTS AT INDONESIA

	Brix	pH	Color	% Color Removal
Carbonated Liquor (Blank)	61.3	8.25	819	
a)Carbonated liquor + MAJ-1 (300 PPM)	61.4	8.19	429	47.7
a) Carbonated liquor + MAJ 1 (300 ppm) filtered with coated filter aid & HPA (0.5 kg/m2)	60.1	7.82	153	81.3 (Total)
b) Carbonated liquor + Maj-1 300 ppm + RFB-SE 200 ppm	60.2	8.16	395.4	51.77
b) Carbonated liquor + Maj-1 300 ppm + RFB-SE 200 ppm filtered with coated filter aid & HPA (0.5 kg/m2)	61.8	7.53	139	83.02 (Total)
c) Carbonated liquor + Maj-1 300 ppm + RFB-SE 500 ppm	61.1	8.03	368.8	55.02
c) Carbonated liquor + Maj-1 300 ppm + RFB-SE 500 ppm filtered with coated filter aid & HPA (0.5 kg/m2)	60.9	7.48	112	86.32 (Total)
d)) Carbonated liquor + Maj-1 300 ppm + RFB-SE 750 ppm	60.9	7.98	322	60.71
d) Carbonated liquor + Maj-1 300 ppm + RFB-SE 500 ppm filtered with coated filter aid & HPA (0.5 kg/m2)	60.8	7.41	102	87.5 (Total)

Results showed high color removal when two stages process are used. In addition to its high power of color removal, HPA technology has the advantage of not generating polluting effluents or high consumption of water and steam in the refining process

CONCLUSION

HPA technology developed by Carbo Solutions International Company has exceptionally high color removal and filterability capacity when it is used in sugar refineries supporting and complementing either carbonatation processes at very low capital investment and operational cost.

The use of HPD allows additional color removals in the carbonation up to 30%; this allows producing fine liquors of a lower color, decrease of operating costs in the stages of secondary discoloration and greater productivity.

Two stages process emerges as an innovative technology for the replacement of traditional color removal processes in sugar refineries, offering low environmental impact, high operating flexibility, low operating cost and very low capital investment

REFERENCES

1. Bennett, M. C. (1974). Physical chemistry of phosphatation and Carbonatation (Parts I, 11) Int Sug J 76: 4044; ibid. 76: 68-73.
2. Egglestone Gillian, Mechanism of removal of undesirable residual amylase insoluble starch from refinery streams by powdered activated carbon, Agribussiness Intelligence, Iforma May 2016
3. F. H.C.Kelly, Mak Fong Keng and Chua Gek Sim, Chemistry Department, Singapore University, Bukit Timah Road, Singapore 1972, Study of grow mechanism of calcium carbonate at carbonatation process for sugar refining
4. James H. Bushong, Carlos Andrés Donado González, Washington Odipo, Julio Mauricio Pineda Marroquín, André Luis Massucato, Emmanuel M. Sarir. Improving daily production capacity and energy efficiency in sugar refineries and sugar mills with CarboUA high performance process aids, 2008
5. Lima I; Eggleston G; Sarir E; Donado CE; Thompson J; St. Cyr E. 2016. How powdered activated carbons can be used to remove undesirable residual amylase, insoluble starch, and select colorants from refinery streams. International Sugar J., 118, 353-362.
6. Moodley, M (2001). Laboratory optimization of the carbonatation process with regard to impurity removal at the Malelane refinery. M-Tech Thesis, ML Sultan Technikon, KwaZulu-Natal.
7. Murray JP and Runggas FM (1975). A study of factors affecting refinery carbonatation. Proc S Afr Sug Technol Ass 49: 90-93.

S.I.T. Paper #1162

Results from the Commercial Operation of a Fractal Shallow-Bed Ion Exchange Decolorizer

S.D. Peacock[1], O. Tzschaetzsch[2], B-C. Schulze[2] & V. Kochergin[1]

[1]Amalgamated Research LLC, 2531 Orchard Drive East, Twin Falls, Idaho 83301, USA

[2]ESCON GmbH, Schlossstrasse 48a, 12165 Berlin, Germany

Abstract

The use of relatively shallow resin beds, in conjunction with efficient fractal fluid distribution, has been shown to reduce both the capital and operating costs of ion exchange systems. The use of highly efficient fractal distributors reduces the amount of waste regenerant produced and minimizes process dilution, which are critical factors in a successful sugar refining application. Regenerant use can further be optimized by means of partial recycle and the membrane filtration of regenerant waste (either using nanofiltration or high pressure reverse osmosis) for brine recovery. The shallow resin depth allows for the use of more kinetically-efficient resins of a smaller size, without resulting in an excessive pressure drop across the bed. Although employed in the beet industry in a number of instances, the first fractal shallow bed system was only installed in a refinery decolorization application in 2008. The principles employed in the design of a more recent system in Malaysia, using an acrylic resin, are reviewed and the results of some data collected during the first two years of commercial operation are presented. Even at higher flow rates and color loadings than originally anticipated, a sustained color removal of more than 75% was achieved, yielding an average fine liquor color of less than 200 ICUMSA units and exceeding the design specification of 70% color removal. The performance achieved on an industrial scale demonstrates the effectiveness of the fractal shallow bed technology.

Conventional Ion Exchange Decolorization Systems

The use of strong base ion exchange resins in the sugar industry has a relatively long history, having been used for the decolorization of syrups from the early 1950's. The creation of styrene / divinylbenzene macroreticular resins in the 1960's led to an enhanced practical capacity for color removal, while the advent of acrylic resins in the 1970's offered effective decolorization and efficient regeneration using a 10% brine solution[1] (Fries and Walker, 1980).

Styrene resins are limited in the level of liquor colorants that can be treated. Beyond a certain color level, brine regeneration is incapable of completely desorbing the color bodies, leading to a loss of decolorization capacity (Getaz, 1988). Over the years, therefore, many ion exchange systems have been built using an acrylic resin for primary decolorization, followed by a styrene resin for secondary decolorization / polishing down to low color levels, with the more effectively-regenerated acrylic resin protecting the secondary styrene resin from organic fouling.

Using an acrylic resin alone for decolorization is usually more cost effective, but yields a lower color removal (Kirkiridis, 1992). Where two-stage decolorization using the same resin is practised, it is usual to cascade the resin from the secondary columns into the primary columns halfway through its life, with new resin then being placed into the secondary decolorization columns. The primary columns containing used resin thereby protect the new resin in the secondary columns from contaminants such as suspended solids (Bento, 1997; Kirkiridis, 1992; Rousset and Bhatt, 2011).

---

1 The selectivity and capacity of styrene-based resins for sugar liquor colorants is ultimately higher, due to tight bonding between the hydrophobic organic backbone of the resin and the aromatic organic groups in the sugar colorant molecules, but for this same reason the color is not as effectively removed from the resin during regeneration, leading to irreversible fouling and a shortened effective resin life.

A review of related publications over the past 20 years in the *Proceedings of the Sugar Industry Technologists* (plus a few other relevant publications) was carried out as an indication of the conventional state of the art in sugar liquor decolorization by ion exchange.

Operating Conditions

Since the earliest use of ion exchange for sugar liquor decolorization, the operating and regeneration conditions employed have not substantially changed, with typical conditions as shown in Table 1 below (Bento, 1997; Bento and Stuchi, 2014; Cheong and Mussebah, 1980; Davis, 2001; de Lataillade, 2001; Dibella, 2000; Cooper, 1976; Fries and Walker, 1980; Getaz, 1988; Getaz, 1990; Hoareau *et al.*, 1977; Inkson and Korea, 2015; Kirkiridis, 1992; Rousset and Bhatt, 2011; Theoleyre *et al.*, 2011; Vawda *et al.*, 2014).

Table 1. Typical operating conditions for sugar liquor decolorization using ion exchange.

Liquor flow rate[2]	1 – 4.5 BV/h most usually 2 – 3 BV/h
Throughput before regeneration	20 – 100 BV/cycle usually 30 BV/cycle (or 23,000 – 30,000 t DS × color IU/m^3 resin)
Regenerant flow rate	1 – 4 BV/h
Expected resin lifetime	200 – 500 cycles

[2] This flow rate, in resin bed volumes per hour (BV/h), is per resin stage volume (primary / secondary). In fact, although flow rates expressed in terms of BV/h are widespread in ion exchange, the specification of a superficial liquor velocity through the bed is more appropriate for design purposes, as flow rates in terms of BV/h are not really comparable for systems in which different bed depths are employed.

Color removals of around 90% are achievable, although color removals between 60% and 80% are more typical. Slower flow rates and longer cycle lengths are required when treating liquors with higher incoming colors (Dibella, 2000; Fries and Walker, 1980) and higher color removals are thus usually only achieved at the lower end of the liquor flow rate range in Table 1. Slightly better color removals are expected when pairing ion exchange as a secondary decolorization technique with carbonatation than with phosphatation (Vawda *et al.*, 2014), as phosphatation and ion exchange tend to target the elimination of the same types of colorant molecules.

Typical resin column depths are of the order of 1.22 – 1.83 m (Fries and Walker, 1980; Getaz, 1988). Using conventional technology, bed depths shallower than this are more prone to color leakage and early breakthrough (as discussed under *Mass Transfer Limitations* and *Liquid Distribution* below), while bed depths greater than this typically result in excessive pressure drop (Getaz, 1988).

The ionic form of the resin during operation has been found to be an important factor in decolorization performance, with resin in the chloride form yielding the best results. Regeneration is thus best carried out using a sodium chloride brine. The quantity of sodium chloride required for effective regeneration ranges from 150 – 300 kg of NaCl per m^3 of resin per cycle, in the form of a 10% brine solution (Bento, 1997; de Lataillade, 2001; Srivastava *et al.*, 2009; Theoleyre *et al.*, 2011). However, most plants consume 225 – 240 kg of NaCl per m^3 of resin per cycle (Cooper, 1976; Dibella, 2000; Fries and Walker, 1980; Getaz, 1990).

Alkaline solutions have been found to be most effective at removing color bodies adsorbed onto the polymer structure of the resin, and the use of a small quantity of sodium hydroxide in combination with the brine solution is more effective than the use of a single regenerant (Getaz, 1988). Adjusting the pH of the brine to 12 is recommended by Cooper (1976) and Davis (2001), while other authors report or recommend the use of a relatively wide range of sodium hydroxide concentrations, as shown in Table 2 below.

Table 2. Sodium hydroxide concentrations in regenerant brine.

	NaOH Concentration *(%)*	**NaOH Consumption** *(kg of NaOH per m^3 of resin per cycle)*
Getaz (1990)	0.1%	9
Bento (1997)	up to 0.7% (usually 0.2%)	–
Dibella (2000)	0.8%	20
Srivastava *et al.* (2009)	2.0%	–
Theoleyre *et al.* (2011)	0.5% – 1.0%	–
Vawda *et al.* (2014)[3]	0.5% 1% – 2%	–

Vawda *et al.* (2014) only recommend the use of an alkaline brine for the regeneration of styrenic resins, with this not being considered necessary for acrylic resins.

The presence of iron (typically derived from the construction materials of the plant) and other inorganic contaminants in sugar liquors results in the formation of complexes with colorant molecules and with the active sites of the resin, leading to reduced performance over time. Periodic acid washes over the life of the plant are thus required to restore decolorization capacity. Acid treatment is usually carried out at a flow rate of 2 – 7 BV/h, using a 4% solution of hydrochloric acid, at a frequency of once every 5 – 23 cycles (Bento, 1997). In most refineries, this is carried out every 12 – 20 cycles (Getaz, 1990; Inkson and Korea, 2015; Kirkiridis, 1992; Theoleyre *et al.*, 2011). Neglecting acid washing can potentially result in a decline in resin performance three times faster than normal (Getaz, 1988).

3 Conflicting recommendations for the quantity of sodium hydroxide to be used for resin regeneration are offered in this publication, probably due to a typographical error.

Acid washing is also effective in dissolving inorganic suspended solids derived from the carbonatation process, as well as removing certain polymeric and irreversibly-adsorbed colorants from the resin structure. Dibella (2000) reports having to carry out an acid wash once every five standard regeneration cycles, using an acid concentration of approximately 5%, but this was primarily in order to eliminate a flow restriction caused by suspended solids contamination.

Regular backwashing (typically once per exhaustion cycle) is also necessary to remove accumulated suspended solids trapped within the bed, as well as any fines created by mechanical damage to the resin beads themselves, thus minimizing any unnecessary pressure drop across the bed (Getaz, 1988).

Regardless of operating conditions or treatment, the performance of ion exchange resins gradually deteriorates over time, due to progressive organic fouling that is the result of inactivation of resin functional groups by non-removable ions and the blocking of the resin pores by polymerized colorants or molecules that are too large to diffuse out of the resin structure during the regeneration cycle (Getaz, 1988).

Mass Transfer Limitations

For decades, it has been recognized that the rate of removal of color bodies from sugar liquors by ion exchange resins is controlled by mass transfer limitations associated with the diffusion of the large colorant molecules to the polymeric adsorption sites within the resin (Fries and Walker, 1980; Jensen *et al.*, 2004). This is partly associated with the high dissolved solids concentration of the liquors conventionally treated in cane sugar refining. Optimization of the resin structure has therefore taken place over time to improve the rate of colorant diffusion (for example, the change from gel resins to macroreticular resins).

A typical plot of the liquor color exiting a resin decolorization system, as a function of throughput over time, is presented in Figure 1 below. Based on classical ion exchange theory, an S-shaped breakthrough curve (as shown in the figure) would be expected. However, the actual situation is somewhat different, with a slow but steady increase in exit color over time (Cooper, 1976). This confirms mass transfer kinetics to be controlling, with thermodynamic equilibrium considerations not being of overriding importance under practical conditions.

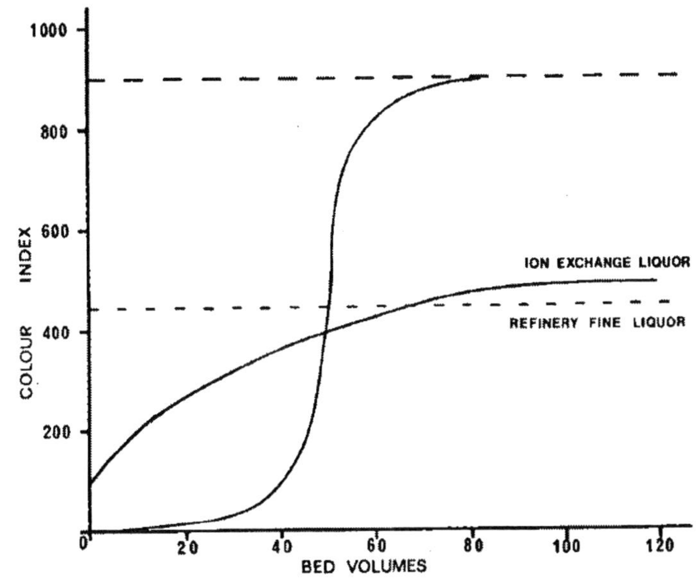

Figure 1. Typical breakthrough curves (Cooper, 1976).

The particular shape of the breakthrough curve in sugar liquor decolorization means that a certain minimum resin bed depth is needed in order to achieve a desired level of color removal. This minimum bed depth has been shown in multiple trials to be relatively shallow, of the order of 0.7 – 1.0 m (Kochergin *et al.*, 2015).

However, the non-ideal shape of the decolorization breakthrough curve is not solely due to kinetic limitations, but is also associated with inefficient liquid distribution (as discussed further under *Liquid Distribution* below). This contributes to large resin bed depths being used in conventional industrial practice, and to the use of two resin beds in series, so as to minimize the leakage of colorants into the final product stream.

A recent development is the more widespread availability of resin beads of a smaller average particle size (300 – 350 microns, rather than the more typical 600 – 800 microns), leading to enhanced diffusion and faster mass transfer characteristics[4]. Based on an analysis of the mass transfer resistances associated with diffusion in an ion exchange system, it is expected that a halving of the average particle size would yield an increase of up to four times in the rate of decolorization (Jensen *et al.*, 2004). Unfortunately, many ion exchange systems of conventional design cannot readily take advantage of resins with a smaller particle size, as decreasing the size of the beads materially increases the pressure drop through the columns (Getaz, 1988).

Liquid Distribution

An additional contributor to the non-ideal shape of the colorant breakthrough curve under industrial conditions is the lack of uniform liquor distribution across the full cross-section of the resin bed. This is usually caused by poor distributor design, sometimes exacerbated by flow restrictions in the bed due to the presence of suspended solids captured within the top layers or trapped air bubbles (Getaz, 1988). It leads to the presence of channeling and "dead spots" within the resin, resulting in premature exhaustion of some parts of the resin bed (which "leak' higher color levels) while other parts of the bed remain less completely utilized. Careful attention to uniform distribution is thus critical to optimum performance.

---

4 In the extreme case, Getaz (1988) reports a thousand-fold improvement in diffusion rate to be possible when using disposable powdered ion exchange material of smaller than 100 microns in size, compared to conventional ion exchange beads. However, Bento (1997) only reports a decolorization capacity improvement of 50 – 100 times due to the reduced mass transfer resistance with powdered resin, when compared to resin over the very wide size range from 300 – 840 microns.

The large bed depths commonly employed in ion exchange decolorization are often an attempt to compensate for poor fluid distribution, thereby increasing capital and operating costs through inefficient resin usage and regeneration (Kochergin *et al.*, 2015). Alternatively, a large liquid headspace is provided above the top resin bed surface, in the hopes that this will make the initial fluid distribution more uniform. However even this approach is not very effective, as *computational fluid dynamics* (CFD) modelling by Foutch *et al.* (2003) showed that the effects of an initial fluid maldistribution will propagate virtually unchanged through a headspace of almost 1 m in depth and into the resin bed itself. Tracer tests carried out using food grade dye have demonstrated that an initial fluid maldistribution will remain intact through resin beds with a depth of 2 m (Kochergin *et al.*, 2015). Additional resin depths are thus ineffective at improving flow distribution and simply increase pressure drop and resin inventory without yielding the desired benefits in terms of performance.

Non-ideal liquid distribution also impacts negatively on the other (non-exhaustion) portions of the operating cycle of an ion exchange decolorizer. For example, during an ideal sweetening-on operation, displacement of the water in the resin bed by liquor should occur without any mixing between the fluids. The column effluent should contain water, followed one void volume later by a step-change to liquor of the same concentration as that entering in the feed. However, this does not happen in practice, with dilute liquor typically starting to leave the column after 0.6 bed volumes, with the concentration steadily increasing over time (Getaz, 1988). This mixing between water and liquor results in an increased energy requirement for evaporation, and hence to an increased refining cost. The optimum flow rate for sweetening-on in conventional ion exchange systems is 1.25 – 2.0 BV/h, with higher flows resulting in an increase in channeling and sweet water dilution (Bento, 1997).

Dilution also occurs during the sweetening-off stage of the process, where mixing results in sweet water of a steadily decreasing concentration exiting the column towards the end of the sweetening-off period. In fact, dilution of the liquor with water is more pronounced during sweetening-off than during sweetening-on, due to the slow rate of diffusion of sucrose out of the resin beads (Getaz, 1988). Flow rates during sweetening-off typically range between 1.5 and 3.0 BV/h (Bento, 1997).

A similar situation happens during the regeneration and rinse portions of the operating cycle, resulting in increased salt consumption, the creation of greater quantities of regenerant waste and increased disposal or treatment costs. This can be a significant problem in various locations, particularly as environmental regulations become more stringent (Getaz, 1988).

Fractal Shallow Bed Ion Exchange Technology

For efficient ion exchange operation (whether in softening, decolorization or any other application), the fluid entering a resin column needs to be immediately uniformly distributed across the entire cross-section of the column area. Conventional liquid distributors in ion exchange systems, consisting of perforated radial pipes (known as "laterals") or nozzles, do not do this very effectively. As a result, fluid flows preferentially (or "channels") through certain portions of the resin bed, while other parts of the bed remain under-utilized. The concentration front within the bed thus spreads, reducing the efficiency of the ion exchange process.

To compensate for the non-ideal initial fluid distribution, conventional ion exchange beds are provided with substantial depth, in an attempt to "even out" the flow distribution using distance and to minimize the phenomenon of early colorant breakthrough. However, the provision of the additional resin depth is costly, relatively ineffective and increases the pressure drop, limiting throughput (Kochergin *et al.*, 2003; Kochergin *et al.*, 2007; Kochergin *et al.*, 2015).

The use of extremely efficient fractal distributors results in a much "sharper" S-shaped plug flow breakthrough curve in ion exchange. As a result, the resin bed depth can be substantially reduced and the exhaustion flow rate substantially increased, without leading to early breakthrough (Schulze *et al.*, 2008). Typically, in softening applications, the resin inventory may be reduced by an order of magnitude, to around 10% of that used in conventional systems, with bed depths of the order of 0.5 m and flow rates ten times those normally employed (Kearney *et al.*, 2001; Kearney and Rearick, 2003). In high-concentration liquor decolorization applications, however, the mass transfer limitations within the system limit the improvement, with flow rates and resin inventories typically only reducing by a factor of two over conventional technology (Kochergin *et al.*, 2003). Although fractal distributors are more costly than conventional low-efficiency distributors, the overall capital cost of the installation, and its lifetime cost of ownership, are reduced compared to the installation of conventional ion exchange decolorization technology.

With the very shallow resin beds employed, fractal ion exchange systems can have much shorter cycle times between regenerations (if used to achieve the same color loading as in a conventional decolorization system). This holds the potential to minimize resin fouling, as longer cycles allow for the diffusion of some solution components into the resin beads that are very slowly adsorbed and cause irreversible fouling. Shorter cycles can lead to reduced resin contamination and better regeneration performance in the long term, leading to a longer resin life (Jensen *et al.*, 2004; Kochergin *et al.*, 2015). However, the uniform fractal flow distribution allows for far more efficient utilization of the whole resin inventory, as compared to standard resin beds where color breakthrough is experienced due to flow channeling, before other portions of the resin inventory are fully exhausted. The achievable color loading in the resin bed is thus increased, extending the usable cycle time of the resin in exhaustion service (sometimes to such an extent as to yield almost the same cycle time as a conventional column – but with a substantially reduced resin inventory). The shallow bed and uniform fractal distribution also facilitate efficient backwashing of the resin.

Due to the efficient distribution from the fractal, the flow rate through the bed can be increased without any turbulence that might disturb the resin bed. This can be a problem with conventional distributor designs, with the poor flow distribution creating an uneven resin level across the vessel surface (Foutch *et al.*, 2003). The fractal distributor (which does not rely on a pressure drop for efficient operation) and the much shallower resin bed reduce the pressure drop across the decolorizer to an almost negligible value, leading to a saving in energy consumption and in capital cost for the manufacture of the resin vessels themselves.

With shallow resin beds and efficient fluid distribution, the sweetening-on, sweetening-off and various regeneration and rinse steps in the ion exchange cycle can be achieved with reduced dilution and mixing. This yields operating cost savings, particularly in terms of refinery energy consumption (Schulze *et al.*, 2008; Kochergin *et al.*, 2015).

Shallow beds are also more tolerant of the presence of a small amount of suspended solids in the feed material, which is a problem commonly encountered in refinery liquors. The frequent cycling and backwash of the shallow bed reduces the risk of solids blockages and channeling within the bed (Kochergin *et al.*, 2003). Should it be required, resin replacement in shallow beds is easier to achieve and less costly.

Perhaps most importantly for the improved cost effectiveness of ion exchange technology in the long term, shallow beds with low pressure drops facilitate the use of ion exchange resins of very small particle size (Kochergin *et al.*, 2007). As discussed above, this would yield substantial benefits in terms of mass transfer kinetics and improved resin productivity. Unfortunately, however, development work in this area by resin manufacturers has focused more on creating small, uniform cation exchange resins for chromatographic applications, rather than on anion exchange decolorization resins. Over time, it is hoped that more manufacturers will produce cost-effective anion resins with smaller average particle sizes.

Numerous fractal shallow bed ion exchange systems have been installed, particularly in the beet sugar industry (for example, see Kochergin *et al.*, 2015). In 2008, the first fractal shallow bed system was successfully installed in a refinery decolorization application in Poland. Recently, a system was installed in cane refinery in Malaysia, employing an acrylic resin to achieve a design decolorization of 70%.

Waste Regenerant Brine Recovery

One of the greatest impediments to the more widespread use of ion exchange technology in the sugar industry is the treatment / disposal problem associated with the waste brine regenerant, which can represent 1 – 2% of the total treated liquor throughput (Cooper, 1976). The waste disposal problem arises from the high color, high COD content and high chloride content of the spent brine (Getaz, 1988; Meadows and Wadley, 1992).

During regeneration, most of the color extracted into the brine stream is released during the initial stages of the regeneration, as shown in Figure 2 below. A saving in brine consumption and effluent generation can thus be achieved by recovering the latter half of the waste brine for re-use in the following regeneration (Getaz, 1988; Inkson and Korea, 2015; Meadows and Wadley, 1992).

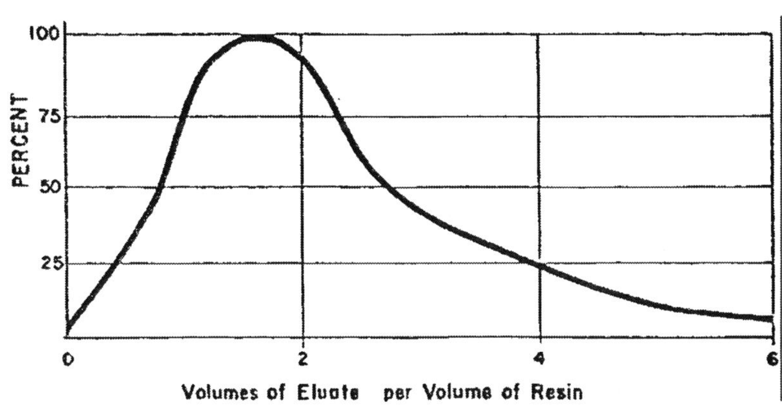

Figure 2. Color profile of waste regenerant brine exiting a resin column (Getaz, 1988).

Over the years, numerous chemical, physical and biological treatment techniques have been investigated for the treatment of waste brine (see, for example, Bento, 1997). Currently, the use of membrane nanofiltration for decolorization of the brine, allowing it to be recycled and re-used for resin regeneration, appears to be the most cost-effective option. This technology is almost considered as standard equipment in an ion exchange decolorization system (Theoleyre *et al.*, 2013).

The sodium chloride and sodium hydroxide molecules in the waste brine pass through the nanofiltration membrane into the permeate stream, while the large organic colorant molecules and other impurities are retained in the retentate, which becomes the final concentrated effluent stream (containing only a fraction of the initial salt concentration and with a substantially reduced volume). Fresh concentrated brine is added to the permeate to make up the desired concentration and it can then be reused. The efficiency of membrane permeate brine for the regeneration of ion exchange resins has been shown to be similar to that of fresh brine, and this process results in a substantial saving in salt consumption, water consumption and effluent generation. Further details may be found in Meadows and Wadley (1992), Verhaeghe *et al.* (1998), McGrath (1998), Theoleyre *et al.* (2011) and Srivastava *et al.* (2013).

Theoleyre *et al.* (2013) and Theoleyre *et al.* (2015) have also presented an alternative solution for waste regenerant brine recovery using a combination of nanofiltration, reverse osmosis and electrodialysis, which is able to achieve an even lower level of environmental impact.

Potential technologies for future use include high pressure reverse osmosis (HPRO) and ultra-high pressure reverse osmosis (UHPRO). These membrane-based technologies allow very dilute effluents (such as rinse water streams with salt concentrations as low as 1 – 2%) to be concentrated and recycled for regeneration use. HPRO allows for concentration of the dilute effluent to a brine concentration of 7 – 8%, while UHPRO can achieve final brine concentrations as high as 10 – 12%.

Design of the Malaysian Decolorization System

During 2014, Amalgamated Research LLC and Escon Gmbh designed and installed an ion exchange decolorization system for a cane refinery located in Malaysia, which was commissioned during April 2015. The system employs fractal shallow bed technology, consisting of a total of three columns, each with a resin depth of 1.22 m, as shown in Figure 3. For efficiency of fluid distribution, the resin columns are toroidal in design (donut-shaped), as shown in Figure 4. They are operated in a "merry-go-round" arrangement, with two of the columns working in a series arrangement at any one time, while the third column is offline for regeneration. After regeneration is complete, the newly-regenerated column is brought online in the secondary decolorization position, while the column that was in the secondary position is moved to primary decolorization duty. The column that was in primary decolorization duty is then taken offline for regeneration.

The system was designed for an average liquor flow of 41.5 m^3/h at a dissolved solids content of 60%(m/m). The expected color range for the feed liquor was 700 – 900 IU, with the design color removal target being 70% (in order to achieve a decolorized liquor color of 200 – 300 IU). A macroporous acrylic resin (CHEMRA Treverlite IXA510) in the chloride form was selected for all columns (*i.e.* only acrylic resin was used, rather than the combination of an acrylic resin followed by a styrenic resin), with a particle size ranging from 400 to 900 microns (and an average size from 600 – 700 microns).

Figure 3. Three-column ion exchange decolorization system installed in Malaysia.

Figure 4. Internal view of toroidal ion exchange column, showing distribution fractals.

A resin volume of 10 m^3 per column was selected, yielding an average flow rate during the exhaustion cycle of 4.2 BV/h. A color loading (or throughput before regeneration) of 50,000 t

DS × color IU per m^3 of resin was chosen, and a resin life of at least 400 – 450 cycles is expected under normal operating conditions. The design color loading for this system is substantially greater than the typical design figures of 23,000 – 30,000 t DS × color IU per m^3 of resin, as shown in Table 1. However, very high color loadings of 50,000 – 60,000 t DS × color IU per m^3 of resin are readily achievable using fractal technology, without any color leakage or early breakthrough to the product liquor. The uniform flow distribution allows for far more efficient utilization of the whole resin inventory, as compared to standard resin beds where color breakthrough is experienced due to preferential flow channeling, before some of the resin inventory is fully exhausted.

The benefit of a high color loading, and the more efficient use of the resin inventory, is that fewer regeneration cycles are needed for each unit of color removed than would be required in a conventional system. In Table 3, a typical example is demonstrated that (for comparison purposes only) uses the same resin inventory for both scenarios, and hence produces the same amount of effluent per regeneration cycle. It is clear that a fractal shallow bed system, with its high color loading, can reduce the quantity of regenerant used (and effluent generated) by 50% or more when compared to conventional technology under the same conditions. This is a very significant benefit, considering that effluent disposal represents the greatest impediment to the more widespread use of ion exchange technology for liquor decolorization. In a similar way, fractal shallow bed technology also reduces liquor dilution due to water rinsing, sweetening-on and sweetening-off.

Table 3. Example comparison between a conventional system and a fractal shallow bed.

	Conventional System	**Fractal Shallow Bed**
Liquor throughput	50 t/h	50 t/h
Liquor dissolved solids %	65%	65%
Liquor color	900 IU	900 IU
Liquor decolorization	70%	70%
Color removal per hour	20,475 t DS × color IU	20,475 t DS × color IU
Color loading per cycle	30,000 t DS × color IU / m^3	60,000 t DS × color IU / m^3

Length of cycle	1.47 h	2.93 h
Regeneration cycles per day	16.4 cycles	8.2 cycles

Resin backwashing, as well as acid washing when required, are carried out within a separate backwash tank containing a fractal fluid distributor. Due to the shallow bed depth employed, the fractal system is more tolerant of a small amount of suspended solids in the feed material and it has been found that backwashing of the resin has typically only been necessary once per month (usually in conjunction with an acid wash).

A GEA nanofiltration system for recovery of waste regenerant brine was also installed, with a design capacity of 0.5 m^3/h of waste brine feed. Regeneration is carried out at a flow rate of 3 BV/h, consuming a total volume of 3 BV of brine per cycle. The first half (1.5 BV) of the brine used consists of recycled brine collected during the second half of the previous regeneration cycle, while the second half (1.5 BV) of the regenerant brine used is obtained from the nanofiltration permeate, topped up with a small proportion of fresh concentrated brine.

Operating Results

The Malaysian decolorization system was commissioned in April 2015. For most of the operating life of the plant thus far, the inlet liquor flow rate has averaged 50 m^3/h, approximately 20% above the design rating for the plant, and the incoming liquor concentration has averaged around 55%. Peak flows have reached 70 m^3/h, or almost 70% above design. Despite these adverse conditions, the color removal achieved following the commissioning of the plant consistently exceeded 80%, with 70% color removal being recorded within the primary resin column alone.

Following the first period of operation, the performance stabilized and the plant continued to achieve an average color removal of more than 75% for an extended period, thereby exceeding the targeted color removal of 70%. The combined color removal achieved by the system (expressed as a percentage of the incoming color) over eleven months of operation is shown in

Figure 5 overleaf. The blue lines represent performance data measured on an hourly basis, while the red line is a 100-point moving average, which shows the underlying trend in the relatively "noisy" factory-collected data.

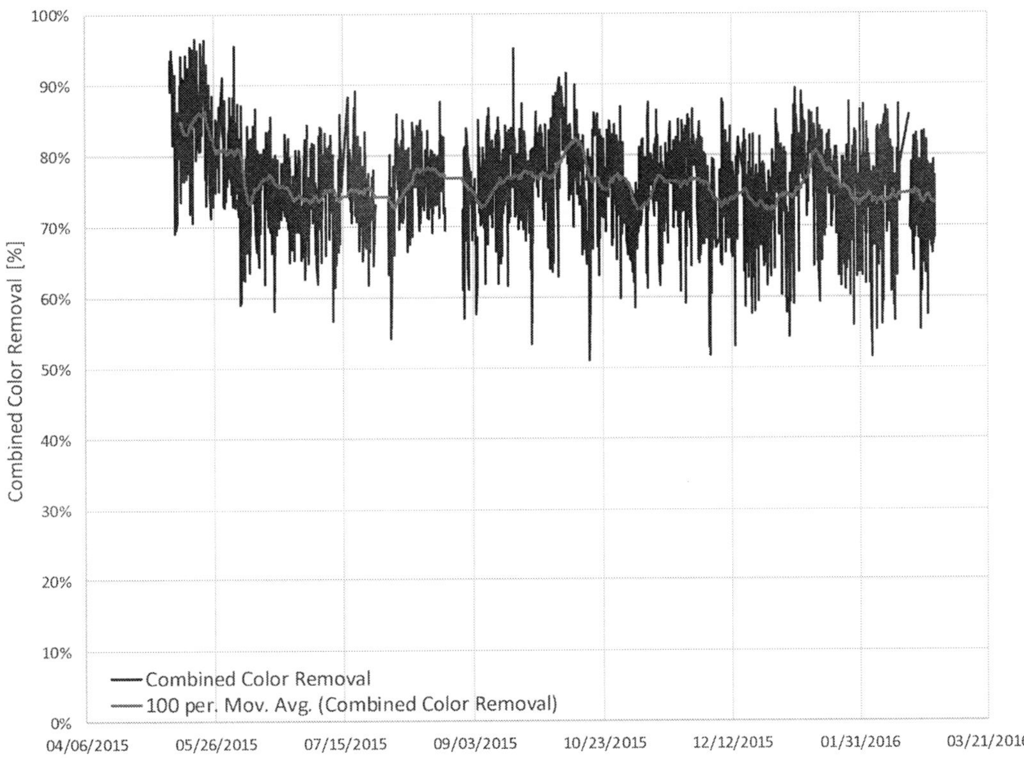

Figure 5. Combined color removal achieved by the plant since commissioning.

Although acid washing of the resin is typically recommended every 10 – 20 regeneration cycles, this was unfortunately not carried out for the first eight months of the plant's life, due to operating conditions at the time. In addition, no routine online cleaning was carried out of the nanofiltration membranes used to recover the waste regenerant brine for reuse. By the time that regular membrane cleaning was implemented, they had been irreversibly damaged. The plant also suffered from serious contamination due to suspended solids, which built up in the resin. However, this was readily removed by backwashing during the regeneration cycle, which was achieved without negatively affecting production. As a result of the non-ideal operating conditions, the resin itself suffered irreversible fouling. However, performance of the system remains satisfactory and replacement of the resin inventory was only carried out after two full years of operation and approximately 400 cycles.

The hourly data presented in Figure 6 show the treated liquor colors leaving the decolorizer during the first eleven months of routine operation. The red lines represent liquor colors measured on an hourly basis, while the blue line is a 100-point moving average, which shows the underlying trend in the relatively "noisy" factory-collected data. The decolorized liquor color averaged approximately 190 IU during this period, which is actually below the target range of 200 – 300 IU required by the customer.

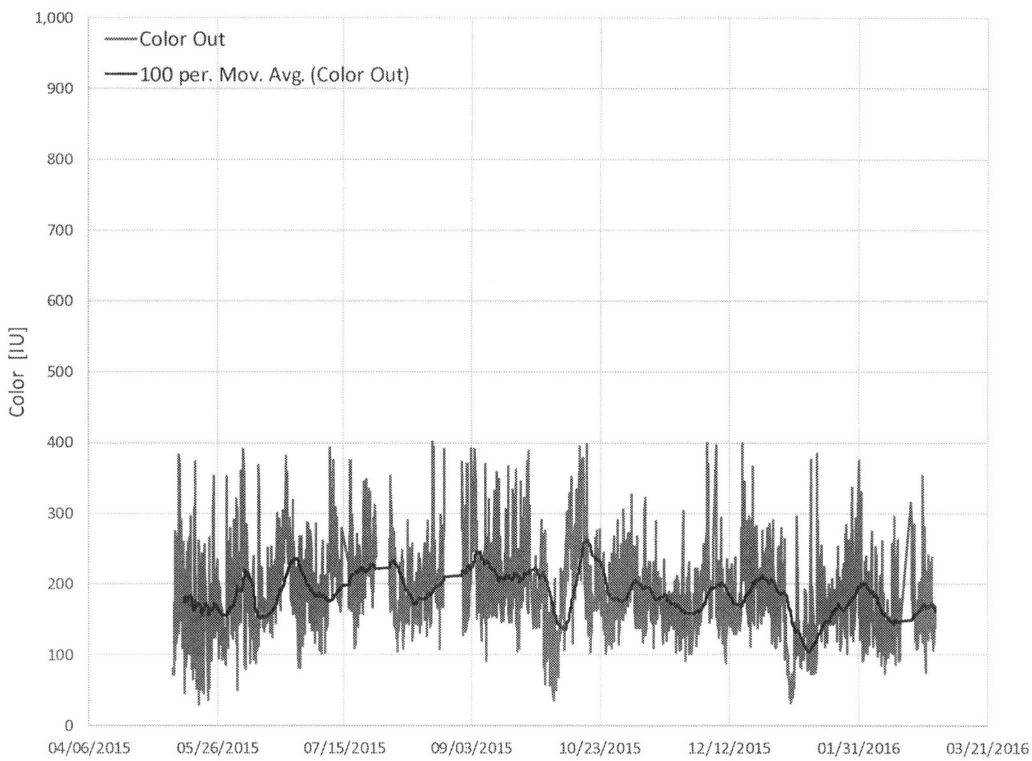

Figure 6. Treated liquor color exiting the plant since commissioning.

Traditional low-efficiency fluid distribution systems in ion exchange vessels suffer from relatively high maintenance costs, due to broken lateral pipes, damaged nozzles and the high cost of handling the gravel and / or inert resin used within the bed distribution system. By contrast, fractals have very low ongoing maintenance costs. In fact, no distributor maintenance has been required at all during the first two years of operation of the Malaysian decolorization system.

Conclusions

Even at higher color loadings and flow rates than originally anticipated, and despite some irreversible damage to the resin due to non-ideal operating conditions, the system achieved excellent results – yielding an extremely satisfied customer. Based on the factory data obtained, some important conclusions may be drawn, namely:

- A consistently high color loading of 50,000 t DS × color IU per m^3 of resin and above is achievable using a fractal shallow bed decolorizer, without significant leakage of color into the product or early breakthrough. Uniform liquid distribution in an ion exchange column eliminates fluid channeling and early color breakthrough, allowing for more effective use of the full resin inventory and thus for installation of a smaller system (in terms of liters of resin per ton of liquor dry solids processed, for example).
- With a much higher color loading, fewer regeneration cycles are required per unit of color removed. This results in a very substantial reduction in regenerant use, effluent generation and liquor dilution (and hence operating cost). As around effluent production is the most significant impediment to the more widespread use of ion exchange decolorization, this is a particularly important benefit.
- The fractal shallow bed technology minimizes mixing and dilution during the sweetening-on, sweetening-off, regeneration and rinse phases of the operating cycle. The total water consumption for sweetening-on, sweetening-off and rinsing was limited to as low as 3 BV per cycle.

- Regular acid washing of ion exchange resins is essential to their ongoing performance. Neglecting to carry out acid washes results in a substantial increase in the rate of irreversible resin fouling and a reduction in performance (Getaz, 1988).
- The partial recycle of regenerant brine, combined with membrane nanofiltration for waste brine treatment, were capable of substantially reducing the final volume of concentrated effluent produced by the decolorization system. The effluent volume was constrained to a quantity small enough to allow it to be blended into the final molasses stream leaving the factory (as feed for an attached ethanol plant).

References

Bento, L.S.M. (1997), "Ion exchange resins for sugar liquor decolourization", *Proceedings of the Sugar Industry Technologists*, **56**, Paper #716 (Symposium contribution), 251-271.

Bento, L.S.M. and Stuchi, A.A. (2014), "Decolourization of refinery sugar liquors with low environmental impact – A semi industrial trial", *Proceedings of the Sugar Industry Technologists*, **73**, Paper #1083, 107-121.

Cheong, O.I. and Mussebah, H. (1980), "Ten years experience with acrylic anion resins as gross decolorizer in cane sugar refining", *Proceedings of the International Society of Sugar Cane Technologists*, **17**, 2389-2397.

Cooper, W.W. (1976), "An investigation into the use of ion exchange resins for decolorising carbonated brown liquor in a sugar refinery", *Proceedings of the South African Sugar Technologists' Association*, **50**, 184-188.

Davis, S.B. (2001), "A review of colour removal mechanisms in sugar refining processes", *Communications from the Sugar Milling Research Institute*, **169**, January 2001, ISBN 186840417-X, 25 pp.

De Lataillade, J. and Rousset, F. (2001), "Ion exchange decolorization: A flexible way to modernization and capacity extension", *Proceedings of the International Society of Sugar Cane Technologists*, **60**, 107-119.

Dibella, P.J. (2000), "Experience in the operation of an ion exchange plant and improvements in the biological treatment of the regenerants", *Proceedings of the Sugar Industry Technologists*, **59**, Paper #774, 67-78.

Fries, W. and Walker, R.W. (1980), "The use of a two bed (acrylic and styrenic) strong base ion exchange system for the decolorization of cane sugar liquors", *Proceedings of the 1980 Technical Session on Cane Sugar Refining Research*, 171-181.

Foutch, G.L., Yi, J., Dhumal, S., Bostic, D.W., Gervais, S. and Gibson, J.E. (2003), "Simulation analysis of Surry NPS condensate polishers", *Proceedings of the 2003 EPRI Workshop on Condensate Polishing*, Electric Power Research Institute, February 2004, 345-359.

Getaz, M.A. (1988), "The application of ion exchange resins in cane sugar refining", *Communications from the Sugar Milling Research Institute*, **146**, April 1988, ISBN 0869806084, 77 pp.

Getaz, M.A. (1990), "A comparative assessment of decolourising resin performance when treating liquor from Huletts and Noodsberg refineries", *Proceedings of the South African Sugar Technologists' Association*, **64**, 223-227.

Hoareau, P., James, D. and Lamusse, J.P. (1977), "Application of ion exchange resins to the decolorisation of melt from very high pol raw sugars", *Proceedings of the International Society of Sugar Cane Technologists*, **16**, 2623-2634.

Inkson, M. and Korea, K.T. (2015), "Bahrain sugar refinery: Effluent issues and solutions", *Proceedings of the Sugar Industry Technologists*, **74**, Paper #1120.

Jensen, C., Pillay, M., Rossiter, G. and Mononen, H. (2004), "The impact of small diameter resin beads on process kinetics and economics in the sugar industry", *Ion Exchange for Today and Tomorrow* (Proceedings IEX 2004), 4 – 8 July 2004, Society of Chemical Industry (UK), 8 pp.

Kearney, M., Velasquez, L., Petersen, K., Mumm, M. and Jacob, W. (2001), "The fractal softener", *Proceedings of the American Society of Sugar Beet Technologists*, **31**, 110-114.

Kearney, M. and Rearick, D.E. (2003), "Weak cation exchange softening: Long term experience and recent developments", *Proceedings of the American Society of Sugar Beet Technologists*, **32**, 257-264.

Kirkiridis, A.P. (1992), "Two-stage decolourisation of refinery liquors by ion exchange resin", *Proceedings of the South African Sugar Technologists' Association*, **66**, 155-158.

Kochergin, V., Jacob, W. and Bornak, W. (2007), "Extending life of decolorization resin by aggressive cleaning procedures", *Proceedings of the Sugar Industry Technologists*, **66**, Paper #918, 47-62.

Kochergin, V., Kearney, M., Schulze, B.-C. and Tzschatzsch, O. (2015), "The use of fractal shallow bed technology for color and ash removal from refinery syrups", *Proceedings of the Sugar Industry Technologists*, **74**, Paper #1108, 70-80.

Kochergin, V., Pryor, T. and Jacob, W. (2003), "Pilot scale thick juice decolorization using fractal equipment", *Proceedings of the American Society of Sugar Beet Technologists*, **32**, 125-131.

McGrath, S. (1998), "The Talosave brine recovery process for treatment of ion exchange regenerant in sugar refineries", *Proceedings of the Sugar Industry Technologists*, **57**, Paper #744, 299-309.

Meadows, D.M. and Wadley, S. (1992), "The recovery of brine from resin regeneration effluent by nanofiltration", *Proceedings of the South African Sugar Technologists' Association*, **66**, 159-165.

Rousset, F. and Bhatt, J. (2011), "Conversion from bone char to ion exchange decolorization at the Lantic Sugar Montreal refinery", *Proceedings of the Sugar Industry Technologists*, **70**, Paper #1009, 233-246.

Schulze, B.-C., Tzschatzsch, O. and Costesso, D. (2008), "Latest developments in liquor decolorization using ion exchange and chromatography", *Proceedings of the Sugar Industry Technologists*, **67**, Paper #948, 245-256.

Srivastava, A.K., Tripathi, S.P., Gopalakrishanan, K. and Rao, G.S.C. (2009), "Development of ion exchange decolourisation and spent brine filtration in India", *Proceedings of the Sugar Industry Technologists*, **68**, Paper #972.

Srivastava, A.K., Jain, D., Babu, R. and Rao, G.S.C. (2013), "Spent brine recovery from ion exchange effluent: The environmentally friendly decolourization option", *Proceedings of the Sugar Industry Technologists*, **72**, Paper #1051, 100-109.

Theoleyre, M.-A., Gonin, A. and Paillat, D. (2011), "Innovative and sustainable process solution for sugar decoloration", *Proceedings of the Sugar Industry Technologists*, **70**, Paper #1013.

Theoleyre, M.-A., Paillat, D. and Lutin, F. (2013), "Innovative and sustainable improvements in salt recovery for sugar decolorization", *Proceedings of the Sugar Industry Technologists*, **72**, Paper #1064, 252-258.

Theoleyre, M.-A., Paillat, D. and Lutin, F. (2015), "Innovative concepts for regeneration of decolorization ion exchange resins in cane sugar refinery: Towards a green and fully integrated technology", *Proceedings of the Sugar Industry Technologists*, **74**, Paper #1105, 26-34.

Vawda, A.S., Sarir, E.M. and Donado, C.A. (2014), "Comparison of different decolourization technologies", *Proceedings of the Sugar Industry Technologists*, **73**, Paper #1088, 189-200.

Verhaeghe, F., Malgoyre, R., Theoleyre, M.-A. and Cartier, S. (1998), "Industrial scale brine recovery by nanofiltration from ion exchange decolorization", *Proceedings of the Sugar Industry Technologists*, **57**, Paper #737, 235-246.

S.I.T. Paper #1163

Recent Advances in Rare Sugars and Potential Applications

Wei-Chun Liu[1*]

[1] Bioprocess Engineering Division, Taiwan Sugar Research Institute, Tainan 70176, Taiwan
*Corresponding Author: a05881@taisugar.com.tw□886-06-2671911 ext. 412

Abstract

Carbohydrates are not only an important source of energy but also mediate a variety of physiological processes in human health and disease. However, increasing the consumption of sugars (such as glucose, fructose, sucrose) added to foods and drinks in the modern diet has been viewed as a major contributor to obesity and health problems. In addition, low caloric alternatives to the present dominant sweeteners such as Aspartame and HFCS have also been questioned about the effects on human health. Therefore, in recent years, rare sugars which taste similar to sucrose but have almost zero-calories and high bioactivities such as scavenging free radical, anticancer, and neuroprotective function have attracted great attention for their applications in many areas ranging from health foods to pharmaceutical and nutrition industry.

Unfortunately, rare sugars rarely exist in nature which restrict their utilizations and chemical synthesis does not satisfy the incresing demand. Nowadays, microbial and enzymatic transformation have become a very powerful tool for the synthesis of rare sugars. "Izumoring tree" developed by Prof. Kan Izumori from Kagawa University illustrates the strategy to design the production processes of all rare sugars (e.g., D-psicose, D-allose, D-tagatose) from readily available materials such as glucose or fructose. This method is more controllable and scalable. Until now, two rare sugars, D-tagatose and D-psicose, have been approved by FDA as food additives in 2003 and 2014, respectively. Moreover, various novel physiological functions of rare sugars have also been discovered, including reduction of body fat and potent inhibitory effect on the production of reactive oxygen species, and thus, may be used for medical purposes. Combined with new fermentation systems and robust biocatalysts for industrial-scale applications, rare sugars are potential game-changers to be the next-generation sweeteners.

Keywords: *Rare sugars; Health food; Food additives; Fermentation; Sweeteners*

Introduction

Increasing the comsuption of sugars, especially high-fructose corn syrup (HFCS), in daily food, beverage, and sweet has raised a major concern about cancer, diabetes, obesity and health problems. According to the *2015-2020 Dietary Guidelines for Americans*, it has removed the restriction on total fat and set limit for added sugars. Moreover, "a new WHO guideline recommends adults and children reduce their daily intake of free sugars to less than 10% of their total energy intake. A further reduction to below 5% or roughly 25 grams (~ 6 teaspoons) per day would provide additional health benefits (press release, March 4, 2015)." However, a five carbon pentitol, xylitol, that has beneficial health properties such as prevention of tooth decay has shown that alternatives to existing dominanat sweeteners are possible. Therefore, in recent years, rare sugars which taste similar to sucrose but have almost zero-calories and high bioactivities such as scavenging free radical, anticancer, and neuroprotective function have attracted great attention for their applications in many areas ranging from health foods to pharmaceutical and nutrition industry. In 2002, the 1st Rare Sugar Congress organized by the International Society of Rare Sugars (ISRS) has defined rare sugars as monosaccharides and their derivatives that are present in small quantities in nature. For example, allulose (psicose), tagatose, xylitol, erythritol and so on are rare sugars that have been well-studied in terms of their efficacy, safety and production from exisitng sources or using microorganism. As shown in **Figure 1**, natural sweeteners now make up only 1% of the total sweetener market by volume. But products like rare sugars, stevia, monk fruit, and sweet proteins are gaining traction and have the potential to gain market share from both sucrose and other alternative sweeteners. Rare sugar is now mainly manufactured by Japanese chemical company, Matsutani Chemical Industry, by British sugar manufacturer, Tate & Lyle (which also makes Splenda), and by South Korean biotech firm, CJ CheilJedang.

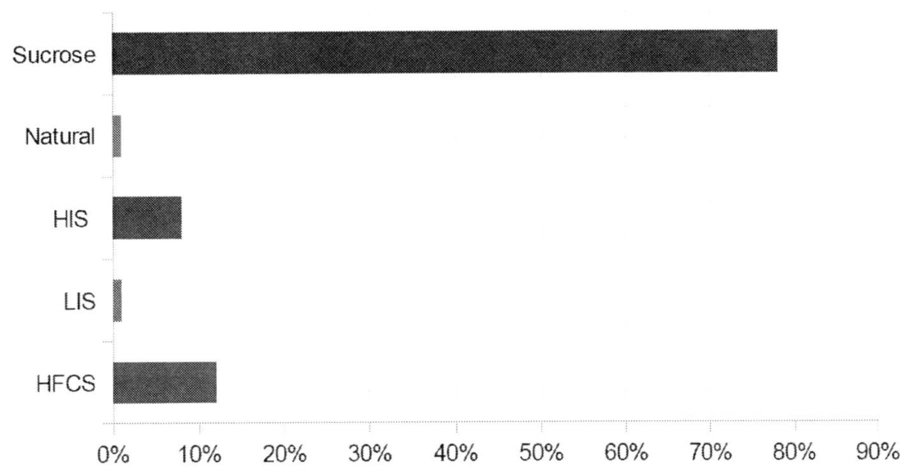

Figure 1. Total sweetener market. High-intensity sweeteners (HIS) include stevia, aspertame, sucralose, etc; low-intensity sweeteners (LIS) include sorbitol, xylitol, mannitol, tagatose, etc. **(Adapted from Life Resource Inc.)**

As mentioned above, rare sugars rarely exist in nature, and are also difficult to chemically syntheize. In this paper, the physiological functions and bioproduction of some rare sugars are reviewed, expecting to open for more discussions of the potentials of rare sugars as the next-generation sweeteners.

Physiological functions of rare sugars

Sucrose is the most commonly-used food sweetener. It has additional properties such as food preservation and is an important ingredient in sweets. Although there is no serious health risk of sucrose, its very high in calories and rising health-related issues such as tooth decay, obesity, ageing and even cancer (Lustig, R. H., *et al.*, 2012) makes sucrose not the so perfect sweetener. But rare sugar that has similar sweetness of sucrose and ultra-low energy is an ideal substitute for sucrose. Until now, only two rare sugars, D-allulose and D-tagatose, have been approved by FDA.

D-allulose (also known as D-psicose) is a rare monosaccharide and a C-3 epimer of D-fructose (shown in **Figure 2**) (Zhang, W., *et al.*, 2016). It has 70% of the sweetness of sucrose and almost zero-calorie. It was reported that the absorption rate of D-allulose is lower than that of other sweeteners. In 2014, it has been approved as GRAS by FDA and is permitted to be used as the food ingredient and dietary supplement. In addition, it shows a competitive inhibition of the influx and efflux transporters of D-fructose and D-glucose and further reduces their absorption. This contributes to several health benefits such as anti-obesity and anti-diabetes (Hayashi, N., *et al.*, 2014). D-allulose also has other physiological functions such as anti-inflammatory effects and neuroprotective effects, especially the inhibition of cancer cell growth. Cancer cells require massive amounts of glucose as an energy source for their dysregulated growth and thus overexpression of various glucose transporters (GLUTs). Due to the competitive inhibition of glucose transporters (GLUTs), D-allulose can suppress cancer growth by reducing both GLUTs expression and glucose uptake (Noguchi, C., *et al.*, 2016). Furthermore, D-allulose also can be applied in agriculture. It can be used to modulate plant growth and control plant diseases. When use of the rare sugar, plant life is halted which might show its potential to be used as the herbicide.

D-tagatose is also a rare ketohexose and an epimer of D-fructose isomerized at C-4 position (shown in **Figure 2**). It has 92% relative sweetness of sucrose but around 2.5-fold decrease in calorie. According to the study, it can control the blood glucose level. In samll intestine, D-tagatose inhibits the degradation of carbohydrate to glucose and in liver, it promotes glucokinase activity which further enhances transfer of glucose to glycogen (Cote, C.L., *et al.*, 2005). It also has been approved as GRAS by FDA in 2003 and as novel food ingredient by EU (EC No. 61) in 2005. Except for the health benefit of glucose level control, D-tagatose is also a tooth-friendly ingredient similar to xylitol.

Other rare sugars like D-sorbose, D-allose and L-form of monosaccarides (*e.g.* L-fructose, L-glucose, L-tagatose, etc.) have also benn applied in virous fields (Li, Z., *et al.*, 2013). D-sorbose can be potentially used as an insect control agent and an intermediate for producing vitamin C (ascorbic acid). L-fructose is a potential inhibitor of several glucosidases which can be used as the drug for treatment of diabetes. L-glucose can be used as an starting material for glycoconjugate vaccines against the bacterium.

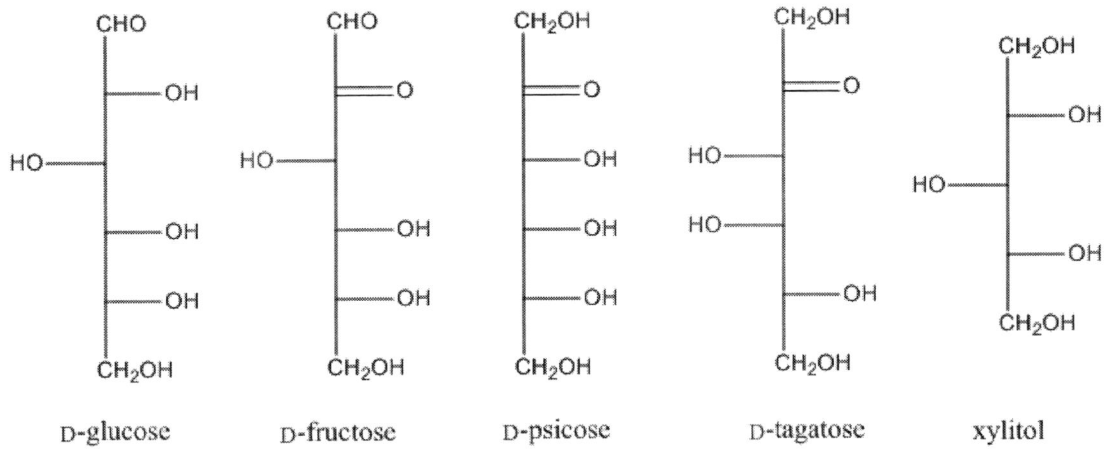

Figure 2. Structures of D-glucose, D-fructose and rare sugars. D-psicose and D-tagasoe are the C-3 and C-4 epimers of D-fructose, respectively.

Biotechnological production of rare sugars

With decades of studies, many potential applications of rare sugars have been gradually disclosed. However, an important factor restricting the utilization of rare sugars is their limited availability and lack of efficient synthetic strategy. In the early years, synthesis of rare sugars was achieved through multiple chemical synthetic steps using either common monosachharides or non-sugar starting materials. Due to the repeated protection and unprotection of glycosides, the method was usually very complicated and the yield was extremely low. Nowadays, the biosynthesis and enzymatic production of rare sugars in large-scale has become a very powerful tool in this field. Especially, Prof. Ken Izumori at Kagawa University in Japan is a pioneer in bioproduction of rare sugars. As shown in **Figure 3**, he created a beautiful “Izumoring tree” to illustrate all possible strategies for the bioproduction of these rare sugars by enzymatic or microbial transformation. In addition, starch, whey or hemicellulose waste can also be used as raw materials to extract starting monosaccharides for massive producing rare sugars. For example, D-glucose can be obtained from starch, D-galactose from whey, and D-xylose from hemicellulose. Combined with using enzymes such as D-tagatose 3-epimerase, aldose isomerase, aldose reductase, and oxidoreductase or even whole cells as biocatalysts, the synthesis of rare sugars can be easily achieved (shown in **Figure 4**) (Van Overtveldt, S., *et al*, 2015).

Bioproudction of rare sugars requires multidiscipline approaches invovled that include fermentation technology, molecuar biology, enzyme technology, and organic chemistry. This strategy offen offers several advantages over chemical synthesis due to highly enantioselective and regioselective properties of enzymes. In addition, enzymatic and microbial reactions usually performed under mild conditions and are eco-friendly. However, there are still few challenges remained to be conquered: 1) enzymes or microorganisms used in the bioproduction pipeline need to be GRAS sources, which may limit some of their applications; 2) a chemical equilibrium (isomerization or epimerization) invovled in biosynthetic methods may lead to low yields and increase difficulties in product purification; 3) enzymatic reactions is relatively more expensive and thus discovery of more robust enzymes is an increasing demand. Therefore, further screening of novel microbial sources and advancements in genomics and biotechnology can improve the properties of existing enzymes and discover new systems for industrial-scale applications, respectively.

Conclusion

Currently, sweeteners are at an intriguing intersection. With the FDA attempting to include a separate "added sugars" listing on nutrition panels, consumers may pay more attentions on those high-calories sugars add. Moreover, many countries intend to introduce "sugar tax" on sugar-sweetened food and drinks. This pushes processors to be more motivated to look for low-caloric or non-nutritive sweeteners for sweetness, and especially "natural" sweeteners is highly demanded. Rare sugars, such as D-allulose, D-tagatose, and xylitol, which show similar sweetness and sensory profile to that of sucrose and bring more additional functional and health benefits, might be potential game-changers to be the next-generation sweeteners.

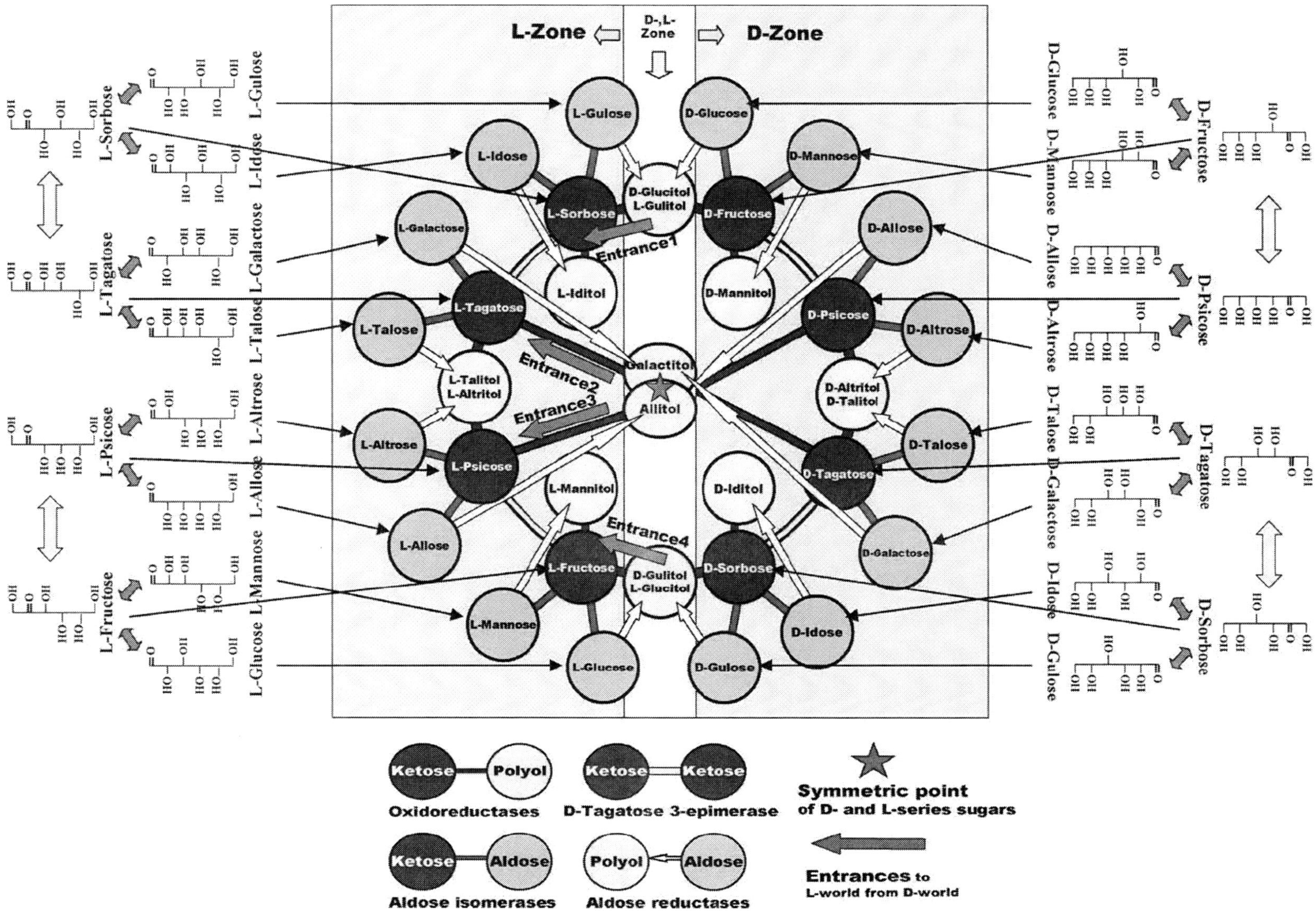

Figure 3. Izumoring tree□the process of all hexoses production. (Adpated from Izumori, K., 2006, Figure 4.)

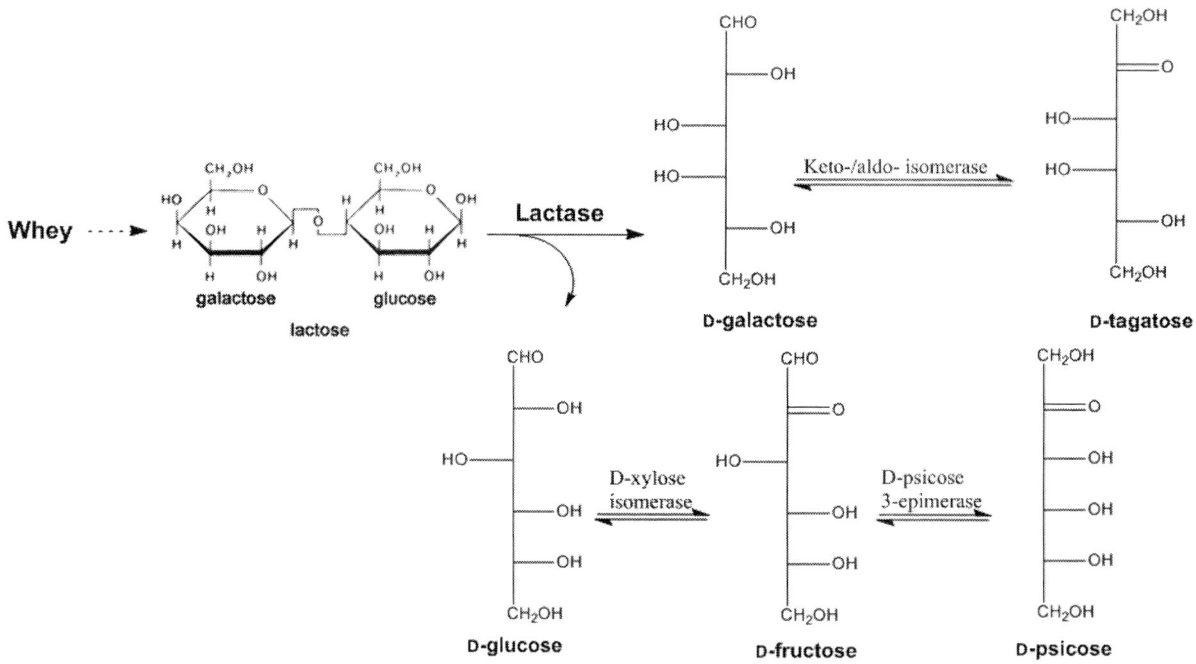

Figure 4. Bioproduction of D-psicose and D-tagatose from cheap natural resources such as starch, whey or hemicellulosic waste.

References

- Cote, G. L., Dunlap, C. A., Appell, M., Momany, F. A. (**2005**). Alternansucrase acceptor reactions with D-tagatose and L-glucose. *Carbohydrate Research*. 340□257-262.
- Granstrom, T. B., *et al.* (**2003**). Izumoring: a novel and complete strategy for bioproduction of rare sugars. *Journal of Bioscience and ioengineering*. 97□89-94.
- Hayashi, N., *et al.* (**2014**). Weight reducing effect and safety evaluation of rare sugar syrup by a randomized double-blind, parallel-group study in human. *Journal of Functional Foods*. 11□152-159.
- Izumori, K. (**2006**). Izumoring: A strategy for bioproduction of all hexoses. *Journal of Biotechnology*. 124□712-722.
- Li, Z., *et al.* (**2013**). Biosynthesis of rare hexoses using microorganisms and related enzymes. *Beilstein Journal of Organic Chemistry*. 9□2434-2445.
- Lustig, R. H., Schimidt, L. A., and Brindis, C. D. (**2012**). The toxic truth about sugar. *Nature*. 482□27-29.
- Noguchi, C. *et al.* (**2016**). D-allulose inhibits cancer cell growth by reducing GLUT1 expression. *The Tohoku Journal of Experimental Medicine*. 238□131-141.
- Van Overtveldt, S., *et al.* (**2015**). A structural classfication of carbohydrate epimerases: from mechanistic insights to practical applications. *Biotechnology Advances*. 33□1814-1828.
- Yang, Y., and Zhang, M.-W. (**2010**). Research progress in rare sugars. □□□□□31□303-

307.

- Zhang, W., *et al*. (**2016**). Recent advances in D-allulose□Physiological functionalities, applications, and biological production. *Trends in Food Science & Technology*. 54□127-137.

S.I.T. Paper #1164

Major New Insights into the Cause of Floc Formation in Alcohol Beverages Sweetened with Refined Cane Sugars

Gillian Eggleston* and Alexa Triplett

*USDA-ARS-Southern Regional Research Center
1100 Robert E. Lee Boulevard
New Orleans, LA 70124, USA
gillian.eggleston@ars.usda.gov

ABSTRACT

The sporadic appearance of floc from refined, white sugar in alcoholic beverages is a large concern to both beverage manufacturers and sugar refiners. With the declining use of high fructose corn syrup as a beverage sweetener in recent years, floc from cane sugars remains a technical problem that is not easily managed. For cane sugars, no definite cause for floc formation has been attributed to any one quality parameter. Silica, polysaccharides, protein, waxes, and color indicator values I.V. (color at pH 9.0/color at pH 3.0) have all been previously implicated, especially with acid beverage flocs. In this study, cane invert sugar mixed with pure alcohol and water to a 60% ABV target) increased light scattering by up to ~1000-fold, i.e., 11.3 to 20.5 NTU, greater than the initial invert sugars (0.018 to 0.024 NTU). Both insoluble and soluble starch, fat, inorganic ash, oligosaccharides, Brix, and pH were found *not* be involved in the prevailing floc formation mechanism. There were strong correlations between haze floc and the I.V. (R^2=0.994) and protein (R^2=0.837) values of the invert sugars, indicating that polyphenol colorants and protein are involved in the floc formation mechanism. Polyphenolic-protein flocs are known to combine and cause floc formation in other alcohol beverages including beers, ciders, and wines. Much lower levels of I.V. ratios and protein are involved with alcohol flocs than those reported for acid beverage flocs sweetened with sugar. More haze formed when select 60% ABV samples were heated to 80 °C because partial or full denaturation of protein exposed hydrophobic polyphenol binding sites. Heating to increase floc formation is consistent with a greater role for hydrophobic rather than hydrogen bonding. By probing the floc formation with nephelometry and a haze (floc) active protein and polyphenol, as well as polar, non-polar, and ionic solvents, major new insights into floc formation caused by refined cane sugars were observed and are discussed.

INTRODUCTION

The sporadic appearance of floc from refined, cane (white) sugars in alcoholic beverages is a concern to both beverage manufacturers and sugar refiners. Although floc is unsightly and is a quality parameter that will not impact health and nutritional status of the beverage, the consumer expects the beverage to be clear and will most likely be deterred from purchasing the beverage. The visible defect may be particulate and sedimentary or "woolly" and suspended in the beverage or precipitated (Edye, 2004). Cohen et al. (1970) reported that, while all acid beverage floc (ABF) forming sugars also develop a floc on the addition of ethanol, not all alcohol haze forming sugars develop ABF. With the declining use of high fructose corn syrup as a beverage sweetener in recent years, floc from either beet or cane sugars, inverted or non-inverted, remains a technical problem that is not easily managed. Typically, beet sugars are less prone to cause haze and the primary cause has been attributed to saponins (Edye, 2004; Iciek et al., 2012). For refined cane sugars, however, no definitive cause has been attributed to any one quality parameter. Silica, polysaccharides (Clarke et al., 1978), protein (Naidoo and Simpson, 2004; Clarke et al., 1978), and fats/waxes have all been implicated (see reviews by Edye, 2004 and Lemos et al., 2013). Polysaccharides in cane sugar can include dextran formed from microbial deterioration (levan and sarkaran polysaccharides are also formed from specific microbial deteriorations in a few parts of the world, particularly dry areas), starch, indigenous cane polysaccharide ISP (arabinoglucan), and small amounts of soluble cellulose and pectin (Edye, 2004). Clarke et al. (1978) put forward a mechanism for acid beverage floc that incorporated a coacervate of a positively charged protein interacting with a negatively charged ISP. It has been reported by McKee et al (2010), for acid beverage floc formation only, that when high (>3.2) color indicator values I.V. (color absorbance ratio at 420 nm of the sugar solution at pH 9.0 to pH 3.0) are present in cane refined sugar there is a strong likelihood that acid beverage flocs will form, but no explanation has been put forward. As high I.V. values are indicators of polyphenolic and flavonoid colorants that naturally occur in sugarcane, these may be involved with floc formation or they may be indirectly involved. Polyphenolic-protein haze are known to form in beer, wine, cider, and fruit juices (Siebert et al., 1996a). The use of I.V. values to predict acid beverage floc formation from cane sugars, however, have not been totally reliable as outliers has been found and the method requires further studies (Iciek et al., 2012).

Lemos et al. (2013) studied twenty-one refined cane sugars from Brazil for their formation of alcohol flocs at 55-89% v/v ethanol, and reported that total starch and dextran positively correlated with alcoholic floc formation. They used a starch method (ICUMSA GS1-16) that is now known to only measure soluble starch and which gives no information of the amount of insoluble starch present, and the Haze dextran method used to determine dextran is not specific to dextran and can measure other haze forming compounds as well. It is most likely that, because insoluble starch has not been previously detected before by commonly used starch methods, its contribution to beverage floc formation has never been realized. Fortunately, a new USDA research method has now been developed by Cole et al. (2016) to measure total, insoluble and soluble starch, utilizing microwave-assisted probe sonication.

Alcohol beverages are high value-added products with an enormous international market. Such products are often sweetened with white, refined sugars sourced from around the world; the sugars can be sucrose or invert sugars. Alcohol beverage manufacturers, can occasionally encounter batches of sugars that cause unwanted floc in their alcoholic products. Unfortunately, there are no rigorous specifications for the prediction of floc from refined cane sugars and, in particular, unwanted false negative results. This may be because there are multiple causes and/or because the underlying reasons for floc formation are still not completely known. At this present time, some alcohol beverage manufacturers estimate floc formation from batches of sugar by creating a 60% ABV sugar base, which is simply invert sugar (or other sugar) mixed with pure alcohol and water to a 60% ABV target. If after 30 min the fortified sugar turns hazy (NTU $\geq$ 2.0) it is considered to be unstable. This fortified sugar must stand for 24 h before it is filtered to remove flocs. Once flocs are removed the now clear, fortified sugar solution can be used to make flavored beverage products.

This study was undertaken to improve understanding of the root cause or causes of floc formation in alcoholic beverages sweetened with refined cane sugars. Another long-term goal of the study is to underpin the development of rigorous specifications for the measurement of alcohol haze formation.

EXPERIMENTAL

Chemicals. All chemicals were analytical grade. Dimethylfromamide (DMF), 1,4-dioxane, tannic acid, gelatin (Type B), tannic acid, triethanolamine, dioxane, and dimethylformamide (DMF) were obtained from Sigma-Adrich (St. Louis, MO, USA). De-ionized water of < 18 ohm resistance) was used in sample preparation. Absolute ethanol (200 proof) was from Koptec (King of Prussia, PA, USA). Sodium chloride was from J. T. Baker.

Beverages. Commercial invert sugars (~75 Brix syrup) from various refinery sources (denoted by upper case letters), and 60% ABV unfiltered and filtered sugar bases produced from the invert sugars, were obtained from a large North American alcoholic beverage manufacturer. The 60% ABV sugar base (1 L) was created by adding invert sugar syrup (260 g) to 780 mL of 77% ethanol plus 30 mL of deionized water. This 60% ABV sugar base (unfiltered) was left for 24 h by the beverage manufacturer to approximately estimate if any floc will form. The unfiltered 60% ABV was then pressure filtered through diatomaceous earth to form the filtered 60% ABV sugar base.

Stock Solutions. Tannic acid stock solution (5,000 ppm) was prepared by dissolving tannic acid in ethanol (20% of final stock solution volume) and making to the volume with de-ionized water. Gelatin stock solution was prepared daily by dissolving protein (5,000 ppm) in de-ionized water.

Floc Measurements by Haze. Light scattering measurements were undertaken with a Hach Model 2000N nephalometer (Hach, Loveland, CO). Samples containing flocs were gently shaken first before measurement. Results were expressed in nepholus turbidity units (NTU).

Effect of a Haze-Active Protein and Polyphenol. Floc was produced by adding gelatin (5,000 ppm; haze-forming protein) or tannic acid (5,000 ppm; haze-forming polyphenol) to the test 60% ABV solution (20 mL in a 100 mL beaker). The samples were incubated for 30 min at 25 °C.

Heating Experiments. The unfiltered 60% ABV samples were incubated for 30 min in a water bath at 80 °C; the 80 °C samples were cooled to 25 °C before measuring the haze. The haze of the samples at room temperature (25 °C) were also measured and compared.

Effect of Solvent Experiments. To 20 mL of unfiltered 60% ABV samples, 0, 2, 4, and 8 mL of either de-ionized water, absolute ethanol, dioxane, DMF, or 20% NaCl solution were added and gently mixed. This resulted in solutions containing 0, 10, 20, and 40% solvent. The haze of the resulting solution left at 25 °C (room temperature) for 30 min was then measured.

Insoluble and Soluble Starch. Samples were analyzed for total, soluble, and insoluble starch using the microwave-assisted sonication/iodometric USDA Research method (Cole *et al.*, 2016). Invert syrups were first adjusted to ~15 Brix before analyses and starch was assayed in triplicate; concentrations are quoted as average ppm on a Brix basis.

Brix (% dissolved solids), pH, Ash, and Fat. Brix was measured using an Index Instruments (Kissimmee, FL) TCR 15-30 temperature controlled refractometer accurate to ± 0.01 Brix, and results expressed as an average of triplicates. The pH was measured on a Metrohm Brinkman 716 DMS Titrino (Riverview, FL, USA) with a Mettler Toledo (Columbus, OH, USA) xerolyte electrode. Inorganic ash was measured according to method AOAC 900.12; the weighed samples was ashed at 600 °C. Fat analysis followed MWL FO 08 method based on AOAC 922.06 method. The homogenized sample was treated with HCl and washed twice with petroleum ether and diethyl ether. Both ash and fat were expressed as % on a wet weight basis and results are averages of triplicate analysis.

Color and Indicator Values. Color was measured as the absorbance at 420 nm and calculated according to the official ICUMSA method GS2/3-9 (1994). All the invert syrup samples and 60% ABV unfiltered and filtered samples were not initally Brix adjusted. Samples (~5 mL) were diluted in triethanolamine/hydrochloric acid buffer (5 mL; pH 7.0) and filtered through a 0.45 μm filter. Color was also measured at pH 3.0, 4.0, and 9.0 by first adjusting the buffer with HCl and NaOH solutions, respectively. The Indicator Value (I.V.) was measured as both the ratio of color at pH 9.0/color at pH 4.0 as well as color at pH 9.0/color at pH 3.0. Results are expressed as an average of triplicates.

Protein. Protein was first quantitatively measured using MWL FO 014 method which is based on AOAC 992.15 method. Weighed samples were combusted to release nitrogen; nitrogen was determined and converted to the protein value. Results of triplicate analyses were expressed as a percent on a wet weight basis; the lower limit of analysis was 0.1%. Protein values from this method were too high and dubious. Thus the presence of protein was next confirmed qualitatively by scanning the invert sugar in a Beckman DU-64 UV/Vis Spectrophotometer between 220 – 350 nm. Protein was present as a peak at 280 nm due the presence of select amino-acids (tyrosine and particularly tryptophan and phenylalanine). Once protein presence was confirmed it was assayed using a Bi-Rad protein colorimetric assay (Bio-Rad Laboratories, Hercules, CA, USA) based on the Bradford dye-binding method (Bradford, 1976). Bovine serum albumen (BSA) standard was used to create the standard curve at 595 nm absorption. This method is based on the dye binding primarily to basic (particularly arginine) and aromatic amino-acids. Results are expressed as an average of duplicates.

Dextran. The Rapid Dextran Test (Ecolab, KS, USA) was used to measure monoclonal antibody dextran. A conversion factor was calculated for each batch of antibody used. Results are expressed as an average of duplicates.

Oligosaccharides measured using ion chromatography with integrated pulsed amperometric detection (IC-IPAD). An oligosaccharide fingerprint chromatogram (up to 12 DP) of each sample was obtained by using a strong NaOH/NaOAc gradient over 40 min (Eggleston and Borges, 2015), using CarboPac PA1 analytical and guard columns (Dionex Corp., Sunnyvale, CA, USA) at 25 °C. All samples were adjusted to the 7.0 Brix prior to comparison.

Statistics. Pearson correlation coefficients were calculated to investigate relationships among the different chemical and physical parameters using Microsoft Excel 2013™.

RESULTS AND DISCUSSION

Floc Formation by Cane Invert Sugars

Four invert sugars produced from sugarcane refined sugars and known to cause alcohol floc formation, were first studied. Floc (haze) formation was measured by light scattering nephelometry, which is used routinely in the quality control of beverages, including alcoholic beverages such as ciders and beers (Chapon, 1993). The haze of the invert sugar ingredients were very low, i.e., between 0.018 to 0.024 NTU, and bore no relationship to the floc formed on the addition of ethanol (Table 1). Flocs formed in the unfiltered 60% ABV samples (invert sugar mixed with pure alcohol and water to a 60% ABV target) increased light scattering by up to ~1000-fold, i.e., 11.3 to 20.5 NTU greater than the initial invert sugars. The flocs formed were white and fluffy and eventually precipitated to the bottom of the bottle as a sediment. As expected, when the 60% ABV samples were filtered the NTU values markedly decreased (Table 1).

Table 1. Haze (NTU), Brix (% dissolved refractometric solids), and pH of initial invert sugars, non-filtered, floc formed alcohol solutions, and filtered alcohol solutions.*

Sample	Flocs Present	Brix %	pH	Ash %	Fat %	Haze NTU
Invert Sugar A	x	77.36	3.85	<0.1	0.5	0.020
Invert Sugar B	x	72.22	3.55	<0.1	0.5	0.018
Invert Sugar C	x	71.80	5.77	0.3	0.3	0.020
Invert Sugar D	x	76.47	4.93	<0.1	0.4	0.024
Unfiltered 60% ABV Solution A	✓	33.42	5.33	<0.1	<0.1	20.50
Unfiltered 60% ABV Solution B	✓	32.64	5.09	<0.1	<0.1	19.30
Unfiltered 60% ABV Solution C	✓	32.4	6.28	<0.1	<0.1	14.84
Unfiltered 60% ABV Solution D	✓	33.22	5.60	<0.1	<0.1	11.34
Filtered 60% ABV Solution A	x	33.92	5.34	<0.1	<0.1	0.024
Filtered 60% ABV Solution B	x	32.52	5.14	<0.1	<0.1	0.030
Filtered 60% ABV Solution C	x	32.41	6.28	<0.1	<0.1	0.023
Filtered 60% ABV Solution D	x	33.96	5.70	<0.1	<0.1	0.024

*All of these samples were provided by the North American alcoholic beverage manufacturer

The pH values of the initial invert sugars varied from 3.85 to 5.77 (Table 1), and were all acidic. The alcohol manufacturer who provided the samples does not adjust the pH of the sugar when producing beverages, but many alcoholic beverages are between pH 3 and 4. As seen in Table 1, the addition of ethanol to the sugars increased the pH, but the filtering of these solutions to remove the flocs had negligible effect on the pH. The initial pH of the invert sugars strongly affected the pH of the final filtered ABV solutions (R^2=0.975; y=0.484x + 3.424). Neither the pH nor the Brix values were found to have any relationship with haze, i.e., floc formation.

Chemical Composition Analyses

At the beginning of this study, the cause of floc formation in alcoholic beverages sweetened with refined cane sugars was not known, so initial research focused on a few possibilities, including starch, dextran, oligosaccharides, fat, ash, protein, color, etc. In this study, none of the invert sugars contained any soluble or insoluble starch (Table 2), which strongly indicated that no starch forms are involved in the floc formation mechanism. Nevertheless, this does not preclude that starch molecules could not be entrained or co-precipitated with the precipitated floc. Lemos et al. (2013) showed that soluble starch can cause the formation of flocs in ethanol solution although the effect of dextran was stronger. High molecular weight dextran was measured using a specific monoclonal antibody method and results are listed in Table 2. All the invert sugars contained dextran but there was not a strong relationship between dextran and haze floc formation. Nevertheless, the dextran content was markedly reduced in the 60% ABV samples suggesting they became insoluble in the ethanol and also that they may be associated with the floc although not with the initial mechanism of floc formation. Fat concentrations of 0.3 to 0.5% were measured in the initial invert sugars which may reflect wax from the surface of sugarcane stalks (Table 1). The fat concentrations were reduced to negligible levels (<0.1%) in the unfiltered 60%ABV samples, which are less than the 0.1% limit of the method of analysis used. Thus fat was dissolved in the ethanol and not involved in floc formation (Table 1). Additionally, inorganic ash concentrations were found to have no relationship with floc formation (Table 1). Fingerprint oligosaccharide chromatograms of the initial unfiltered 60% ABV samples (A to D) are illustrated in Fig. 1. The samples contained very few oligosaccharides which were not related to floc formation. The chromatograms of the initial invert sugars and filtered samples (not shown) were very similar to the unfiltered samples.

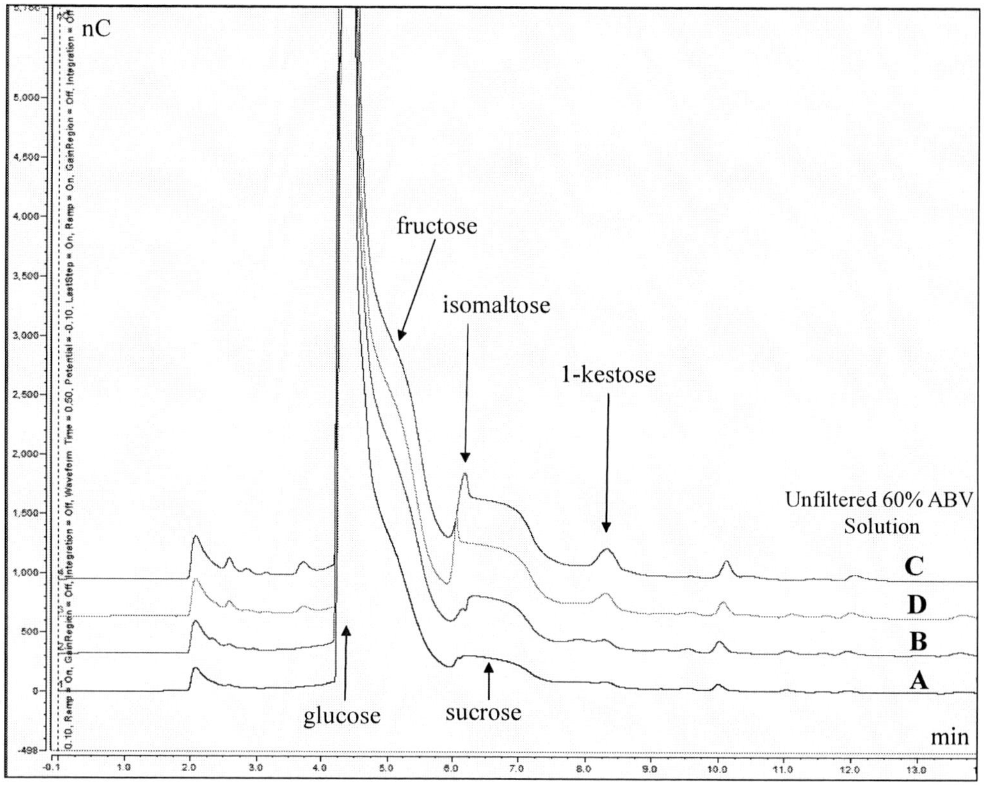

Fig. 1. IC-IPAD oligosaccharide chromatograms of unfiltered 60% ABV solutions.

Table 2. Compositional analyses of initial cane invert sugars A to D, unfiltered, floc formed alcohol solutions, and filtered 60% ABV solutions.*

Sample	Insoluble Starch (ppm/Brix)	Soluble Starch (ppm/Brix)	Dextran (ppm/Brix)	Qual. Protein	Quant. Protein (µg/mL)	Color at pH 3.0 (ICU)	Color at pH 4.0 (ICU)	Color at pH 7.0 (ICU)	Color at pH 9.0 (ICU)	I.V. 9/3	I.V. 9/4
Invert Sugar A	0	0	165.6	✓	0.60	244.4	248.7	277.3	370.7	1.52	1.50
Invert Sugar B	0	0	n.d.□	✓	n.d.□	220.9	222.6	255.7	322.5	1.46	1.45
Invert Sugar C	0	0	186.7	✓	0.11	238.8	237.5	252.7	300.9	1.26	1.27
Invert Sugar D	0	0	102.9	✓	0.13	224.7	229.3	247.9	290.2	1.27	1.27
Beet Invert Sugar (control)	0	0	13.7	✓	0.04	159.9	167.5	167.5	175.5	1.10	1.05
Unfiltered 60% ABV Solution A	0	0	5.2	✓	29.67	306.1	308.3	309.9	350.2	1.14	1.14
Unfiltered 60% ABV Solution B	0	0	n.d.□	✓	n.d.□	297.4	300.0	310.6	335.8	1.13	1.12
Unfiltered 60% ABV Solution C	0	0	4.5	✓	29.73	319.4	317.7	325.3	349.5	1.09	1.10
Unfiltered 60% ABV Solution D	0	0	0	✓	25.79	290.2	296.4	300.5	301.2	1.08	1.07
Beet Invert Sugar (control)	0	0	1.9	✓	25.69	291.5	297.1	302.4	302.0	1.04	1.02
Filtered 60% ABV Solution A	0	0	n.d.†	n.d.	n.d.	298.7	299.8	306.6	341.3	1.14	1.14
Filtered 60% ABV Solution B	0	0	n.d.	n.d.	n.d.	296.7	295.2	307.8	331.5	1.12	1.12
Filtered 60% ABV Solution C	0	0	n.d.	n.d.	n.d.	319.8	n.d.‡	350.8	358.6	1.12	n.d.‡
Filtered 60% ABV Solution D	0	0	n.d.	n.d.	n.d.	321.4	299.2	297.9	319.9	1.07	1.07

*All of these samples were provided by the North American alcoholic beverage manufacturer

□n.d.=not determined because not enough sample; †n.d.= not determined; ‡n.d.=not determined as anomalous

The color of all the invert samples was measured at pH 3.0, 4.0, 7.0, and 9.0 according to ICUMSA protocols, and the color Indicator Value I.V. was calculated as the ratio of color at pH 9.0/color at pH 4.0 as well as pH 9.0 color/color at pH 3.0. As expected, colors at pH 3.0 and 4.0 were lower than at pH 7.0 and 9.0, because color intensity depends on the solution pH due to changes in molecular structure, ionization, and association-dissociation equilibria (Dietz et al., 1959); the changes are not linear and color intensity changes most steeply in the pH range 6.0 to 8.0.

The colors measured at pH 3.0, 4.0, and 7.0 did not exhibit any significant relationships with floc formation (NTU), although a weak linear correlation (R^2=0.404) existed for color at pH 9.0. It is well known that cane derived colorants, e.g., flavonoids and phenolics, are sensitive to the higher pH values while process derived colorants are not (Smith et al., 1981). In strong contrast, there was a strong, positive polynomial relationship with color I.V. values of the invert sugars with haze formation (see Fig. 2). When we received other floc forming cane invert sugars (N=7) from the alcohol beverage manufacturer this relationship was consistent. The two I.V. values were strongly related to each other since a very strong linear correlation (R^2=0.996; y=1.0028x) existed between them. The I.V. 9.0/3.0 values, however, correlated more strongly with haze formation (R^2=0.969) than the I.V. 9.0/4.0 values (R^2=0.816), which indicated that the former is more preferable for use. This is in agreement with findings of McKee et al. (2010) for acid beverage flocs. It must be noted that the I.V. 9.0/3.0 values that were correlated strongly with floc formation in this study, i.e., 1.20 to 1.50, were much lower than the I.V. 9.0/3.0 range reported to predict acid beverage floc formation with white, refined cane sugars, i.e., >3.2 (McKee et al., 2010). Moreover, McKee et al. (2010) observed that I.V. 9.0/3.0 values <2.1 indicated a low floc potential for acid beverage flocs. The differences in results are most likely because of the ethanol addition in alcohol beverages and/or possibly because this study used invert sugars. Overall, the results in this study suggest that much lower concentrations of polyphenolic colorants are involved in the formation of floc in alcoholic beverages. Sugarcane juice contains a complex mixture of phenolic colorants mainly polyphenolic and flavonoid colorants which are monomeric and sensitive to pH. The major polyphenols are caffeic, chlorogenic, and coumaric acids (Duarte-Almeida et al., 2011), and the predominant flavonoids are tricin, naringenin, luteolin, and apigenin (Smith and Paton, 1985; Duarte-Almeida et al., 2011). Flavonoids are water soluble and at acid pH are in the non-ionized form.

All the invert sugars A to D and their corresponding 60% ABV samples were first scanned between 220 and 350 nm to qualitatively determine the presence of protein, and contained protein (Table 2). The invert samples were subsequently quantitatively analyzed for protein using a dye-binding method (Bradford, 1976), and results are also listed in Table 2. The protein levels in the invert sugars were very low ranging from 0.11 to 0.60 μg/mL (ppm). Naidoo and Simpson (2004) reported that higher levels, i.e., >2.00 μg/mL, of protein in refined cane sugars assayed with a protein method also based on the Bradford dye-binding method, predicted that acid beverage flocs would form. Similar to the I.V. values (Fig. 2), the protein had a polynomial relationship (R^2=0.837) with haze (floc) formation suggesting that protein was somehow involved in floc formation. When we received other floc forming cane invert sugars (N=7) from the alcohol manufacturer this relationship was found to be consistent (see Fig. 3).

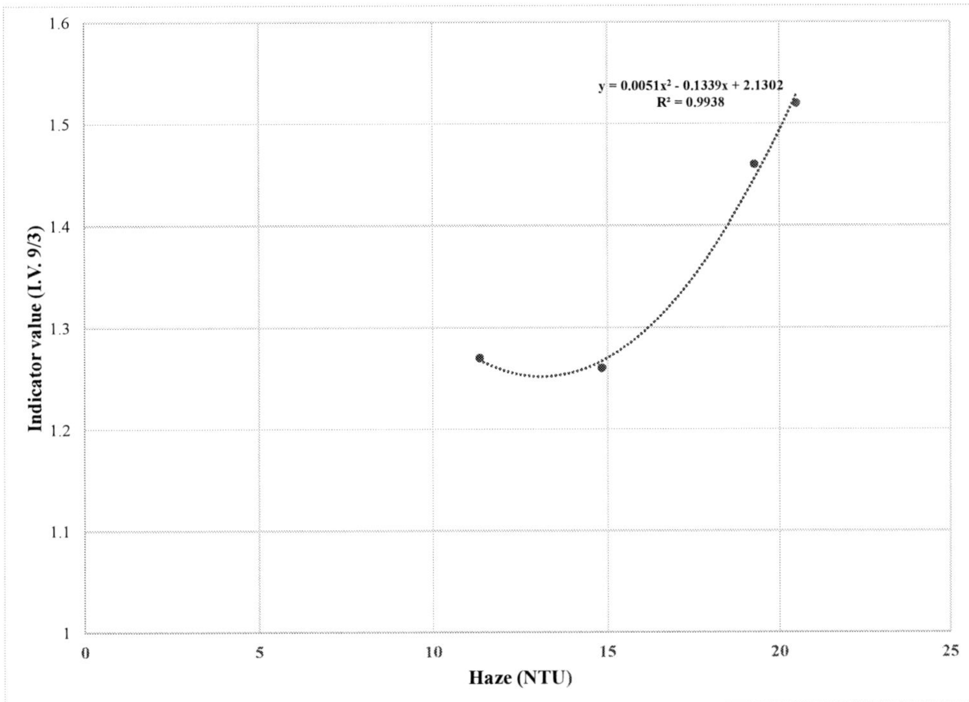

Fig. 2. Relationship between floc formation (haze) and the color indicator values (color at pH 9/color at pH 3) for initial invert sugars A to D (see Table 2).

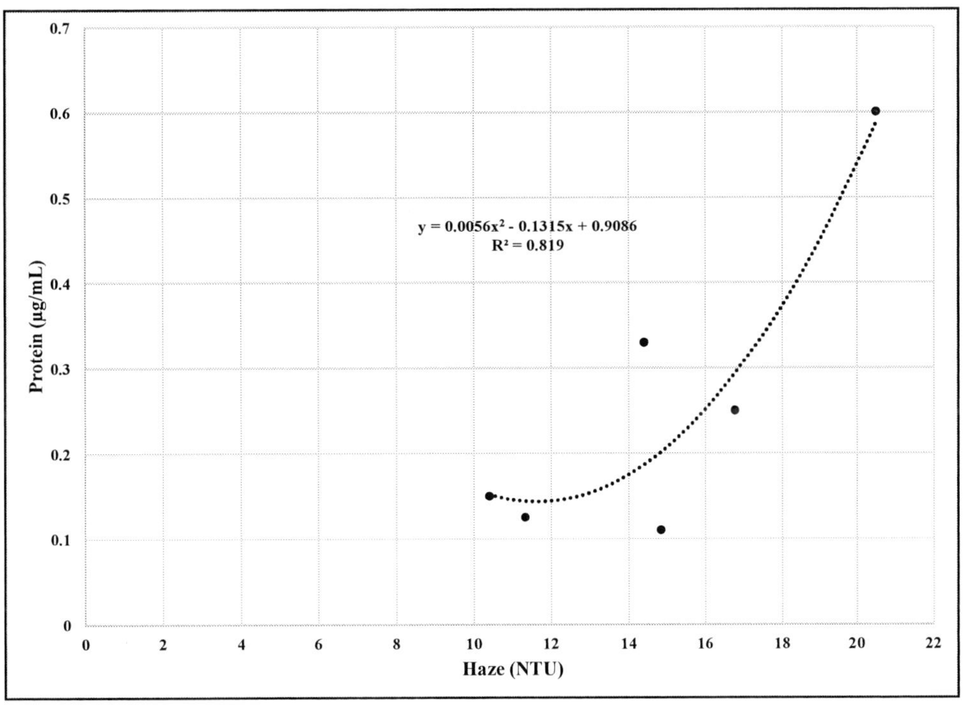

Fig. 3. Relationship between floc formation (haze) and the protein concentrations for initial unfiltered 60% ABV solutions produced from invert sugars A to G. Not enough sample from invert sugar B was available for protein analyses.

Small amounts of protein naturally occur in sugarcane (Clarke et al., 1978) and enzyme proteins are also added during sugar manufacture. In particular, amylases are often routinely added during sugar manufacturing to degrade starch. Dextranase is occasionally added to degrade dextran when it is present due to sugarcane deterioration. Floc formation in other alcoholic beverages, i.e. beer, has also been reported to be affected by proteins such as proteases added during beer processing (Seibert et al., 1996a). Not all proteins, however, are haze-active, i.e., contribute to floc formation. Proteins which readily interact and precipitate with polyphenols are generally large, have high proline (an amino acid) content, and lack secondary or tertiary structure (Asano et al., 1982). The protein assay used in this study is based on dye-binding primarily to basic (particularly arginine) and aromatic amino-acids, which means it is not sensitive to proline. A protein assay which is more sensitive to proline may have provided protein values that had an even stronger relationship with floc formation than what was found in this study (Table 2 and Fig. 3). To the best knowledge of the authors no such proline-sensitive protein assay has been developed yet.

Sugarcane protein in refined sugars has been reported to contain proline, although aspartic acid and alanine (both amino acids) are the major constituent amino acids (Clarke et al., 1978). Proline is also one of the major amino acid constituents of dextranase enzyme from *Chaetomium* gracile fungi (Hattori et al., 1981). Proline is also contained in amylase proteins produced from *Bacillus subtilis* and *Bacillus licheniformis* bacteria (Kuhn et al., 1982) which are added in sugar manufacture. Furthermore, Clarke et al. (19878) reported that "it has been observed, in attempts to treat floccing sugars with enzymes, that the addition of any protein, e.g., amylase, pectinase, dextranase, to a sugar solution enhanced floc formation".

Relative Levels of Haze-Active Proteins and Polyphenols in Cane Invert Sugars

Since the compositional analyses results suggested polyphenolic colorants (indicated by the color I.V. value) and proteins may be involved in floc formation these were further investigated. Estimates of the relative amounts of haze-active proteins in the 60% ABV samples were undertaken by adding tannic acid (TA), a very haze-active polyphenol that is known to combine with haze-active protein to form a floc that can be measured by light scattering nephelometry (Chapon, 1993). Conversely, estimates of the relative amounts of "sensitive" or haze-active polyphenolics in the 60% ABV samples were made by adding gelatin, a very haze-active protein (Siebert et al., 1996b) which is known to combine with haze-active polyphenolics. No pH adjustments were undertaken in these experiments which were undertaken at 25 ˚C. The results for the invert sugars and the control (beet invert sugar) are listed in Table 3. The beet invert sugar that did not form any visual haze (Table 2), showed a very small increase in haze as a result of the addition of both tannic acid and particularly protein (Table 3). These results indicated that the beet invert sugar must contain a very small amounts of haze-active protein and slightly more haze-active polyphenols. The I.V. value of the beet invert was 1.10 which was low compared to values for the cane invert sugars (see Table 2), but as the value was greater than 1.0 it indicated that the beet invert contained a very small amount of polyphenol colorants. Such natural colorants most likely came from the beet leaves and tuberous root. The beet invert sugar also contained a small amount of protein (0.04 μg/mL) detected by the Bradford dye-binding protein assay (Bradford, 1976), which confirmed that haze-active protein was present in the beet invert sugar (Table 2).

Table 3. Haze (NTU) resulting from the addition of haze-active gelatin protein or haze active poly-phenolic tannic acid to the initial unfiltered 60% ABV samples incubated at 25° C for 30 min.*

Additive concentration (ppm)	Cane Invert Sugar (A)	Cane Invert Sugar (C)	Cane Invert Sugar (D)	Beet Invert Sugar (control)
		Gelatin B□		
0	16.9	13.6	9.8	0.101
25	16.5	15.7	8.5	0.125
100	16.3	17.4	8.2	0.125
500	13.4	11.3	7.3	0.116
		Tannic Acid†		
0	16.9	13.6	7.6	0.068
25	15.4	12.8	9.2	0.098
100	14.7	12.7	9.2	1.210
500	12.2	10.9	7.9	0.089

*The A, C, and D 60% ABV samples were re-created in the USDA laboratory from the original invert sugars

□ Preliminary results showed that the gelatin alone caused negligible contributions to haze

† The stock solution of tannic acid (5000 ppm) used to create the solutions was 0.063 NTU

Results listed in Table 3 also showed that cane invert sugar D contained some haze-active protein since haze was formed on the addition of tannic acid. The reduced value at 500 ppm of tannic acid is due to dilution because the additive was added by volume. In comparison, the addition of gelatin to sugar D caused a decrease in floc formation indicating this sugar had negligible amounts of polyphenols or that the polyphenols were all already bound to sites on the sugar's protein. Invert sugars A and C did not seem to contain any available haze-active protein that could bind with tannic acid because the addition of this haze-active polyphenol had no effect (Table 3). Cane invert sugar C had increased haze formation on the addition of gelatin (the reduced value at 500 ppm of gelatin is due to dilution because the additive was added by volume) indicating not all of its polyphenolic colorants were bound. Beers are known to contain a considerable amount of haze-active protein and very little haze-active polyphenol, whereas commercial apple juices are just the opposite (Siebert et al, 1996a). However, the highest hazes in alcohol beverages have been reported to depend on the ratio of haze-active protein to haze-active polyphenol (Siebert et al., 1996b). Moreover, different polyphenols and tannins can also show variations in interaction with a given protein (Bennick, 2002).

Nature of the Floc Interactions

The nature of polyphenolic-protein interactions has been the subject of many studies because they are involved in beer floc formation. It is generally accepted that the polyphenol-protein interaction in solution is a reversible interaction leading to an equilibrium between the soluble protein/polyphenol complexes and the reactants (Bennick, 2002). These soluble complexes may reach a size where they are no longer soluble and this creates a floc. In turn, the flocs then may aggregate or they may undergo changes resulting in precipitation/sedimentation (Bennick, 2002).

It must be noted that in this study the flocs caused by the addition of cane invert sugars eventually precipitated to the bottom of the bottles.

Heating of unfiltered 60% ABV samples recreated in the laboratory from the initial invert sugars, was undertaken at 80 ˚C to evaluate if hydrophobic interactions were involved in floc formation. As shown in Table 4, there was a small but definite increase in the turbidity of the A to D samples on heating, which indicated that partial or full denaturation of the protein exposed additional polyphenol binding sites. Heating to increase floc formation is also consistent with a greater role for hydrophobic rather than hydrogen bonding (Siebert et al., 1996b). This is because hydrophobic groups tend to be concentrated in the interior of water-soluble proteins, it is likely that more hydrophobic (polyphenol) bonding sites were exposed by the heating as protein hydrogen bonds are broken with the heat (Siebert et al., 1996b). Heating has similarly been shown to increase the hydrophobicity of ovalbumin and lysozyme proteins (Kato et al., 1986). In this study, the heating of the flocs (60% ABV samples) to 80 ˚C not only consistently increased floc formation but also tended to create some larger floc particles (results not shown), which is further strong evidence that protein is involved with the formation and aggregation of flocs. During sugar processing at both the factory and refinery, proteins have been exposed to temperatures >85 ˚C for prolonged periods and are presumed to be thoroughly denatured. The results in Table 4, however, suggest proteins in refined sugars are not always completely denatured as previously considered, but further studies are needed to confirm this.

Table 4. Haze (NTU) resulting from the incubation of the initial unfiltered 60% ABV samples at 25 and 80 ˚C

Unfiltered alcohol solution	Haze at 25 ˚C	Haze at 80 ˚C
Cane Invert Sugar (A)*	12.3	14.2
Cane Invert Sugar (B)*	13.6	16.7
Cane Invert Sugar (C)*	10.7	11.5
Cane Invert Sugar (D)*	7.16	7.90
Beet invert sugar (control)	0.059	0.066

*The A to D 60% ABV samples were re-created in the USDA laboratory from the original invert sugars. Initial Haze (NTU) values at 25 ˚C were lower because they had not been left as long to form as the original samples (see Table 1).

The type of bonding involved in the floc formation was further probed by investigating the effect of adding polar, non-polar, and ionic solvents (Table 5). Since the initial protein-polyphenol interaction is considered reversible (Siebert et al., 2016b) it should be possible to dissolve already-formed floc in DMF (a hydrogen bond acceptor) and dioxane (non-polar) solvents. As shown in Table 5, all the flocs in the initial 60% ABV samples were progressively dissolved by dioxane and particularly DMF, although not all the flocs were dissolved in 40% of the solvent. Dissolution in DMF suggests that hydrogen bonds play some role in floc formation. Bennick (2002) observed that hydrogen bonds as well as hydrophobic effects are involved in polyphenolic/protein interactions.

Water aliquots added in the same proportions of the other solvents also caused a decline in haze most likely because of dilution as the flocs dissolved in this polar solvent. Thus the components of the floc are water soluble. The solubility of proteins in aqueous solutions depends on the distribution of hydrophilic and hydrophobic amino acid residues on the protein surface. As stated earlier, hydrophobic residues predominantly occur in the globular protein core.

Table 5. Effect of various concentrations of polar and non-polar solvents on Haze (NTU) formation or dissolution in the initial unfiltered 60% ABV samples incubated at 25° C for 30 min.*

% Final Volume Added	Water	Ethanol	Dioxane	DMF	20% Sodium Chloride Solution
Polarity:	*Very Polar*	*Low polar*	*Non-polar*	*Polar (hydrogen bond acceptor)*	*Ionic*
Cane Invert Sugar (A)					
0	16.2	16.2	16.2	16.2	16.2
10	11.3	16.7	14.4	13.0	13.2
20	11.5	21.1	13.9	12.5	9.8
40	7.2	21.1	12.9	9.8	5.3
Cane Invert Sugar (C)					
0	12.6	12.6	12.6	12.6	12.6
10	10.2	12.2	11.2	11.3	10.0
20	9.4	12.5	10.7	10.2	9.3
40	5.9	12.9	9.2	8.4	5.1
Cane Invert Sugar (D)					
0	7.4	7.4	7.4	7.4	7.4
10	6.2	7.4	6.8	6.2	4.6
20	4.8	8.0	6.3	5.1	4.4
40	3.7	8.4	6.0	5.0	2.7
Beet Invert Sugar					
0	0.072	0.072	0.072	0.072	0.072
10	0.065	0.086	0.068	0.075	0.087
20	0.021	0.091	0.087	0.093	0.088
40	0.022	0.116	0.082	0.088	0.070

*The A, C, and D 60% ABV samples were re-created in the USDA laboratory from the original invert sugars. Initial Haze (NTU) values are lower because they had not been left as long to form as the original samples (see Table 1).

In comparison to the solvent water, all the cane invert sugars dissolved even further in the ionic sodium chloride solution. Hagerman and Butler (1978) previously reported that ionic interaction between proteins and polyphenols is considered to be of little or no importance. Furthermore, Asano et al. (1982) reported that, in acidic conditions, such as pH 4.5 in alcoholic drinks sweetened with sugar, hydroxyl groups of polyphenols have no charge and, therefore, ionic bonding is not involved in the floc formation mechanism. Therefore, the sodium chloride most likely allowed more floc to dissolve because it altered the dielectric constant of the solvent. In strong contrast, ethanol, a miscible solvent, is known to decrease the dielectric constant of water which in turn allows proteins to come close together and eventually precipitate, i.e., desolubilize the protein. Furthermore, at high concentrations such as in this study, ethanol accelerates the denaturation of protein by disrupting hydrophobic interactions between amino acids. These reasons most likely explains why the cane invert sugars produced even more haze on the addition of 10 to 40% ethanol (Table 5). This was also further evidenced by cane invert sugar A which contained more protein than the other sugars, exhibiting the greatest increase in haze with ethanol addition (see Table 2).

Solution pH is also known to often influence how proteins behave by affecting the net charge of the protein molecules. This can lead to more or less ionization, with a change in ionic attraction or repulsion which may also alter the protein solubility or the molecular conformation (Siebert et al., 1996a). This in turn may influence the accessibility of the sites on the protein to which the polyphenols can bind. Generally, it has been found that precipitation of polyphenol complexes is pH dependent, with the greatest precipitation occurring near the isoelectric point of the protein due to the decreased electrostatic repulsion of protein molecules at this pH (Hagerman and Butler, 1978). The cane invert sugars provided by the beverage manufacturer were acidic (see Table 2) but not adjusted for pH. At such acidic pH, the proteins in the cane sugars will be positive charged (Clarke et al., 1978), thus more studies are now warranted on the effects of pH.

Critical Concentrations and Ratios of Components for Floc Formation

Polyphenols are thought to be multivalent ligands and may form crosslinks with two or more protein molecules. Thus the stoichemetry and size of the polyphenol-protein complexes depends on the concentrations as well as the reactants and protein/polyphenol ratio (Siebert et al, 1996). As the beet invert sugar (control) in this study contained very low amounts of protein and polyphenols (see Table 2) but still did not form floc, there are most likely critical concentrations of protein and polyphenols both individually and in relation to one another. An excess of one component may change the interaction, but further studies are needed on this.

Solutions to Floc Problems

Although most sugar refinery processes reduce the amount of protein, it is still often found in refined sugars (Clarke et al., 1978), although at very low concentrations which have not always been detected. Dunsmore et al. (1978) reported that acid beverage floc forming compounds found in refined sugars could be removed with select refining processes, but they were unable to explain why they worked. For example, clarification by carbonatation removed floc while phosphatation (with and without flocculants) clarification did not. Proteases (enzymes that break down protein) were not suitable because pH and temperature optima for such enzymes are not practical in a sugar refinery. Refinery liquors filtered (0.45, 0.60, 0.80, and 1.20 µm pore sizes) at temperatures < 60 °C rendered them floc negative irrespective of the filter pore size. Liquors filtered through the smallest filter pore size of 0.45 µm were always floc negative even when filtered at 80 °C, suggesting the components involved in floc formation are very small, i.e., $<$ 0.60 µm. Thus, overall the higher the temperature the greater was the tendency to floc and the larger the filter pore size the greater was the amount of floc formed. The results presented in this paper can now explain the temperature results. As shown in Table 4, on heating floc solutions to 80 °C there was a definite increase in floc (haze) formation because heating exposes hydrophobic groups concentrated in the interior of the water-soluble protein (that may already have been partially denatured during processing) which are polyphenol bonding sites, causing more flocs to form (Siebert et al., 1996b).

Dunsmore et al. (1978), also reported that refinery liquors filtered through filter aids used in the refinery, e.g., Kieselguhr, Celites 505 and 535 and fine filter aid Celite Filter Cell, at 60 ˚C did not floc, further indicating that the exposure of the protein to higher, denaturing temperatures, to expose more polyphenol binding sites is critical to floc formation. This is also further evidence that hydrophobic bonding is critically involved in the floc formation mechanism and not ionic bonding as suggested by Clarke et al. (1978). The fact that ionic flocculants used in phosphatation clarification did not remove floc components (Dunsmore et al., 1978) is also another indication that ionic bonds are not involved.

More recently, Eggleston et al. (2015) undertook laboratory studies with powdered activated carbon, and observed that powdered activated carbon (PAC) removed amylase protein from refinery liquors after phosphatation clarification. It is also most fortunate that PAC was also shown to preferentially remove of polyphenolic and flavonoid colorants (Eggleston et al., 2015). This confirmed the research by Paton and Smith (1982), who used chromatography profiles to show that granular activated carbon was capable of removing all polyphenolic and flavonoid colorants from refinery liquor whereas ion exchange and bone char were not. Therefore, granular activated carbon, used in many sugar refineries as a decolorizing process, or powdered activated carbon could be very useful in removing the components involved in floc formation, and more studies are now warranted.

CONCLUSIONS

This study, has shown that the formation of alcoholic beverage flocs at room temperature (~25 ˚C) by the addition of cane invert sugars is *not* due to fat, inorganic ash, soluble and insoluble starch, Brix, pH, and oligosaccharide components of the invert sugars. There were strong correlations between haze floc and the I.V. (R^2=0.994) and protein (R^2=0.837) values of the invert sugars. Since the I.V. value is an indicator of polyphenolic/flavonoid colorants these results indicated that polyphenol-protein flocs were involved in the floc formation mechanism. Polyphenolic-protein flocs are known to combine and cause floc formation in other alcohol beverages including beers, ciders, and wines. Much lower levels of I.V. ratios and protein are involved with alcohol flocs than those reported for acid beverage flocs sweetened with sugar. This is most likely because ethanol at the high concentrations in alcohol beverages accelerates the denaturation of protein and exposes more polyphenol binding sites. More haze formed when select 60% ABV samples were heated to 80 ˚C because further partial or full denaturation of protein exposed hydrophobic polyphenol binding sites. Heating to increase floc formation is consistent with a greater role for hydrophobic rather than hydrogen bonding. By probing the floc formation with nephelometry and a haze (floc) active protein and polyphenol, as well as polar, non-polar, and ionic solvents, further evidence was found that hydrophobic interactions and hydrogen bonding are involved with floc formation and interactions, with hydrophobic interactions being more critical. There was very little evidence for covalent or ionic bonding of protein to polyphenolic colorants. Since I.V. is an indicator of haze-active polyphenolic compounds, more research is now warranted to ascertain the specific polyphenolic compounds involved in floc formation. Further studies are currently being undertaken to determine the mechanism of protein-polyphenol complexes. Dextran concentrations were also found to be weakly related to floc formation but it is not known if this neutral polysaccharide is entrained or co-precipitated with protein-polyphenol complexes or more involved in the floc formation mechanism.

Overall results in this paper suggest that the removal of polyphenol and flavonoid colorants during sugar manufacture and refining now has an additional incentive. Furthermore, the minimum amount of enzyme proteins should be added in sugar manufacture to control starch and dextran. Certainly adding these enzymes without checking if they are working should be avoided. Granular activated carbon, used in many sugar refineries as a decolorizing process, or powdered activated carbon could be very useful in removing both the protein and polyphenolic components involved in floc formation, and more studies are now warranted to confirm this.

ACKNOWLEDGEMENTS

The authors thank Mr. Eldwin St. Cyr for excellent technical assistance and Dr. Barry Hurlbert of USDA-ARS-SRRC for help with the protein assays. Mention of trade names or commercial products in this article is solely for the purpose of providing specific information and does not imply recommendation or endorsement by the U.S. Department of Agriculture. USDA is an equal opportunity provider and employer.

REFERENCES

Asakura T, Adachi K, Schwartz E (1978). Stabilizing effect of various organic solvents on protein. *J. Biol. Chem.*, **253**, 6423-6425.

Asano K, Shinagawa K, Hashimoto N (1982). Characterization of haze-forming proteins of beet and their roles in chill haze formation. *J. Am. Soc. Brew. Chem.*, **40**, 147-154.

Asano K, Ohtsu K, Shinagawa K, Hashimoto N (1984). Affinity of proanthocyanidins and their oxidation products for haze forming proteins of beer and their formation of chill haze. *Agric. Biol. Chem*, **48**, 1129-1146.

Bennick A (2002). Interaction of plant polyphenols with salivary proteins. *Crit. Rev. Oral Biol. Med.,* **13**, 184-196.

Bradford MM (1976). A rapid and sensitive method for the quantitation of microgram quantities of protein utilizing the principle of protein-dye binding. *Anal. Biochem.,* **72**, 248-254.

Chapon L (1993). Nephelometry as a method for studying the relations between polyphenols and proteins. *J. Inst. Brew.,* **99**, 49-56.

Clarke MA, Roberts EJ, Godshall MA, and Carpenter FG (1978). Beverage floc and cane sugar. *Internat. Sugar J.*, *80*, 197-202.

Cohen, M.A., Dionisio, D.G., and Drescher SJ (1970). The isolation and characterization of certain impurities responsible for quality problems in refined cane sugar. *Proc. Sugar Indust. Technol.,* 123-165.

Cole M, Eggleston G, Gilbert A, Chung Y (2016). Development of an analytical method to measure insoluble and soluble starch in sugarcane and sweet sorghum products. *Food Chem.*, *190*, 50-59.

Deitz, V., F. Carpenter, C. Rieger, and H. Rootare (1959). The influence on pH on some properties of char liquors. *Bone Char Research Project Tech. Report*, 1959. No. 53.

Duarte-Almeida JM, Salatino A, Genovese MI, and Lajolo FM (2011). Phenolic composition and antioxidant activity of culms and sugarcane (*Saccharum officinarum* L.) products. *Food Chem.*, **125**, 660-664.

Dunsmore A. Heal MJ, Matic M and Rugggas FM (1978). A practical solution to the acid beverage floc problem. *Proc. Sugar Indust. Technol. Mtg*, 514-527.

Edye L (2004). Sugar quality in soft drink manufacture: The acid beverage floc problem. In: Shaidi F, Spanier A, Ch Ho., Braggins T (eds) *Quality of Fresh and Processed Foods*. Kluwer Academic/Plenum Publishers, New York, 317-326.

Eggleston G and Borges E (2015). The multiple applications of ion chromatography oligosaccharide fingerprint profiles to solve a variety of sugar and sugar-biofuel industry problems. *J. Agric. Fd. Chem.*, **63**, 2841-2851.

Eggleston G, Vawda A, Zatlokovicz J, Lima I, and Eldwin St. Cyr E (2014). Use of activated carbons to remove undesirable residual amylase from factory and refinery streams. *Proc. 2014 Sugar Industry Technol. Mtg*, Canada, **LXXIII**, 330-342.

Hagerman AE and Butler LG (1978). Protein precipitation method for the quantitative determination of tannins. *J. Agric. Food Chem.*, **26**, 809-812.

Hattori A, Ishibashi K, Minato S. (1981). The purification and characterization of the dextranase of *Chaetomium gracile*. *Agric. Biol. Chem.,* **45**, 2409-2416.

Iciek J, Blaszcyk I, Biernasiak J, Lisik K and Wojtczak M (2012). Floc in acidified solutions of white sugar – a literature review. *Zuckerindustrie*, **137**, 449-453.

Kuhn H, Fietzek P, Lampen J (1982). N-Terminal amino acid sequence of *Bacillus licheniformis* α-Amylase: Comparison with *Bacillus amyloliquefaciens* and *Bacillus subtilis* enzymes. *J. Bacteriol.*, **149**, 372-373.

Lemos LR, Nogueeira A, Wosiacki G, Lacerda LG, and Demiate, IM (2013). The influence of different amounts of dextran and starch in crystallized sugar in the formation of floc in acidic carbonated solutions and alcoholic solutions. *Sugar Tech*, **15**, 65-70.

McKee M, Triche R, Godshall MA, Richard C (2010). Floc prediction in refined white cane sugar: The SPRI Rapid Floc Test. *Proc. SPRI Conf*, New Orleans, 28-31.

Naidoo Y and Simpson R (2004). The detection of protein in refined sugar. *Proc S. Afr. Sugar Technol Assoc.,* **78**, 569-572.

Paton NH and Smith P. (1982). A HPLC study of the changes in colorant composition following factory decolorisation of raw liquors with bone char, resin and granular carbon. *Proc. Sugar Proc. Res. Conf.*, 1-23.

Siebert KJ, Carrasco A, and Lynn PY (1996a). Formation of protein-polyphenol haze in beverages. *J. Agric. Fd. Chem.*, **44**, 1997-2005.

Siebert KJ, Troukhanova NV, Lynn PY. (1996b) Nature of poly-phenol-protein interactions. *J. Agric. Food Chem.*, **44**, 80-85.

Smith P and Paton NH. (1985). Sugarcane flavonoids. *Sugar Technology Reviews*, **12**, 117-142.

S.I.T. Paper #1165

COGENERATION ENHANCEMENT USING LOW PRESSURE VAPOUR FOR CRYSTALLISATION

Vivek Verma† and Dinesh Saini‡
vivek.verma@sprayengineering.com

ABSTRACT

Crystallization and power generation in sugar factory have very close association with each other. Synergically integration of crystallization within system has direct effects on the factory Cogeneration enhancement. Cogeneration is actually thermodynamically efficient utilization of fuel where thermal energy is used to generate electricity and the remaining heat is utilized to meet the process requirements in sugar industry. This paper gives an insight to augment power generation by use of low pressure vapour from fourth effect falling film evaporator using innovative approach & efficient technologies and gives live example of Dhanalakshmi Srinivasan Sugars Pvt. Ltd., Tamil Nadu, India. This plant is exporting surplus power @ 100kW/Ton of Cane using efficient and sustainable technology. The plant is having Spray Continuous Pan (SCP®) for crystallization and utilizes vapour from fourth effect falling film evaporator.

Keywords: Cogeneration, Evaporation, Continuous Crystallization, Refined Sugar, Colour Value, Massecuite Boiling, Flashing System.

INTRODUCTION

In the last two decades many sugar mills started practicing co-generation and contributing in reducing the energy deficit that India is currently facing. But many mills are unable to realise the benefits of co-generation for which it is intended for. There are several factors contributing to the issue. However apart from the non-technical factors there exist several technical factors which need to be addressed. The time has come for the implementation.

KEY FACTORS

The key areas which have major impact on enhancing power export are:

1. Reduction in use of process thermal energy with optimized vapour bleeding scheme.
2. Reduction in in-house electrical power consumption.
3. Recovery of waste heat from all streams for use in process or power generation.
4. The installation of more efficient cogeneration systems.

† Managing Director, Spray Engineering Devices Limited.
‡ Team Leader, Spray Engineering Devices Limited.

5. Installation of energy efficient equipment in the process house.
6. Optimization of process layout and operational parameters.

So one's thrust should be on minimization of the steam requirement for sugar processing through efficient vapour bleeding by most suitable selection of evaporator type and size.

SYSTEM CONFIGURATION AND ACHIEVEMENTS

A live example on embracing various technologies which are implemented in Dhanalakshmi Srinivasan Sugars Pvt. Ltd. (DSSPL), (3500 TCD Plant), Tamilnadu, India is reflected next, that led this plant such efficiency. The current steam demand of the plant is 31-32% on cane and achieved highest exporting power range i.e.100-105kW/Ton of cane.

Steam is consumed directly or indirectly mainly in two areas of the sugar processing i.e. evaporation and crystallization. This paper presented different techniques to enhance cogeneration with minimum steam consumption in sugar factories in order to make the sugar factory more profitable by increasing the power export.

This plant operates at 110 bar 540^0C cycle with one 110 TPH boiler and 23 MW Power Generation with Sulphur-Free Refined Sugar production. Apart from the steam economy and power generation milestones of this plant has few more special features.

- A modern 2 roller 4-mill tandem plant with electrical drives and an inline shredder.
- The clarification is Sulphur free with only defecation followed.
- A modern boiling house with a combination of Falling Film Evaporators (FFEs), Direct Contact Heaters, Spray Continuous Pan (SCP®) for A, B, C, R1 and R2 Massecuite boiling with Melt Clarification system.
- No Ion-Exchange Column in the plant for any clarification.
- No Deep Bed Filter is in operation.
- CIP system with full automation in line FFEs.
- Refined Sugar is produced in Spray Continuous Pan (SCP®); there is no Batch Pan in the plant.
- Steam and vapour balance done to achieve 31-32% process steam consumption on cane.
- 40% savings on power consumption by the advanced AC-VFD Motors in the process plant over the conventional systems.
- Fully automated Sugar and Co-Generation units with DCS controls.

- State of art cooling and condensing system with gravity inflow cooling towers.

The achievement of the plant lies in the process house especially the evaporation and crystallization, in this plant the systems were designed and implemented to ensure its supremacy in achieving the performance.

A. EVAPORATOR STATION

Concentration of cane juice to saturation point alone consumes major portion of total energy demand. The quantum of steam used for this purpose is very significant for co-gen. Optimized selection of total temperature difference and number of effects reduces a lot the sinking of the energy in this section.

Making availability of more numbers of streams for process heating and that too at required temperature & pressure, do phenomenal improvements in the sugar process.

Typically, more is the number of evaporator effects; greater will be the steam economy. The evaporator station in this plant was designed to deliver 0.65-0.70 absolute bar pressure to pans and that too from 4th effect vapours. The whole thing that lies behind all this happening is by crystallization with fourth effect of Falling Film Evaporators (FFEs). Falling Film Evaporator contrary to Rising Film and Robert type evaporator can be easily operated with very low ΔT i.e. as low as 3-4°C. FFEs are acting as the backbone for the efficiency levels that plant has achieved.

Extensive vapours coming out of the different effects of the evaporators are used for juice heaters and vacuum pans. FFEs making whole evaporation station as amalgam of efficient bleeding system along with flashing system with enhancing flash recovery are making the plant high energy efficient & reduced steam demand.

The efficient distribution of liquid for constant wetting of heating tubes with juice is well taken care for a trouble-free operation of a Falling Film Evaporator. The FFE with hi-tech cascade distributor design forms a thin and even film of feed solution at inner surface of the tube maintaining positive juice circulation. The competent design modification of falling film evaporators and its distributor design helped in achieving marvelous consequences in this plant. The evaporator station setup at this plant is as under:

- Quintuple Evaporator station with seven identical Tubular Falling Film Evaporators (7 Nos. x 1500 m^2), Total Heating Surface Area = 10,500 m^2 [9000 m^2 heating surface is in operation and 1500 m^2 in the standby].

- Condensate Flash Vessel System with flashing off vapours from hot condensate to be further utilised for juice evaporation. Parallelly it eliminates the condensate pumps of individual steam/vapours consumers. Direct Contact Juice Heaters (DCH) for Raw, Defecated and Clear Juice & Syrup heating are used in very close approach.

- Single Floating Evaporator for cleaning all effects.

- Working temperature range of the evaporator is 125°C to 75°C.

Sr. No.	Evaporator Effects	Operating Pressure (bar A)	Operating Temp. (^{0}C)	Operating Temp. Difference (ΔT)
1	1st Effect	1.8–1.9	118-120	6-8^{0}C
2	2nd Effect	1.3-1.4	110-112	8-9^{0}C
3	3rd Effect	0.9-1.0	98-102	7-10^{0}C
4	4th Effect	0.6-0.65	93-96	5-7^{0}C
5	5th Effect	0.35-0.40	75-78	12-15^{0}C

Table 1: Present operating parameters of evaporator station

Figure 1: Evaporator Station

The bleeding scheme is the heart of evaporator station where lot of energy can be saved for enhanced power generation capacity specifically by operating crystallization part using fourth effect vapours of falling film evaporator.

DSSPL entire juice heating is only with direct contact heaters (DCH) (no Tubular Juice Heater or Plate Heat Exchanger).

a. Mixed Juice is two stages heating, first with Pan Vapors followed by 5th effect vapour. These two stages heating raise the juice temperature to 70-74°C.

b. Limed juice heating is in three stages all through DCH using 4th, 3rd & 2nd effect vapour.

c. Clear juice heating is single stage using 2nd effect vapour.

d. All crystallization part is taken by Spray Continuous Pan (SCP®) which are operating with 4th effect vapour of falling film evaporator.

Sr. No.	Description	Configuration	Operating Temperature
1	MJ 1st Heating	Pan Vapour	54-57 ^{0}C
2	MJ 2nd Heating	5th Effect Vapour	74-77 ^{0}C
3	DJ 1st Heating	4th Effect Vapour	86-89 ^{0}C
4	DJ 2nd Heating	3rd Effect Vapour	96-99 ^{0}C
5	DJ 3rd Heating	2nd Effect Vapour	107-110 ^{0}C
6	CJ 1st Heating	2nd Effect Vapour	107-110 ^{0}C
7	All Pans (A, B, C & R)	4th Effect Vapour	86-89 ^{0}C

Table 2: Vapour bleeding arrangement

The said bleeding scheme has reduced the steam consumption to 31.0-32.0% on cane and enhanced the power generation capacity.

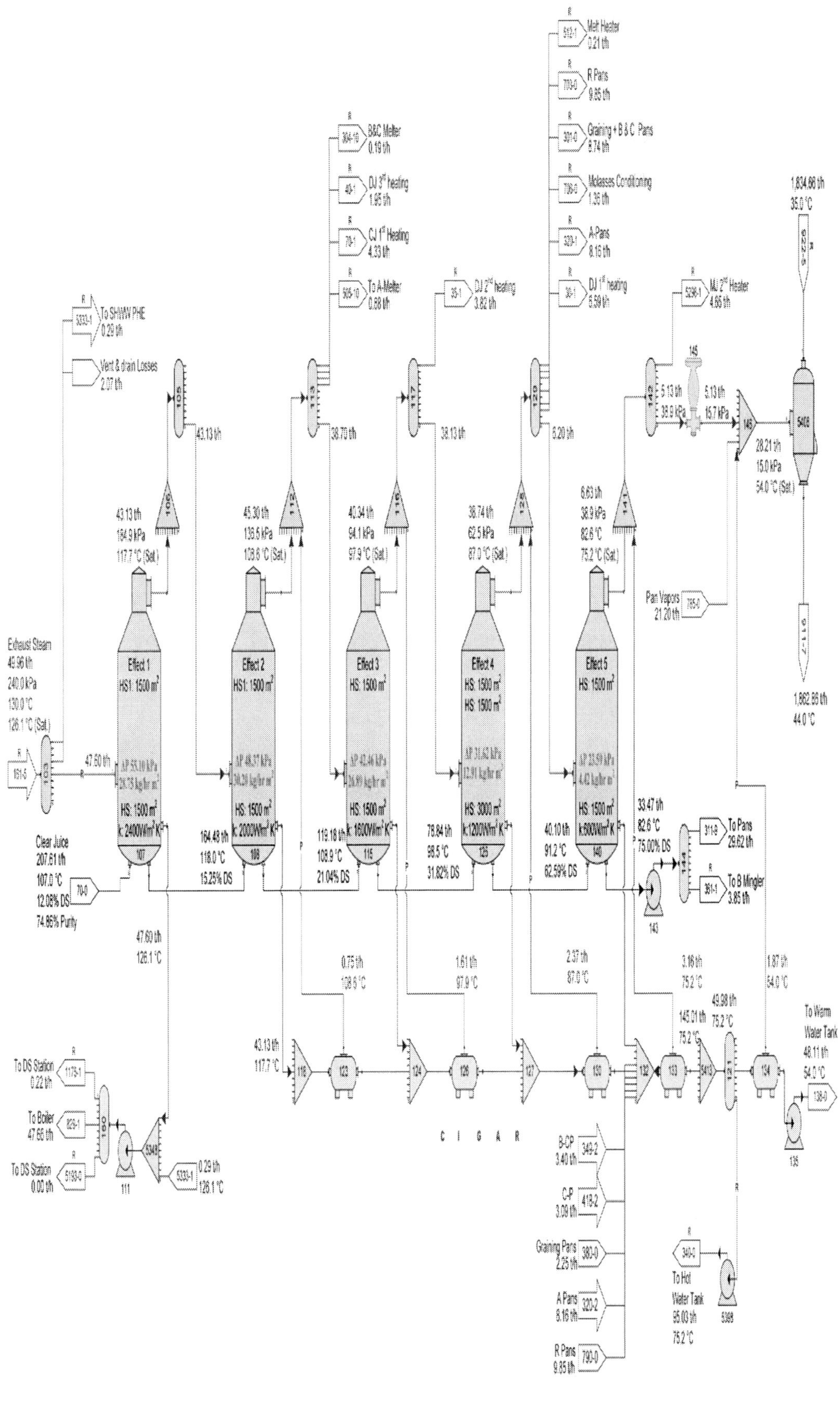

Melt Heater 0.21 t/h
R Pans 9.85 t/h
Graining + B & C Pans 8.74 t/h
Molasses Conditioning 1.36 t/h
A-Pans 8.16 t/h
DJ 1st heating 6.69 t/h
B&C Melter 0.19 t/h
DJ 3rd heating 1.95 t/h
CJ 1st Heating 4.33 t/h
To A-Melter 0.68 t/h
DJ 2nd heating 3.82 t/h
MJ 2nd Heater 4.65 t/h
To SHWW PHE 0.29 t/h
Vent & drain Losses 2.07 t/h
1,834.66 t/h 35.0 °C
1,862.86 t/h 44.0 °C
Exhaust Steam 49.96 t/h 240.0 kPa 130.0 °C 126.1 °C (Sat.)
Effect 1
Effect 2
Effect 3
Effect 4
Effect 5
Clear Juice 207.61 t/h 107.0 °C 12.08% DS 74.86% Purity
Pan Vapors 21.20 t/h
To Pans 29.62 t/h
To B Mingler 3.85 t/h
To Warm Water Tank 48.11 t/h 54.0 °C
To DS Station 0.22 t/h
To Boiler 47.65 t/h
To DS Station 0.00 t/h
C I G A R
B-CP 3.40 t/h
C-P 3.09 t/h
Graining Pans 2.25 t/h
A Pans 8.16 t/h
R Pans 9.85 t/h
To Hot Water Tank 95.03 t/h 75.2 °C

Figure 2: Process simulation diagram of actual operating Evaporator Station

B. CRYSTALLIZATION

Crystallization is another major energy intensive area. This section further has significant effect on the efficiency of the plant.

But in DSSPL all Pans are Spray Continuous Pan (SCP®) so the demand is of minimum possible steam pressure for massecuite boiling. Lower pressure requisite for this plant for pan bleeding has made higher impact on the co-gen.

Continuous pans are characterized by constant boiling point and constant heating surface to volume ratio in contrast to batch pan boiling where temperature increases with boiling. In low pressure operation pans tube temperature does not increase much and also improves sugar quality beside thermal energy benefits there are quality benefits as well.

The effect of low pressure boiling on sugar quality of DSSPL is that their refined sugar colour is only 33-34IU. The SCP® configurations of pans are efficiently making optimizing evaporative crystallization of sugar solutions. SCP® has shown high efficiency without any compromise on the process parameters.

Spray Continuous Pan (SCP®) is easily operated at very low ΔT i.e. as low as 20-25°C. Spray Continuous Pan played a significant role in the reduction of energy consumption and to enhance power generation that plant has achieved.

The installed vacuum pan for various massecuite boiling is as under:

- All three Spray Continuous Pans (SCP®) with 9 chambers each for A, B, C, and R1 & R2 massecuite boiling with complete automation.
- Capacity of Spray Continuous Pans: A – 50 TPH, B – 20 TPH, C – 15 TPH, R1 – 25 TPH and R2 – 20 TPH.
- Molasses Conditioner for AH, BH, CL and R1 molasses of direct contact heating type.
- 2 nos. of Seed Crystallizer of 20T for A and B Seeds.
- 2 nos. of Grain Crystallizer of 30T for B and C Grain.
- 5 nos. of supply tanks of 2 for syrup & melt and 3 for molasses storage.

Figure 3: Pan Station

The process engineering design of the plant have made significant impact on various aspects of factory operation and equipment usage and finally to power generation and export. This design together reduces the utility requirements to a greater level in the sugar process house. Middle pressure steam requirement is eliminated. Furthermore, all these together alter the co-gen in an encouraging way such that the total electricity generation of the plant gets increased by 9-12% if compared with other factories operating in the vicinity.

Other remarkable parameters of this plant are as under:

Sr. No.	Particulars	Unit	SEASON WISE			
			2015-16	2014-15	2013-14	2012-13
1	Plant Crushing Rate	MT	2700	3200	3000	2000
2	Power Consumption per ton of Cane	kWh/T	30.61	29.40	30.84	30.79
3	Cane Crushed (Lakh)	MT	6.56	5.00	3.88	1.73
4	Crushing Days	Days	243	160	130	90
5	Sugar Colour (Season average)	IU	33.93			

Table 3: Season wise parameters

Figure 4: Cogeneration Unit

Parameters	UOM	Jan-2016	Feb-16	Mar-16	Apr-16
Power Generation per Ton of Cane	kWh/T	135.96	128.11	135.85	135.35
Power Consumption per Ton of Cane	kWh/T	32.50	29.81	31.86	30.14
Power Export per Ton of Cane	kWh/T	103.46	99.3	103.99	105.21

Table 4: Month wise cogeneration details

RESULT AND DISCUSSION

A lot of opportunities for energy saving exist in the sugar industry and there is need to exploit these for better performance. It is imperative for cogeneration to be adopted in the industry to increase electricity output and to make better use of process steam through optimised bleeding arrangement supported by efficient evaporators and continuous pans. The deployment of energy efficient equipment and technology for the improvement of co-generation can entirely change the present scenario of sugar industry.

Increasing power export beyond present level of 70-80 kWh/Ton to 100-110 kWh/Ton, becomes very important for profitability of this industry.

This concludes that 12% of total electricity generation can be enriched by correction in energy management system as an ultimate measure in the efficiency improvement of sugar plant. Adopting such technology and idea, the sugar infrastructure will have enough power to cater the energy demand of themselves and their vicinity.

Other performance parameters of the Dhanalakshmi Srinivasan Sugars, Tamilnadu, India is as under:

Sr. No.	Descriptions/Parameters	Value
1.	Crushing Capacity	3500 TCD
2.	Cane Crushed (Lakh MT)	5.0
3.	Bagasse % Cane	27.5
4.	Fibre % Cane	13.10
5.	Mixed Juice Purity (%)	81-82
6.	Mixed Juice Brix (%)	13.8
7.	Clear Juice Purity (%)	81.9
8.	Clear Juice Brix (%)	12.5
9.	Syrup Purity (%)	81.75
10.	Syrup Brix (%)	70-75
11.	Steam Consumtion of the Plant	31-32% on cane
12.	Total Heating Surface Area of FFEs	10,500 m^2
13.	A Masseciute (%) cane	29
14.	B Massecuite (%) cane	11
15.	C Massecuite (%) cane	8.5
16.	Refinery Massecuite (R_1) (%) cane	28
17.	Refinery Massecuite (R_2) (%) cane	21
18.	Sugar Colour	< 35 IU
19.	Power Export	100-110 kWh/Ton

Table 5: Performance parameters of the plant

FURTHER SCOPE

There is further scope of improvement in power cogeneration with addition & balancing of energy efficient equipments & technologies like bagasse drying, flue gas heat recovery, exhaust condensate heating etc.

S.I.T. Paper #1166

OPTIMIZATION OF WHITE SUGAR COLOR MANAGEMENT THROUGH THE UTILIZATION OF ON-LINE COLORIMETERS WITH VIDEO MONITORING

Catherine Bouché*, Norbert Duc
ITECA SOCADEI, 445 rue Denis Papin, 13592 Aix en Provence, France

Abstract

The priority for any factory operation is to produce sugar that meets its customer's specification while minimizing the financial impact of such compliance. To ensure low color sugar, the default position tends to be to over wash it, thus leading to substantial water volume and energy consumptions, with a high amount of remelted sugar which should instead be reduced. The over washing at the sugar end process can also mask significant variations occurring within the beet end operation.

The paper describes the installation of two online ITECA color camera systems at ERAWAN SUGAR, a Thai factory founded in 2006. It presents the experience feedback over a one year period. The equipments are installed above two large vibrators conveyors and measure moist white sugar discharged from every batch of two sets of four batch centrifugals. The digital cameras send images and videos to a computer in control room that processes the images in Real time. Based on the correlation with the laboratory measurements, ITECA software calculates the sugar color in ICUMSA units. The on-line measurement is used to optimize the individual centrifugal washing time and keep track of the different massecuite qualities. The Colobserver® software automatically detects any non conformity to avoid contaminating the dryer with out of spec products.

The paper shows how the factory rapidly and significantly stabilized the final sugar color production around the color set point of 30 IU. It also shows that a close monitoring of the sugar color brings a considerable improvement of the global operation efficiency, especially when comparing the on-line measurement to conventional laboratory sampling methods.

Introduction

Now more than ever before, in a very competitive international landscape, it is essential to produce good quality sugar at lower cost. Among all the parameters a factory must track to produce white sugar meeting high quality standards, color is a key factor.

The standard approach recommended by ICUMSA to measure sugar color is a method based on the analysis of samples taken on the sugar conveyor at variable time intervals. The number of analysis performed is always limited by the speed of the sampling and the means implemented by the laboratory to carry out the work.

We found at ERAWAN that, at best, the laboratory could analyze one sample every 1 or 2 hours, therefore leaving several discharges of centrifugals unseen, with possible color variations between two measurements. With so little information it is very hard for the operators to adjust the process properly and there is a high risk to send out of spec product to the dryer.

The implementation of ITECA colorimeters, taking more than 2 images per second over the full width of each conveyor, not only provides a constant tracking of the whole production with high precision,

but the measurements are also delivered in ICUMSA Units as the equipments are first correlated to the standard method.
The paper shows that the operators can now detect any fluctuation, react in real time by optimizing the individual washing time of each centrifugal and feel confident that the production complies with the specification.

ERAWAN SUGAR application description

After the crystallization, the massecuite coming out of six batch pans flows into two consecutive sets of four FIVES CAIL D412 batch centrifugals feeding two separated vibrating conveyors (Lines 1 and 2 - Schematic 1). The 1,000 tons of refined sugar produced per day are conveyed to a dryer or may be occasionally diverted back to a melter in case the quality produced is out of spec.

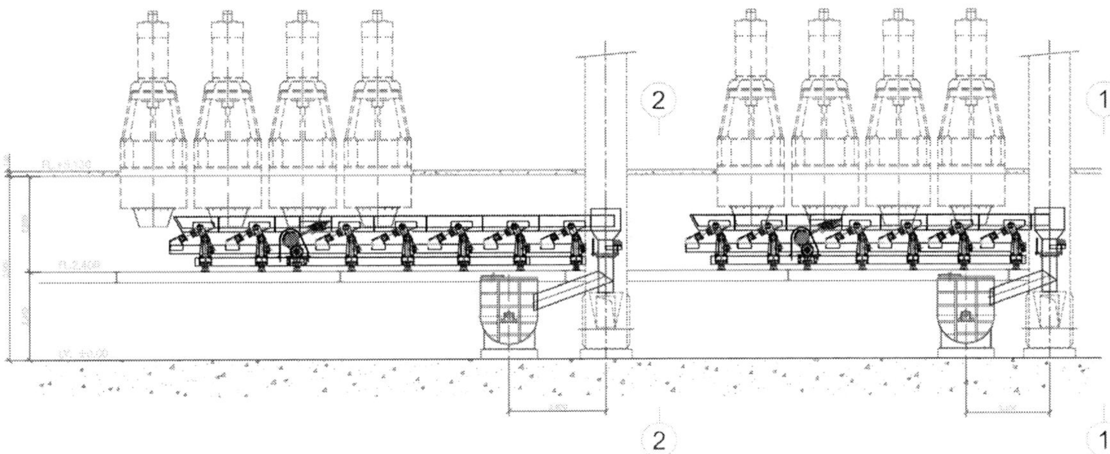

Schematic 1: Configuration of the two sets of four FIVES CAIL centrifugals

ITECA installation consists in two colorimeters type Colobserver® mounted at the centrifugals floor on top of each conveyor (Photo 1, 2 & 3 below). They measure moist white sugar of 30 ICUMSA maximum in color, discharged from every batch of eight centrifugals. Powerful stroboscopic Xenon light sources illuminate the sugar surfaces without being affected by the ambient light. Besides, the use of broadband light sources, from the UV to the Near IR, ensures a measurement across a wide range of sugar colors with high precision and enables the automatic detection of brown lumps using image processing techniques. High resolution color calibrated cameras measure the reflected light and send the images to the same computer in control room where they are processed in real-time by dedicated and customized ITECA software.
To closely monitor the overall performance of the installation, internet remote monitoring and secured screen sharing were set up. This provides ITECA with real time access to the installation from our facility in France or from anywhere in the world in case of need.

Photo 1, 2 & 3: Colobserver® ITECA at the centrifugal floor, above vibrating conveyors

The data is filtered and a mathematical algorithm applied to calculate the coloration values in ICUMSA units after correlation with the laboratory. The video of the sugar is continuously displayed in real time on the main screen with the color trends of each centrifugal and the global trend of the production. Videos are registered according to a predefined registration strategy: it can be done for instance whenever the software detects a non conformity and an alarm is triggered. As the colorimeters communicate with each centrifugal, operators know from which one the sugar they are watching is coming from and they always find the images very helpful to identify and trace the origin of the problems faced.

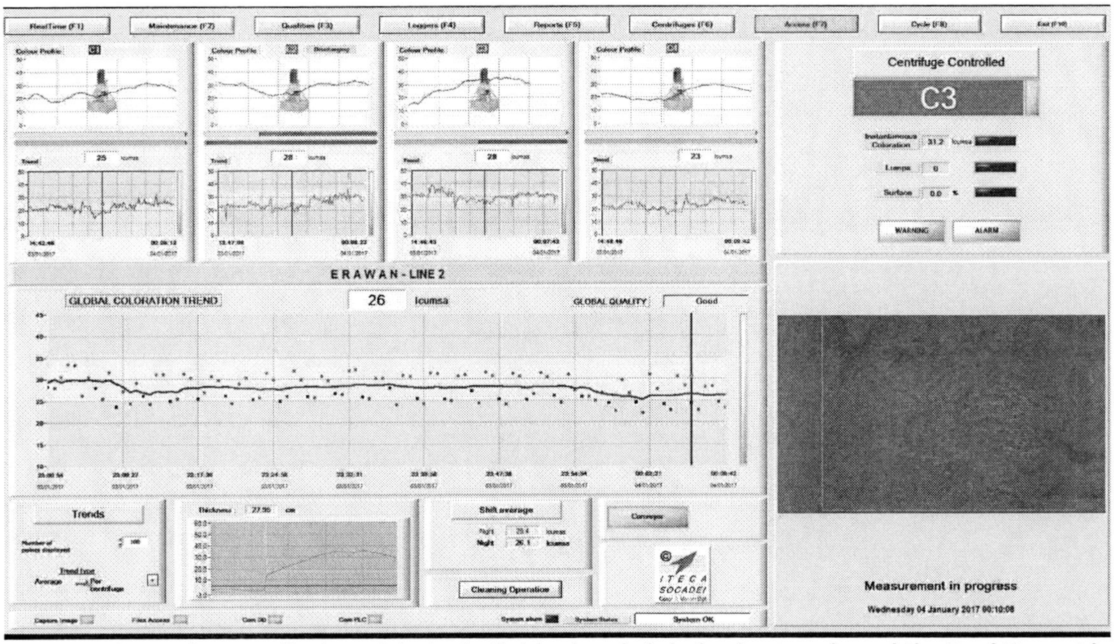

Figure 1: Control room display with the individual and global color trends and the Real time video

On the main screen, the individual color trends of the four centrifugals of each line are displayed as well as the color global trend over 24 hours of production. The video displayed is the sugar coming from the centrifugal under control, being C3 of Line 2 in the case presented in Figure 1, with a color target of less than 30 IU. A set of colored banners used as visual support: green when the measurement is close to the target, blue when it is too low (over washing) or red when it is too high, makes it easier to read the measurements from afar. In Figure 1 above, operators could see that the global color was slowly drifting down to 20 IU because of centrifugals C1 and C4 that were over washing and they could take necessary actions, in this case reducing the washing time, to go back to the color set point of 30 IU. Those adjustments were done manually so far but the plan for the campaign 2017 is to do it automatically through FIVES's Smart Control^{TM} that can alter the amount of sprayed water according to both the basket load up to 1250 Kg and the color measured by the Colobserver® for more precise and immediate results.

Results and discussions

Correlation with the laboratory standard measurements

In March 2016, the critical first step that was carried through commissioning by ITECA's engineer was the correlation of the colorimeters analysis with the laboratory standard ICUMSA color measurements.

A predetermined set of sugar samples discharged by each of the centrifugal were taken across the full range of color produced at the plant, following a rigorous procedure [1] to minimize measurements biases. Each sample was then analyzed by both the laboratory and ITECA's colorimeters and their results compared to establish the correlation equation, which, once integrated into the software, allows the colorimeters to calculate and display the sugar color in ICUMSA units without having to recalibrate the instruments over the full campaign.

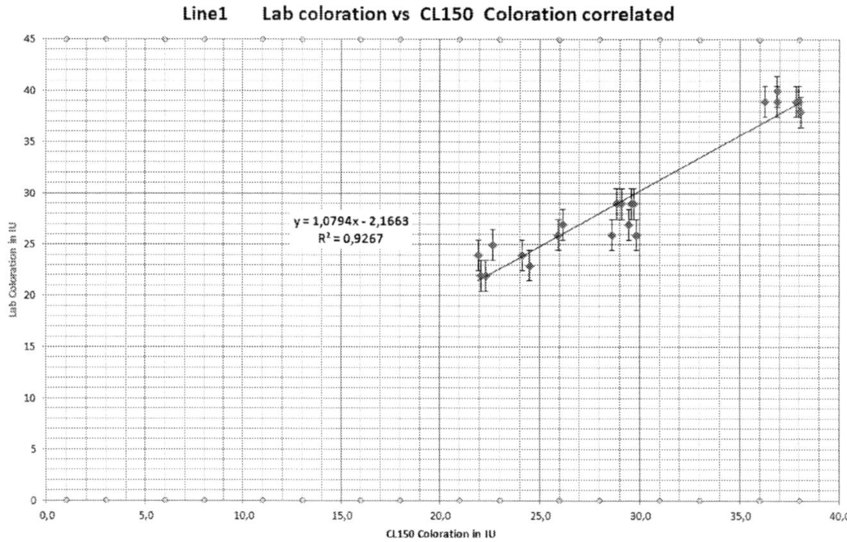

Figure 2: Sampling procedure Line 1 during the commissioning before setting the Correlation

It is important to note that in addition to providing continuous measurements, the colorimeters ensure very stable and repeatable precise measurements, which is far more challenging when using sampling methods.

Importance of individual tracking

When using the traditional measurement methods, it is more likely that the color taken into account is the average color of several discharges of various centrifugals. The average value is a good indicator of the overall production but it rather smoothes artificially the value of the color measured, when the reality can be that only one centrifugal with some default is causing the average color to increase or even exceed the acceptable limits.
ITECA colorimeter communicates in real time with each centrifugal and therefore can track the color profile of each discharge individually.

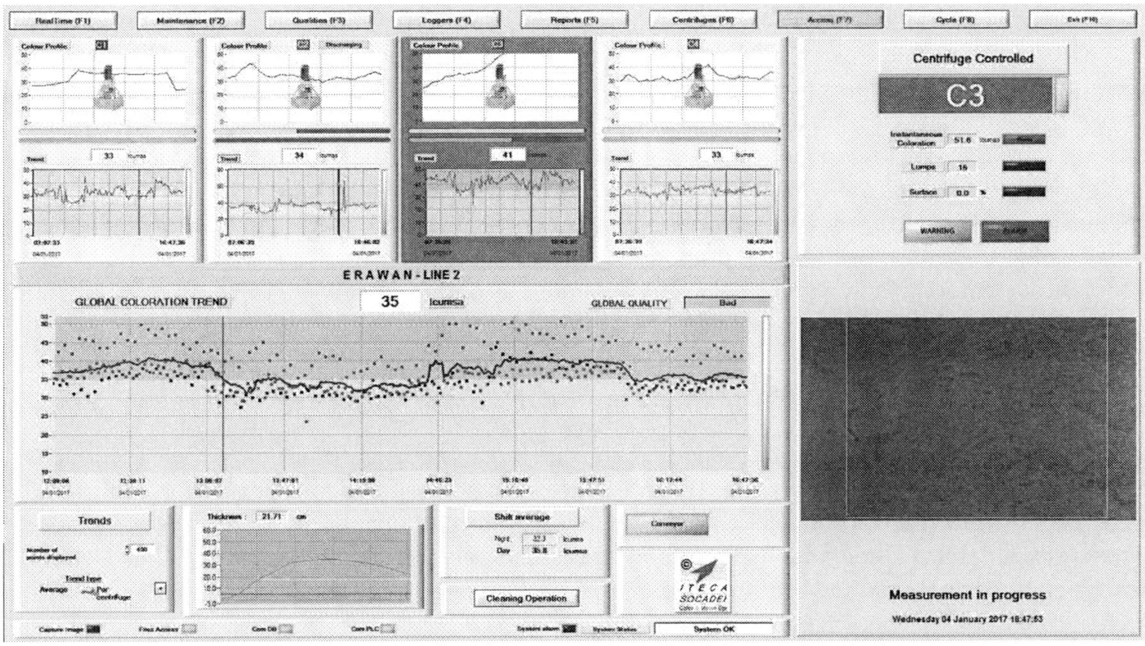

Figure 3: Main display showing centrifugal C3 exceeding the 'Alarm' threshold

In the case presented in Figure 3 above, centrifugals C1, C2 and C4 were producing sugar with color within the color target level whereas C3 was discharging sugar with a too high average color of 41 IU, causing the overall production color to get out of specification.

Clearly in this case, the solution to regain an acceptable color level was to address the issue on centrifugal C4.

Follow up of the massecuite qualities

The color of the sugar discharged by the centrifugals strongly depends upon the quality of the massecuite feeding them.

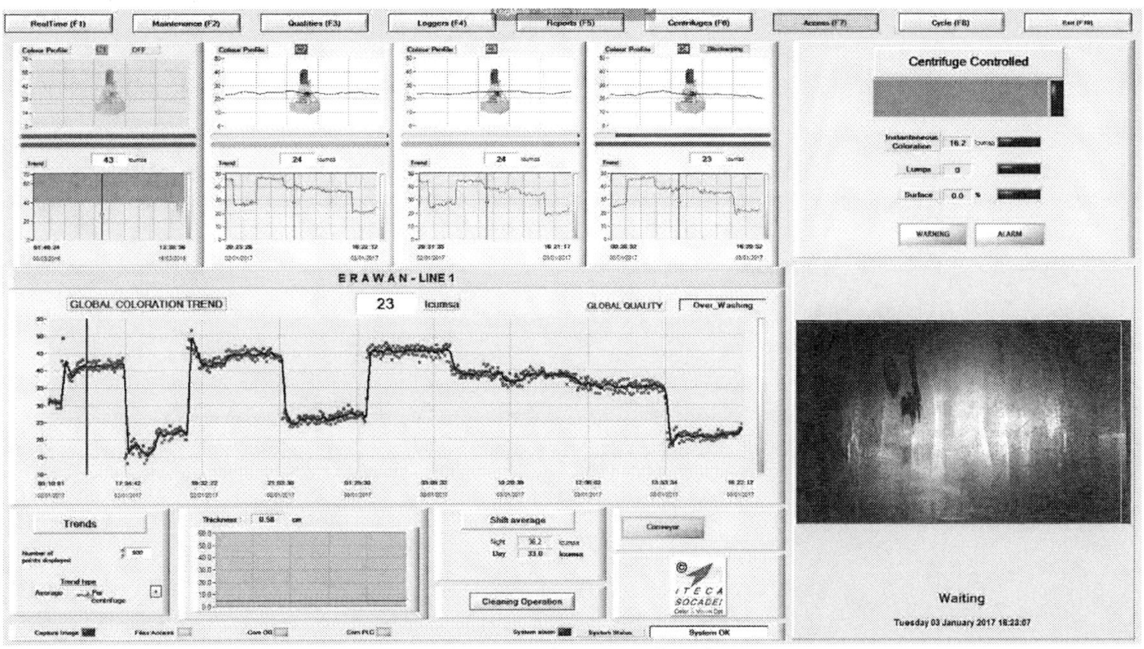

Figure 4: Impact of the massecuite qualities at the centrifugal output

This was highlighted with a short trial that was set up keeping the centrifugals washing time constant, and looking at the sugar color at the centrifugal output: it shown abrupt drops of 20 IU for every change of massecuite quality (Figure 4).
It is therefore necessary to anticipate any sudden change in the process parameters, hence considering the massecuite quality when adjusting the washing time of each centrifugal independently: the better the massecuite quality, the lower the wash water volume needed for the overall color to remain within the predefined limits. Again, this fine-tuning can be performed automatically entering the on-going color measurement into the centrifugal PLC so it can modify the water volume sprayed on the next batch.

Monitoring of the process failures

We have seen that an individual tracking of each centrifugal is critical to optimize their sugar color. Thanks to real time image processing, ITECA cameras provide an additional detection tool, totally decorrelated from the color measurement. Any non conformity appearing on the sugar surface is automatically detected, counted, its surface measured and alarms can be triggered if their number, dimension or color exceed pre-set thresholds.

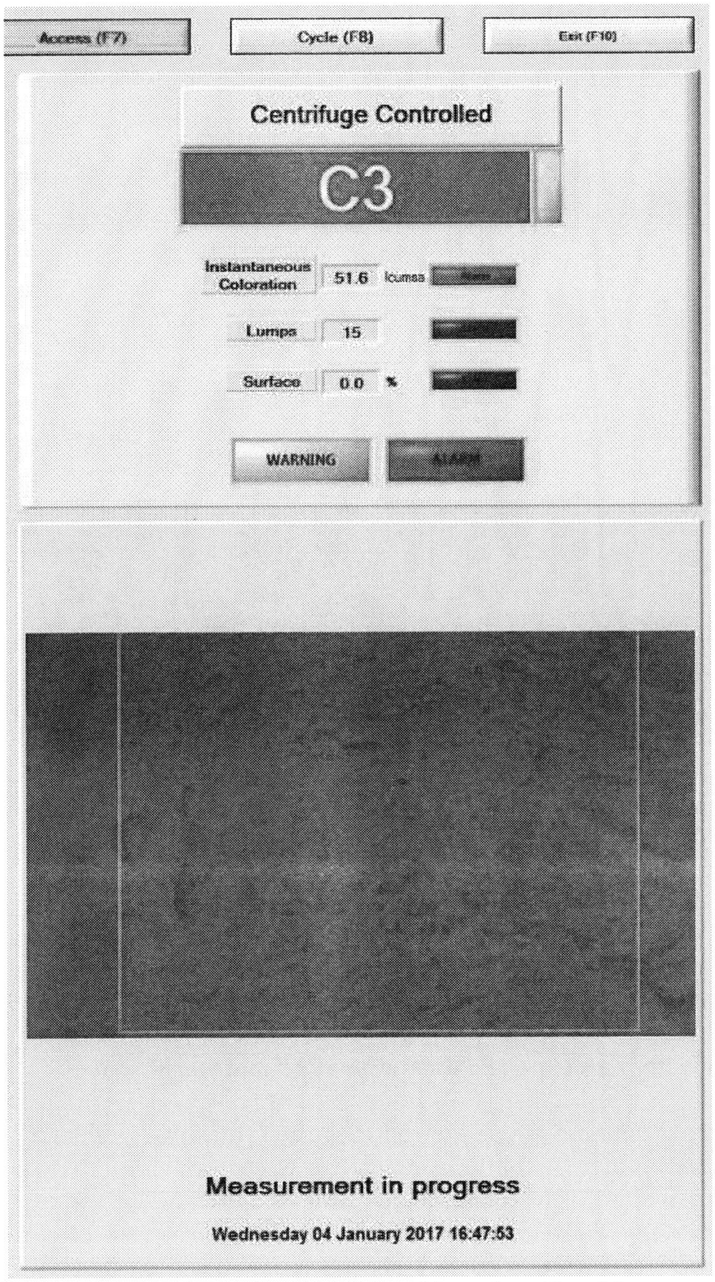

Figure 5: Very small brown lumps detection highlighted in red on the video

A zoom on the case described above (Figure 3), shows that centrifugal C3 was producing too high color levels because of the presence of brown lumps coming out of each batch. However they were very small, the camera was still able to detect them and trigger an alarm based on their number (15 lumps as shown on Figure 5 above). It enabled the operators to identify and fix nozzle blockages issues causing the formation of brown lumps.

An additional capability of ITECA control software is to communicate with the plant PLC any time a critical threshold is exceeded which enables the camera to initiate an action: out of spec product can be automatically deviated to avoid contaminating the dryer. During that first one-year period at Erawan Sugar, operators would manually divert out of spec sugar - mainly on 'lump events', into the melter but the plan is to set up the automatic version later this year.

Each event was recorded as well as the date and time of its occurrence.

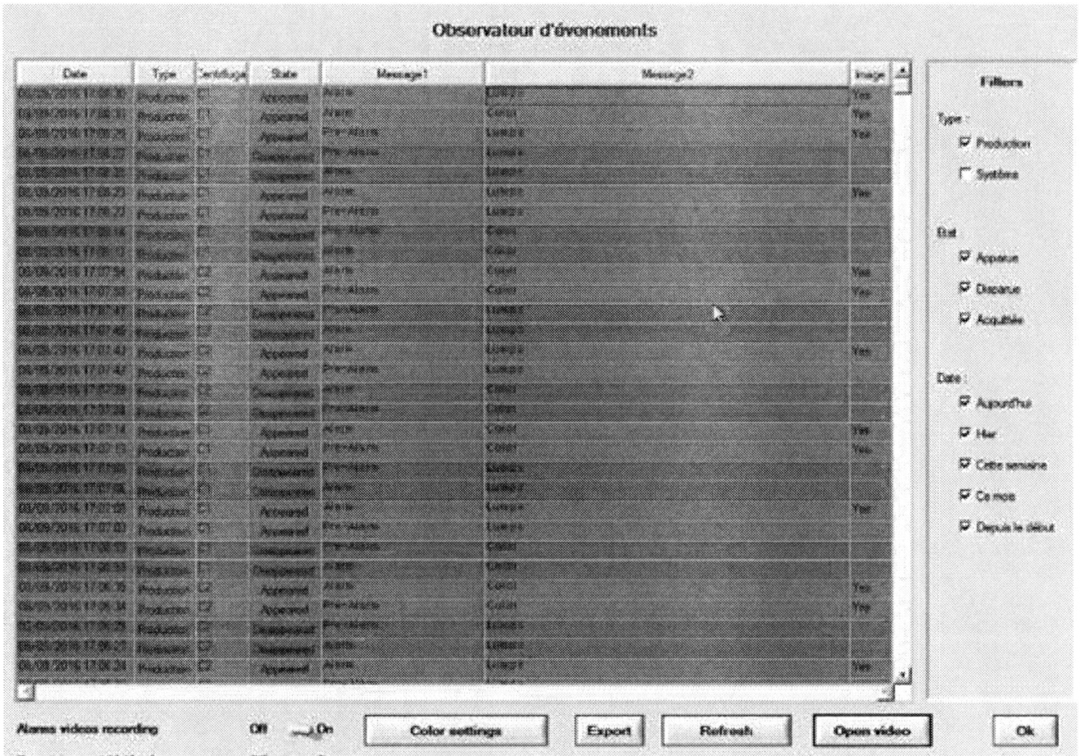

Figure 6: Example of Colobserver® Alarm logbook

It is possible to click on any of the alarms and watch the corresponding video as 24 hours of taping are registered around the detection time. It can be done at any time, without interrupting on-going measurements, and is very useful for comfortably analyzing root causes of the identified issues.

Interest of the wash water reduction and the automatic "auto wash"

The basic concept to evaluate the interest of reducing the centrifugal washing time, and as a matter of fact, the wash water amount, is that one liter of water can remelt up to 3 kg of good sugar. It is therefore essential to avoid over washing the centrifugals and only use the exact amount of water required to keep the color as close to the color set point as possible. If we consider an average water consumption of 2.5 liter/s/centrifugal, only 1 second reduction of washing time **avoids remelting more than 16 tons of good sugar per day. It represents a daily production loss of 26 min.** When less sugar is remelted during the crushing season of 120 days, in addition to huge time savings avoiding reprocessing it, there is also a significant reduction of the water and energy consumption as well as an increase in production capacity.

During the first year of service of ITECA cameras at Erawan Sugar, the washing time was adjusted manually after analyzing the individual color trends of each centrifugal. This period was necessary for the operators to make it a routine to use the color cameras as an efficient diagnostic tool and understand the impact of the washing time adjustment on the sugar color discharged from the centrifugals. As can be seen on Figure 3, the time elapsing between the time a color drift is detected and the time the color goes back to the color set point can be relatively high, as it depends on the operator's reactivity to adjust ITECA Colobserver allowed the centrifugal washing times.
For a much greater reactivity and accuracy, the washing time optimization must be done automatically: this is what will be performed this campaign using FIVES's control system

improvements that can alter the amount of sprayed water according to both the basket load and the color measured by ITECA cameras. Regardless of the disparity between each centrifugal or the massecuite quality, the exact amount of wash water will be fine-tuned to maintain the color at the predefined set point. Furthermore, as it is able to perform accurate cost analyses based on detailed records of machine usage, FIVES Smart Control™ web application should be very helpful to calculate the exact profit obtained by the washing time adjustment.

Conclusion

After only a one year period, significant improvements were achieved to the stabilization of the sugar color at the centrifugal output using ITECA color cameras. Both production shifts (night/day) totally trust and rely on ITECA Colobserver® measurements to take immediate corrective actions on the process whenever there is a color drift or a non conformity is detected.
The automatic individual autowash and the out of spec product automatic diversion that will soon be set up will certainly improve considerably the overall efficiency. It will also ensure an optimal level of quality whatever the process fluctuations encountered.
In order to go further in the improvement strategy, another ongoing project at Erawan Sugar is the installation of ITECA pan microscopes that would be used to enhance the crystal yield and optimize the massecuite quality feeding the centrifugals.

References

[1] C. Bouché, B. Gaillac and A. Ferrete. Sampling theory applied to indirect methods of analysis.

[2] C. Parker, B. Howe, C. Haynes and C. Bouché, British Sugar plc and ITECA SOCADEI. Optimization of white sugar colour management through the utilization of on-line colour cameras. SIT Proceedings. New York, 2016

S.I.T. Paper #1167

A detailed view on the dextran induced processing problems

Abraham, K. [b, a]*; Rohde, A.*[b]*, Flöter, E.* [a]

[a]*Technical University Berlin, Berlin, Germany,* [b]*Sternenzym, Ahrensburg, Germany*

Abstract

The presence of polysaccharides in sugar cane and beet raw juices causes several negative effects during the sugar production process, which are usually reduced by enzymatic decomposition. These process effects are not only a function of the existing level but also of the molecular weight distribution. Different dextran fractions have consequently specific process effects to different extents. An accurate process control hence requires the detailed knowledge about the existing dextran level and ideally about the molecular weight distribution. Hence is the determination and characterization of dextran in raw juice a prerequisite to assess occurring processing problems and particularly to apply appropriate enzyme levels resulting in a controlled decomposition process. Low dextran levels and broad molecular weight distributions in raw juices make this still challenging. A precise and simultaneously time saving analytical method is hence still lacking. To improve on this situation, new approaches for the determination of dextran in sucrose solutions have been developed and compared to the commonly used but rather unspecific and inaccurate Haze method. The latter is additionally restricted to high molecular weight dextran which is therefore not suitable to completely assess the dextran-related processing problems. The two alternative methods both based on polarisation appear to be superior with respect to precision and capture of a broader molecular size range, which seems to be a major advantage over the Haze method. The improvement of analytical tools will allow a more targeted and hence optimised mitigation of dextran induced processing problems due to controlled enzyme applications with known substrate to enzyme ratios.

Keywords: dextran, determination, molecular weight, dextranase

1 Introduction

The presence of polysaccharides in sugar beet and cane juices causes as well-known several negative effects during sugar processing. This especially relates to microbiologically produced dextrans, which are mainly composed of α-(1→6) linked glucose units. Furthermore, branching points of up to 5% can occur in the chain which are mainly represented by α-(1→3) glycosidic linkages. Dextran is extracted along with the juice and can basically affect every following sub-process during the sugar production process. The extent and type of these process effects not only depend on the polymer concentration, but also on the molecular weight distribution (Promraksa, 2008).

A higher viscosity due to the presence of high molecular weight dextran is a well-known fact, which leads to worsened filtration, evaporation and crystallization rates. Furthermore, is the presence of dextran supposed to cause crystal shape and size modifications, which certainly concerns the sucrose crystallization (Abdel-Rahman, 2007) but also the formation and growth of calcium carbonate crystals during the juice purification process (Sarka et al., 2008). It is not fully understood whether these effects are primarily caused by either high or low molecular weight dextran. Consequently, are the specific effects of the different dextran fractions still not clarified at all. Nevertheless, it is known that both high as well as low molecular weight dextran play a decisive role in such effects. These effects are usually mitigated by enzymatic decomposition, which results in a gradual decrease of the molecular weight. To better understand the extent of the dextran related problems and to enzymatically decompose dextran in a controlled manner, it is necessary to completely determine and ideally to characterize the existing polymer levels. This intention is especially hampered by relatively low levels and broad molecular weight ranges in raw juices, which still makes an accurate dextran determination a great remaining challenge (Day & Sarkar, 1986).

Therefore, the present work tries to make a step towards the determination of a total dextran value comprising the whole molecular size range while developing, improving and comparing different methods to determine dextran in synthetic thin juices. Improved analytical tools combined with a subsequent detailed analysis of the specific process effects of the different dextran fractions and enzymatically decomposed dextran is supposed to finally allow an overall optimised and controlled mitigation of dextran related problems.

2 Material and Method

2.1 Materials

In the present work the dextran-degrading enzyme Sugazym DX L (SternEnzym) was used for the enzymatic decomposition of dextran with varying molecular weights. High (Sigma-Aldrich, T2000), middle (Carl Roth, T500) and low molecular weight dextrans (Carl Roth, T40) were used for all dextran determination methods applied.

2.2 Determination of Dextran

2.2.1 Haze Method

Haze method according to the ICMUSA GS1/2/9-15 (2011) was adapted to analyse 15% (w/w) sucrose solutions and additionally modified with regard to the calibration procedure. The calibration was performed by using middle molecular weight dextran (T500) and optimised by implementing the filtration step as a necessary step to equalize measuring and calibrating procedure. The subsequent procedure was done in accordance with the ICUMSA description, which involves a starch and protein removal followed by filtration. Dextran is then precipitated with ethanol and the turbidity is measured in a spectrophotometer using a wavelength of 720nm (ICUMSA, 2011) .

2.2.2 Enzyme method

The optical rotation of 15 % sucrose solutions with defined dextran levels of different molecular weights were measured before and after a complete enzymatic decomposition. Therefore, samples were divided into two parts, of which one was incubated with dextranase in surplus (1000 mg/kg juice of Sugazym DX L by SternEnzym) for 60 minutes at 55°C. The difference in optical rotation was used to calculate the dextran level. Measurements of optical rotation were carried out by using a high-performance circle polarimeter provided by Schmidt & Haensch. Middle molecular weight dextran (T500) was used to prepare the calibration curve. Calibration procedure was completely adjusted to the measuring procedure.

2.2.3 Membrane method (Collaborative work with Schmidt & Haensch)

The optical rotation of 15 % sucrose solutions with defined dextran levels of different molecular weights were measured before and after the separation of dextran via membrane filtration. The difference in optical rotation between the feed and the permeate was used to calculate the existing dextran level. Standard industrial equipment has been used for performing the membrane filtration and measuring the optical rotation (Autofilt and high-performance circle polarimetry, Schmidt & Haensch).

The measurement was performed at room temperature and pressure as a driving force for the membrane filtration of 3 bar was applied. The first aliquot of the filtrate was discarded before the actual sample was collected. Middle molecular weight dextran (T500) was used to prepare the calibration curve and again, the calibration procedure was completely adjusted to the measuring procedure.

3 Results and Discussion

3.1 Details on the respective methods

3.1.1 Haze method

A fundamental view to the respective analytical principles is a necessary prerequisite to interpret the determined dextran values exactly. This mainly refers to the analysis of potential disruptive factors and subsequently the adjustment of the calibration procedure.

The calibration standard used is a key factor to understand the obtained total dextran values. In most methods the response signal is dependent on dextran molecular size and therefore the dextran fraction used for calibration should preferentially resemble the size, or size distribution, of dextran in samples to be measured. This is difficult to ensure. Assuming a broad dextran molecular weight distribution in sugar beet and cane raw juices, middle molecular weight dextran (T500) seems to be the most appropriate choice for preparing the calibration curve and has therefore been used for this purpose for all methods applied.

This is especially important regarding the commonly used Haze method, which is based on alcoholic precipitation. The degree of alcoholic precipitation differs greatly with the molecular weight present, which directly correlates with the determined dextran values. The knowledge of the dextran fraction used for calibration is hence an essential need to interpret the Haze data appropriately. The calibration procedure of the Haze method entangles another source of systematic error. The measuring procedure according to the ICUMSA description involves a step to remove starch and proteins. The latter requires a filtration step which is not included in the ICUMSA instructions for the preparation of the calibration curve. It seems likely that some dextran adheres in the filter pores due to wetting, which results in underestimated dextran values. The need to make the filtration step a part of the calibration procedure is illustrated in Figure 1 (A), which shows the difference of absorption before and after the filtration of dextran solutions containing the corresponding amounts of trichloracetic acid and filter aid. Obviously some dextran remains in the filter pores due to adhesive forces. This clearly supports the need of inclusion of this filtration step in the calibration procedure. The different calibration curves are compared in Figure 1 (B), which additionally underlines the necessity to completely adjust the calibration to the measuring procedure. In the following, Haze data consequently refers to the calibration procedure including filtration.

3.1.2 Enzyme method

The Enzyme method uses the change of the optical rotation due to the enzymatic hydrolysis of dextran. Dextran is completely hydrolysed by application of Sugazym DX L provided by SternEnzym. The difference before and after the hydrolysis is used to calculate the dextran level. This enzyme method is based on a relatively new principle to determine dextrans, whereat the main idea was at first mentioned by Singleton et al (Singleton et al., 2002).

To apply this method, it was necessary to find the incubation time necessary for complete dextran hydrolysis. Therefore, the change of optical rotation was monitored during an ongoing enzymatic hydrolyses process. Figure 2 (A) shows the difference of optical rotation during an enzymatic decomposition of high and low molecular weight dextran directly in the polarimetric tube at an incubation temperature of 20°C. This reveals that the change of optical rotation in pure dextran solutions at this very high enzyme level remains constant after 20 min incubation time, indicating complete decomposition. Earlier work has shown that the presence of sucrose has an inhibitive effect on the enzyme reaction (Abraham et al., 2014), which is probably caused by sucrose acting as a non-competitive inhibitor. Despite the relatively low sucrose level of about 15% in raw juices, slightly changed hydrolysis rates may arise. Therefore, the hydrolysis experiment in the polarimteric tube was also performed for 15%

sucrose solutions containing the same amount of 5000 mg high as well as low molecular weight dextran per kg sucrose. Results are shown in Figure 2 (B). In this case complete degradation is achieved after 30 minutes. High as well as low molecular weight dextran show in this case very similar changes in optical rotation, which is an important requirement for the determination of dextran values irrespective of molecular size. To be absolutely sure that dextran is completely decomposed, an incubation time of 60 minutes and the optimal incubation temperature of 55°C were used for the final measuring procedure.

3.1.3 New Membrane method

The new Membrane method works in a similar manner as the enzyme method but uses a membrane filtration step to remove dextran from the solution. And again, the difference of optical rotation of a solution with and without dextran is used to calculate the dextran value. The difference of optical rotation is in this case much higher in comparison to the enzyme method, which makes it less susceptible to fluctuations. This method is a completely new approach and represents a promising and expandable possibility for the determination of dextran of different molecular weights. The separation of dextran via membrane filtration needs also some fundamental investigations. This includes the investigation of the separation efficiency of the membrane used, which is shown in Figure 3 (A). Solutions of different molecular weight dextrans were membrane filtered and the separation efficiencies were calculated as a percentage retention. In pure dextran solutions high as well as middle molecular weight dextran seem to be completely separated, while just 60 to 70% of low molecular weight dextran is separated by the membrane used. It is worth to mention that the molecular weight cut off (MWCO) of a membrane represents a loosely line not representing a sharp separation limit. Many factors can have an influence on the retention efficiency of a membrane such as molecular shape, sample composition and concentration as well as operating conditions. Especially linear, water soluble molecules such as dextran can still snake through the membrane pores while using a MWCO clearly below the molecule to be separated. Exactly this seems to be the case for low molecular weight dextran with an average molecular weight of 40 kDa, which is obviously not completely retained in pure dextran solutions by the ultrafiltration system used.

As already mentioned, the sample composition is an important factor influencing the membrane process. Interactions between sucrose and membrane pores as well as between sucrose and dextran molecules cannot be ruled out and can also have an effect on the membrane separation process. The first step to analyse these interacting effects is made by the analysis of sucrose retention occurrences due to adhesion effects. Therefore, pure 15% sucrose solutions were membrane filtered and the changes in optical rotation were investigated, see Figure 3 (B). Similarly to the Haze method, effects due to adhesion occur. The resulting optical rotation indicates that a small amount of sucrose remains in the membrane pores. Understandably, this effect decreases over the filtration time assuming the development of a saturated wetted pore surface. To minimize this influence of the sucrose retention on the measurement, the first aliquot of the filtrate is discarded. However, a minute difference in optical rotation caused by the sucrose retention remains even after that discarded permeate volume and remains hence unavoidable.

This fact necessitates the implementation of that membrane filtration step such that samples contain identical sucrose concentrations. Same as for the other two methods, the calibration procedure is completely representing the actual measurement procedure to eliminate systematic errors caused by the sucrose retention during filtration. And again, middle molecular weight dextran was used for calibration, due to size considerations. The separation efficiency of the different dextrans as well as the sucrose retention are the main factors influencing the measurement results. Higher dextran retention as well as increases in the sucrose retention cause higher differences in optical rotation, which consequently results in higher determined dextran values.

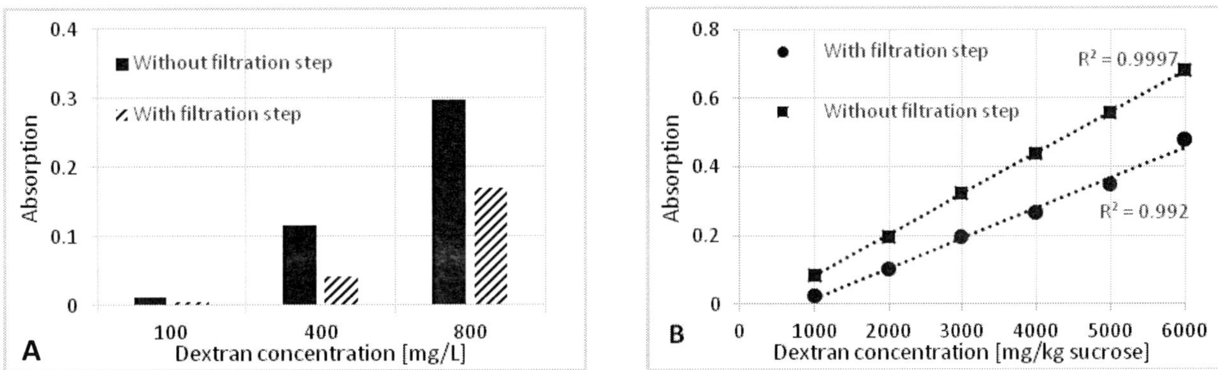

Figure 1 Fundamental analysis of the commonly used Haze method (A) Filtration loss of dextran solutions containing TCA and filter aid (B) Calibration curve in- and excluding the filtration step for 15% (w/w) sucrose solutions

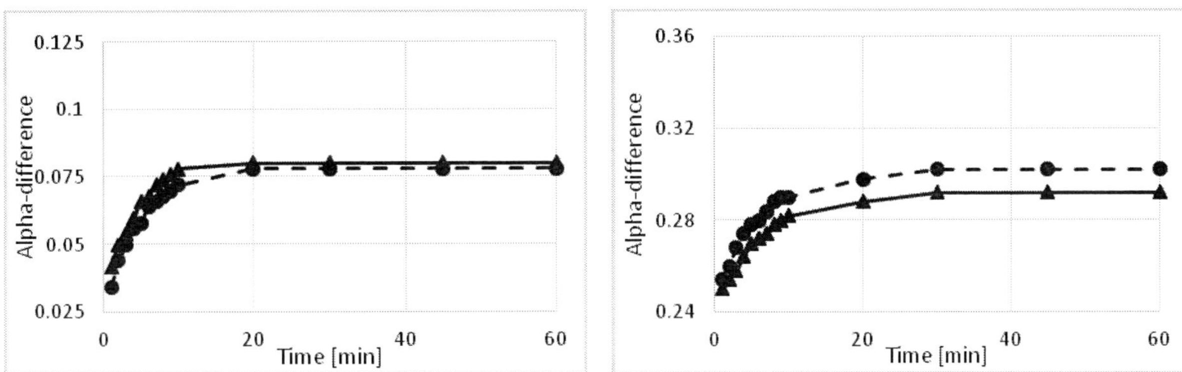

Figure 2 Fundamental analysis of the enzyme method – Incubation of dextran in the polarimeter tube at 20°C (A) ▲-T2000 and •-T40 dextran without sucrose (B) ▲-T2000 and •-T40 dextran in a 15% sucrose solution

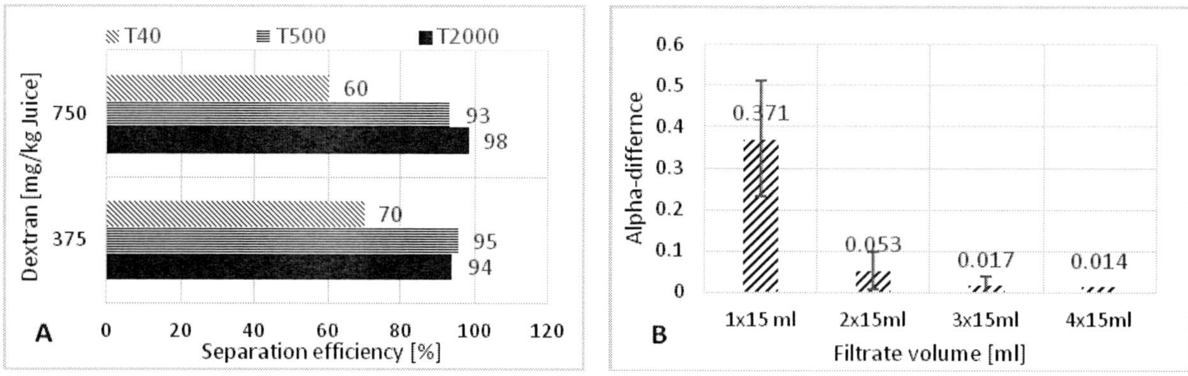

Figure 3 Fundamental analysis of the membrane process - (A) Separation efficiency of dextran with varying molecular weights (B) Membrane filtration of 15% sucrose solutions investigated over 60 ml filtrate

3.1.4 Comparison of different methods for the determination of dextran

After describing the fundamentals of the respective methods, the determination of dextran with varying molecular weights and concentration levels in 15% sucrose solutions can be discussed. As mentioned before, the determination of dextran irrespective of molecular weight is essential to adequately assess the processing problems and also to apply appropriate enzyme levels. This is still a problem to solve. Therefore, the methods explained above have been applied to samples containing high as well as low molecular weight dextran at varying levels. Data shown in Figure 4.

The determination of high molecular weight dextran (T2000) is shown in Figure 4 (A). First of all, it should be noted, that all methods used were calibrated with middle molecular weight dextran (T500) with the purpose to possibly cover all sizes within a total dextran value while assuming a broad molecular weight distribution in realistic raw juices.

As already explained, the degree of alcoholic precipitation differs with molecular size, which explains the overestimated Haze values for the determination of high molecular weight dextran shown as red dots in Figure 4 (A). The main disadvantage of the commonly used Haze method is however its restriction to high molecular weight dextran, which was confirmed by the underestimation of the levels of low molecular weight dextran, see Figure 4 (B). This illustrates impressively how sensitive the Haze method is to variation of molecular size of the dextran studied. Nevertheless, the Haze method currently appears to represent best practice to determine dextran in the industrial reality. Thus is this method the most appropriate reference for evaluation of Enzyme and Membrane method discussed. However, the new methods have also to be judged on their trueness and precision.

The applicability of these two methods for the determination of high molecular weight dextran is also shown in Figure 4 (A). Both methods clearly perform superior to the Haze method. The values determined appear to be fairly close to the known values of the synthetic samples. Data points on the grey line would mean a perfect match. This clearly indicates the two methods well suited for the determination of high molecular weight dextran in 15% aqueous sucrose solution.

In order to overcome the traditional problems of the molecular size dependency of dextran determination also low molecular weight dextran has to be detected at similar accuracy by any improved method. Results of the methods are shown in Figure 4 (B). As mentioned above, the Haze method determines expectedly too low values for low molecular weight dextran. Here again, both the Enzyme and the Membrane method show much better results almost reproducing the known values of the synthetic juices. However, the scattering of the data is a problem which needs to be addressed. The data shown in Figure 4 illustrate that in contrast to the Haze method the two new methods do not show systematic deviations from the known values due to differences in dextran molecular size. The major advantage of the Enzyme method is its independence of dextran molecular size. Irrespective of the starting dextran molecular size change of optical rotation due to the full enzymatic hydrolysis should be identical. This makes the method particularly interesting to determine total dextran levels. The change in optical rotation caused by the complete enzymatic hydrolysis is unfortunately limited so that inaccuracies of the polarisation assessment propagate directly into the accuracy of the dextran determination. To this end an improved polarisation measurement could benefit the quality of the Enzyme method.

The determination of low molecular weight dextran in sucrose solutions by the membrane method reveals higher dextran values than expected from the data on low molecular weight dextran in pure dextran solutions, see Figure 3 (A). As already explained, too low values were determined due to incomplete retention of the dextran. This finding in sucrose solution supports the assumption that sucrose adhesion to the membrane possibly causes hampering of the processes that originally linear T40 molecules snake through the membrane pores in significant quantities. Higher retention results in an improved quality of the measurement because systematic deviations are reduced.

This interaction of dextran and sucrose in membrane filtration of aqueous solutions certainly deserves more attention. Prior to any detailed analysis of the nature of this interaction this beneficial effect has to be verified. To do so the membrane filtration will be combined with highly accurate chromatographic analytical methods to determine the compositions of feed and permeate solution in more detail.

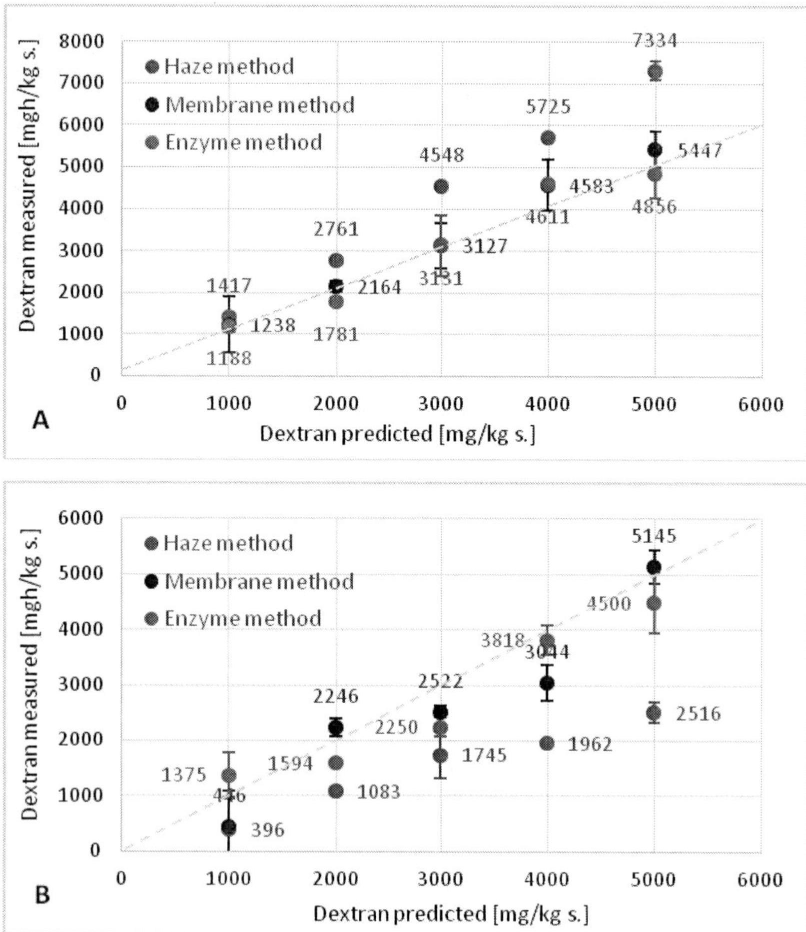

Figure 4 Determination of different molecular weight dextrans (A) High molecular weight dextran (B) Low molecular weight dextran

Compared to the Enzyme method, which actually offers only a limited number of targets for improvement, the Membrane method has more variables that could be varied. Consequently, further optimisation of that process is still possible. These are at first variation of the operating conditions such as pressure or time. Secondly changes of the sample composition or preparation due to targeted addition of processing aids or changes of the membrane are important factors to consider.

Nevertheless, both methods seem even in their current state of infancy to be viable options to determine dextran levels in sucrose solutions irrespective of dextran molecular weight. It is comforting to realize that the differences between the dextran values determined by these two methods are rather small. The improved determination of dextran levels marks a major step towards a controlled mitigation of dextran-induced processing problems.

Next to the reliable determination of total dextran levels, the characterization of the dextran present with respect to the molecular weight distribution is of importance as well. It is hence desirable to also differentiate between high and low molecular weight dextran fractions because the effect of their presence during processing is known to vary. Exploration of the two methods discussed regarding this target consider the rate of enzymatic decomposition, the application of multiple membranes and alike.

4 Conclusion

Two new methods to determine dextran concentrations in synthetic juices were discussed. Critical review of the Haze method, which as industrially respected method serves as reference method, yielded an improved calibration for that method. In comparing the Haze method with the Enzyme and Membrane method the correct reproduction of total dextran levels in the juices irrespective of the molecular weight of dextran was considered a key target. To this end, different dextran fractions were used (T40, T500, and T2000).

The Haze method is generally acknowledged to be rather unspecific, inaccurate and also restricted to high molecular weight dextran. The data obtained by the Haze method show an overestimation of high and an underestimation of low molecular weight dextran, which is in line of expectation when calibration is performed with a medium range dextran. This confirms earlier findings. The Enzyme and the Membrane method differ in essence in the sample preparation while both are characterized by polarisation measurements. In the Enzyme method full hydrolysis of dextran yield a difference in polarisation between feed and treated sample. In the Membrane method, feed and permeate differ in the polarisation. In both case the differences in polarisation are converted to dextran levels in synthetic juices. Both method appear superior to the Haze method. They show practically no variation with the molecular size of dextran considered. However, both methods suffer from some scattering due to the accuracy of the polarisation measurement applied.

Even though both methods are subject to further improvement, this work marks a major step towards better analytical tools for the determination of dextran. This is a prerequisite for adequate control of the enzymatic hydrolysis processes in order to reduce dextran-induced processing problems.

5 References

- Abdel-Rahman, E.-S. (2007). *Investigations on the influence of dextran during beet sugar production with special focus on crystal growth and morphology*. PhD. Berlin.
- Abraham, K., Hagen, S., Schlumbach, K., & Flöter, E. (Eds.) (2014). *Enzymatic Decomposition of Dextran in Sucrose Solutions*.
- Day, D. F., & Sarkar, D. (1986). Methods of Analysis for Dextran in Sugar Molasses and Jucie. *Proc. Int. Soc. Sugar Cane Technol*, 748–754.
- ICUMSA (Ed.) (2011). *ICUMSA Method Book*. Berlin: Dr. Albert Bartens KG.
- Promraksa, A. (2008). *Reduction of Dextran in Raw Sugar Production*. PhD. Suranaree.

- Sarka, E., Bubnik, Z., Kadlec, P., & Vesela-Trilcova, A. (2008). The particle size of carbonation mud, and possibilities for influencing it. *Journal of Food Engineering*, 45–50.
- Singleton, V., Horn, J., Bucke, C., & Adlard, M. (2002). A New Polarimetric Method for the Analysis of Dextran and Sucrose. *American Society of Sugar Cane Technologists*, *2002*, 112–119.

S.I.T. Paper #1168

Application of Molecular Biotechnology in Detecting Sugarcane Pathogens

Yu-Hsuan Chen[1], Li-Shyang Chen[2], Chin Sheng Hu[2], and I-Jen Huang[3]

[1] Taiwan Sugar Research Institute

[2] Taiwan Sugar Business Division

[3]Taiwan Sugar Biotechnology business

Abstract

Sugarcane was infected with *Sporisorium scitamineum*, *Peronosclerospora sacchari*, *Xanthomonas albilineans*, *Leifsonia xyli. Xyli.*. that can cause systemic disease. It has severely reduced the yield of sugarcane and impacts the economy of sugar-refining. When the infection cannot be early eradicated the pathogen will transmit through the sugarcane and harm the entire farm. Nowadays, many highly sensitive and specific diagnostic techniques have been developed to detecte and track the pathogens.

In this study, we designed a diagnostic primer set which can be used for multiplex PCR and real-time PCR simultaneously. The primer set has an amplification efficiency of 90%~110%. The sensitivity of Multiplex PCR and real-time PCR technique is 1,000 copies and 10 copies respectively. For this reason, we use real-time PCR to detect the sugarcane might be susceptible to the pathogen infection in the field. The result showed that *Xanthomonas albilineans* existed in all fields was scattered widely and unevenly in the leaf. The population of *Xanthomonas albilineans* in the leaf was higher than *Sporisorium scitamineum* and *Leifsonia xyli. Xyli.*. In addition, the distribution of pathogens in sugarcane leaf is inconsistent. The population of *Xanthomonas albilineans* in the sample was higher at the bottom of the stalk than the upper of the stalk. Overall the result display in this study could be guidelines for detection of sugarcane pathogens and meanwhile reduce the error of sampling and enhance the accuracy of detection.

Keyword: S*ugarcane Smut; Downy Mildew; Leaf Scald; Ratoon Stunting Disease; Sporisorium Ccitamineum; Peronosclerospora sacchari; Xanthomonas albilineans; Leifsona xyli subsp; Real-time PCR; Multiplex PCR*

Introduction

Sugarcane as an important world crop, it is grown in tropical and sub-regions. Sugarcane not only can make sugar, but it also classified into the energy crops. The cultivation of sugarcane are reached more than 8,000 hectares. The cane stalk is harvested about once 1.5 year after planting. During the growing season, sugarcane is susceptible to invasion of pathogens. It severely reduces the yield of sugarcane and impacts the economy of sugar-refining. There are many pathogens will be hazardous to sugarcane, the most common and more serious damages are caused by *Sporisorium scitamineum* (Us), *Peronosclerospora sacchari* (Ps), *Leifsonia xyli. Xyli.* (Lxx), and

Xanthomonas albilineans (Xa). The infected plant is leading to pathogen disease, including smut, downy mildew, ratoon stunting and white streaks.

Sugarcane smut is caused by *Sporisorium scitamineum.* The most recognizable characteristic of this disease is a black or gray growth that is referred to as a smut whip. Sugarcane smut is disseminated via teliospores that are produced in the smut whip. These teliospores located either in the soil or on the plant, germinate in the presence of water. The spores of the whip are scattered by the air and very quickly spread throughout the plant. The prevention of spread disease was spending a lot of time and workers.

Downy mildew as an obligate parasite caused by *Peronosclerospora sacchari* which is a systemic fungal disease in sugarcane. The symptoms on plant differ, it depends on the time infection occurs. When the pathogens are infected and germinated the young plant stage, the plant will lead to die after 4 to 5 month. Infection at a later stage cause pale green to yellow longitudinal stripes, which increase in length after each leaf unfolds. The leaf stripes may become a chlorotic mottling and then will lead to necrosis. In Taiwan, downy mildew is controlled though the planting resistant varieties of sugarcane.

Leaf scald is a systemic disease caused by *Xanthomonas albilineans.* There are 3 clinical phases, called acute, latent, chronic. In the acute phase of the disease, sudden plant death occurs with withered leaf tip or no symptoms. Later stage, affected stalks may be stunted with the development of axillary buds bearing symptoms of the chronic phase. The recognizable characteristic of the leaves were chlorosis, necrosis or inward curling. A longitudinal section of the stalk shows a reddish discoloration of the vascular bundles, particularly at the nodes. The most important way that disease can spread widely from plant to plant on tools of planting.

Ratoon stunting disease caused by the bacterium *Leifsona xyli* subsp. It can cause 14% crop yield losses. Disease expression can be enhanced by drought stress, it can cause more than 50% yield losses. The disease has no easily recognized external symptoms, only stunting of growth that may not always be apparent in the field. Internally there is usually an orange or orange-yellow discoloration of the vascular bundles, not easily observed the symptom.

Furthermore, the four major pathogens were described above. There are many pathogens that can cause diseases in sugarcane such as chlorotic streak disease and white leaf disease, the sympotms all similar to leaf scald. It shows that similar symptoms can also be caused by a number of other maladies. A primary method to identify pathogen investigation in the fields has often relied on disease symptoms. This method is more reliant on visual assessment, but sometimes it is difficult to confirm infection by phenotypic symptom observation due to a latent infection of pathogens. Although the best disease prevention method is cultivate the resistant varieties of sugarcane. But there still not exist a sugarcane variety can produce highly sugar accompanied by defense all pathogens and suitable for all environment. The resistance of sugarcane varieties in

Taiwan were shown in Table 1. Nowadays, the use of resistant sugarcane varieties is the most effective method of prevention and control. However, an appropriate and early detection method for the sugarcane pathogen is critical in early warning, prevention and control of those disease.

Table 1 The varieties of sugarcane resistance

Disease resistance (Incidence rate)	Leaf scald	Smut	Downy mildew	Ratoon stunting disease
High resistance (0~10%)	F160, ROC5, ROC12, ROC15-16	F146, ROC9-10, ROC13-15, ROC18-20, ROC22-25	Nco310, F134, F160, ROC1, ROC2, ROC4, ROC6, ROC8, ROC9-10, ROC13~15, ROC20~25	F161, F162, F165, F172
Middle resistance (20.1%~30%)	F177, ROC5-7, ROC12, ROC19-20	F160, ROC1, ROC7-8, ROC11, ROC16-17, ROC21, ROC26	ROC3, ROC7, ROC11, ROC12, ROC16, ROC26	F159, F160, F166, F169-170
Susceptibility (more than 30%)	ROC8-11, ROC13	Nco310□F134□F177	F155□ROC5□ROC18	N310, F153, F158, F167, F171

Results and discussion

Primer efficiency and specification of Real-time PCR

In the past decade, polymerase chain reaction (PCR) assay has become a standard method for the detection of a wide range of pathogens and biomarkers in diagnostics. Lately, laboratories are progressively shifting from conventional PCR toward a method referred to as kinetic, real-time, or quantitative PCR (qPCR), which allows monitoring of the progression of the amplification during the cyclic reaction. Ct is the intersection between an amplification curve and a threshold line. It is a relative measure of the concentration of target in the PCR reaction. Good design of the primers is critical for successful results. All primer designs must also be experimentally validated with important quality control assays. The criterion to pass is an amplification efficiency 90%~110%.

In this study, we designed 5 diagnose primer set. One primer sets (2.5S-F/R) were specific for ribosomal gene as internal control gene in sugarcane and four primer sets were UbE-F/R, Ps252-F/R, Lxx-F/R and Xa-F/R, specific for *Sporisorium scitamineum, Peronosclerospora sacchari, Leifsonia xyli xyli.,* and *Xanthomonas albilineans,* respectively. All primers result in efficiency between 90% and 110% at annealing temperature 30~60°C (Table 2) and are typically accepted. Optimizing of an annealing temperature is about 55.6°C~59.5°C by melt curve (Fig. 1.). The sensitivity of each primer detected is about 10 copies per reaction.

Table 2. The efficiency of Real-time PCR primer

	2.5S-FR		UbE-FR		Xa-FR		Lxx-FR		Ps252-FR	
Annealing Tm (°C)	Efficiency (%)	R^2	Efficiency (%)	R^2	Efficiency (%)	R^2	Efficiency (%)	R^2	Efficiency (%)	R^2
62.0	114.9	1.000	99.5	0.997	103.6	0.998	101.3	1.000	121.1	0.991
61.7	116.6	0.998	107.4	0.998	102.7	0.997	110.4	0.999	-	-
60.8	115.6	0.992	97.3	0.996	102.3	0.998	111.3	0.999	116.9	0.997
59.5	109.4	0.999	101.3	0.999	106.6	0.997	110.7	0.999	-	-
57.8	107.9	0.998	98.2	0.999	104.6	0.998	100.2	0.998	105.2	0.996
56.5	106.7	0.999	102.7	0.980	100.3	0.997	102.1	0.999	-	-
55.6	109.7	0.998	94.5	0.995	106.5	0.998	100.9	0.999	-	-
55.0	105.2	0.998	105.5	0.998	104.0	0.999	107.3	0.995	108.7	0.999
30.0	102.0	0.999	97.3	0.998	99.4	0.992	93.2	0.995	110.3	0.992

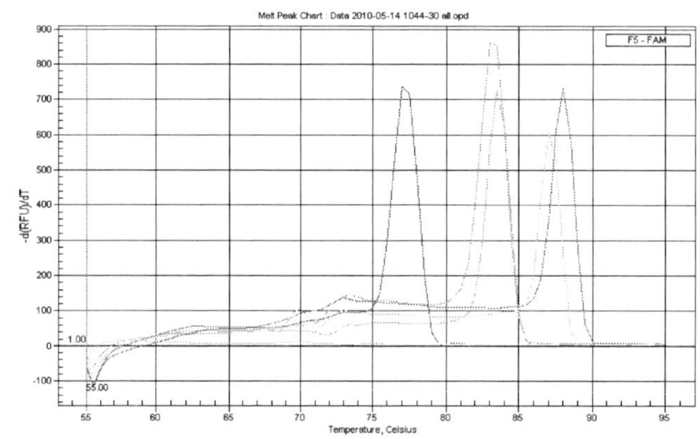

Fig. 1 Real-time PCR melt curve of the 5 specific primer pairs. Blue color encoded the melt peak of internal control gene (2.5S-F/R). Green color encoded the melt peak of the Us gene (UbE-F/R). Purple color encoded the peak of Ps gene (Ps-252 F/R). Orange color encoded the melt peak of Xa gene (Xa-F/R). Red color encoded the melt peak of Lxx gene (Lxx-F/R).

Optimization of Multiplex PCR

Considerable time and effort can be saved by simultaneously amplifying multiple sequences in a single reaction, a process referred to as multiplex polymerase chain reaction (PCR). Multiplex PCR is becoming a rapid and convenient screening assay in both the clinical and the research laboratory. Annealing temperatures for each of the primer sets must be optimized to work correctly within a single reaction, and amplicon sizes should be different enough to form distinct bands when visualized by gel electrophoresis. It is important to consider the specificity of designed primers to the target sequences, while preparing a multiplex assay, especially since competition exists when multiple target sequences are in a single reaction vessel.

Primers designed from 2.5S, Us, Ps, Xa, and Lxx gene amplified fragments of 171 bp, 215 bp, 252 bp, 145 bp and 116 bp, respectively. The multiplex PCR had sensitive detected at 1E+03 copies of target, except the primer set of Ps gene had sensitive detected at 1E+04 copies. Hence, we optimized conditions for 5 primers under a single set of reaction.

The optimization of $MgSO_4$ was designed from the result of Ps252 primer in single PCR. Then, the condition was applied in the duplex, triplex and quadruplex PCR. In the triplex PCR, the 3 mM $MgSO_4$ was successfully amplification which differently detected of 3 combination genes (Fig. 2A). The result of quadruplex PCR was shown in Fig. 2B. The concentration of 4 mM $MgSO_4$ was successfully detected of 4 combination genes in a single reaction. After adjust Ps252 primer, $MgSO_4$ and dNTP concentration. The optimized condition was applied in multiplex PCR, which the limit detected at 1E+03 copies. We simulate exposure risks under field conditions to estimate the probability of dose. The result was shown in Fig. 3. In addition, we found that the difference in concentration of target at 100 fold was interfered the interpret result.

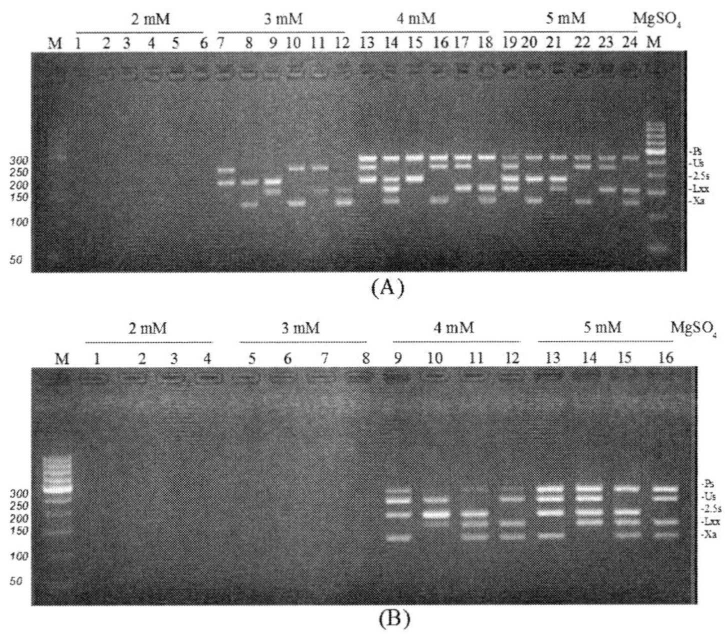

Fig. 2. The effect of $MgSO_4$ concentration in multiplex PCR. (A) Triplex PCR of different 3 primer pairs combination. (B) quadruplex PCR of different 4 primer pairs combination. The amplified fragments of 2.5S, Ps, Us, Lxx and Xa were 171, 252, 215, 145 and 116 bp respectively. All amplified PCR products were analyzed by 3% agarose gel electrophoresis.

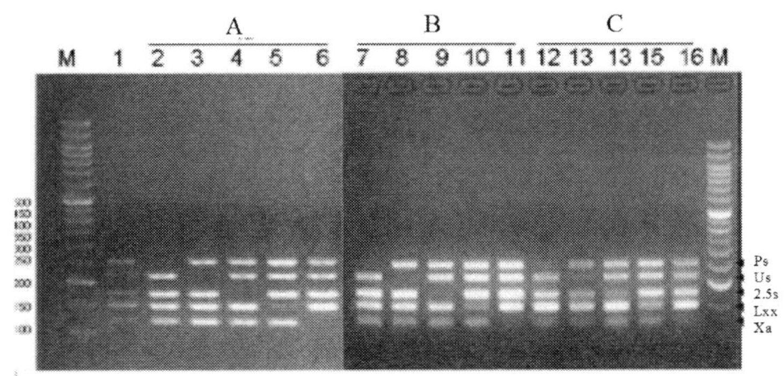

Fig. 3. The efficiency of 5 primers with different concentration of target gene in the multiplex PCR. (A) 1E+06 copies of 4 genes with one 1E+03 copies. (B)1E+05 copies of 4 genes with one 1E+03

copies. (C) 1E+04 copies of 4 genes with one 1E+03 copies. Lane 1, the limit detected at 1E+03 copies of five target gene (2.5S, Ps, Us, Lxx and Xa). The 1E+03 copies gene from leaf to right of each group was PS, Us, 2.5S, Lxx and Xa. The PCR products were analyzed by 3% agarose gel electrophoresis.

Detection of pathogens in the agricultural field samples

The crops field study was performed during the growing season from 2011 to 2013. Leaf samples (n=1030) differing in sampling sites, cropping years, and planting variety were chosen and used for detection of four target pathogen. In the first crop on 2011, the infected plant was screened and recorded of the symptomatic disease. Then, diagnose the pathogens by multiplex PCR and real-time PCR simultaneously and confirm the detection specificity.

After verify the detection technology, the sample from 2012 and 2013 crop years were chosen and detected using the high sensitivity and specificity real-time PCR. The result showed that the four pathogens were detected in field crop with the diversity of pathogen populations. The population of *Xanthomonas albilineans* (21.2%~30.2%) was higher than *Sporisorium scitamineum* (0%~1.3%), *Peronosclerospora sacchari* (2.2%~12.3%) and *Leifsonia xyli. Xyli.* (3.7%~4.1%) (Table 3.). Due to the population of *Xanthomonas albilineans* was shown to be extremely high, consequently the scald disease was high occurred in the field. Among of sugarcane varieties in the field, the ROC10 was highly resistant to smut and downy mildew disease but susceptibility to scald disease (Table 1.). The result was agreed to previously reported that resistance to disease is associated with limited colonization of sugarcane. Because of the sugarcane variety in field can not resistance all the pathogens, thus need use other method to assist prevent pathogen infected. However, some of the scald symptom plant were not detected using real-time PCR, because the distribution of pathogens in sugarcane plant was irregular. The distribution of *X. albilineans* in sugarcane has affected to accurately detect.

Table 3. Detection of pathogen population in sugarcane field crop

Year (sample amounts)	*Sporisorium scitamineum*		*Peronosclerospora sacchari*		*Xanthomonas albilineans*		*Leifsonia xyli. Xyli.*	
	Multiplex PCR	Real-time PCR	Multiplex PCR	Real-time PCR	Multiplex PCR	Real-time PCR	Multiplex PCR	Real-time PCR
2011 (84)	0%	3.5%	0%	27%	12%	42.8%	0%	29%
2012 (309)	-	0%	-	12.3%	-	30.2%	-	3.7%
2013 (637)	-	1.3%	-	2.2%	-	21.2%	-	4.1%

Distribution of *X. albilineans* in sugarcane

The susceptibility of leaf scald will need to prevented by early pathogen detection and early warning. Although real-time PCR is a highly sensitive technology for nucleic acid detection, there still some samples with leaf scald symptom detect with negative result. Realize the distribution of *X. albilineans* in sugarcane will reduce the error of sampling and enhancing the accuracy. Hence, quantitative detection of *X. albilineans* in field-grown sugarcane with different degree of leaf-scald to make sure the distribution of *X. albilineans* in sugarcane.

The sugarcane with withered leaf tip symptom differing in part from the stalk (bottom) to shoot tip (top) include cane stalk, bud, leaf and tiller were chosen. Detection of samples within the sugarcane plant of withered leaf symptoms, the *X. albilineans* pathogen was highly detected in stalk and bud. The pathogen population in samples taken from the bottom of the plant was higher than the top part of the plant (Fig. 4 (A)). In addition, the pathogen distribution in leaf was unevenly distributed (data not showed).

The sugarcane with withered leaf and tiller (half of main stalk in length) both symptoms was detected and shown in Fig. 4 (B). The result of main stalk appeared that highly pathogen concentration can be detected in stalk and bud and reduce from bottom to top. The stalk and bud of tiller was pathogen negative. In addition, we also detected the pathogen in leaf samples. The result showed that the pathogen distribution in leaf was unevenly distributed (data not showed).

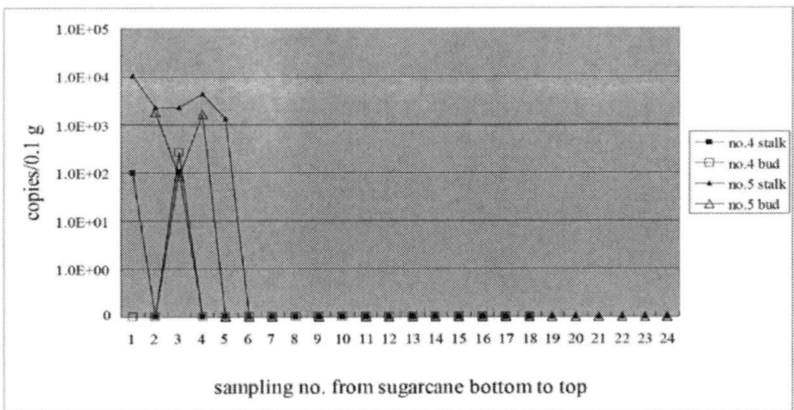

(A)

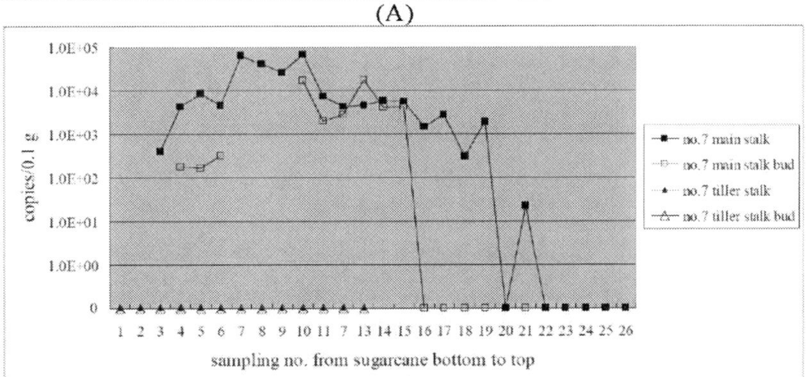

(B)

Fig. 4. Quantitative detection of *X. albilineans* in the sugarcanes field crop with different

degree of leaf-scald symptoms. (A) the withered leaf tip sugarcane. (B) the withered leaf and tiller symptoms sugarcane.

Conclusion

Even though the detection of pathogen in leaf and bud were convenient, but the good sampling is critical for successful results. From our result was suggested that sample from the bottom part (especially in stalk and bud) was high efficiency and accuracy to detect the *X. albilineans* in sugarcane. In addition, the distribution of pathogens in sugarcane leaf is inconsistent. The population of *Xanthomonas albilineans* in the sample was higher at the bottom of the stalk than the upper of the stalk. Overall the technology we established in this study can specificity detection the sugarcane pathogens. And the result display in this study could be guidelines for detection of sugarcane pathogens and meanwhile reduce the error of sampling and enhance the accuracy of detection.

S.I.T. Paper #1169

The use and development of image analysis on massecuites

Arend Wittenberg[1], René Daniels[2], Daniel Nouws[2]

Abstract

Since 2003 Suiker Unie used an off-line image analysis system (IAS) to measure the size distribution of crystals in industrial massecuites. This paper describes the technical set-up, the method of sampling, analyzing and distribution calculation. It gives an overview of the potential use of IAS, such as optimization of full-seeding-systems, tracing the development of CSD during a pan strike and the influence of crystal morphology on run-off-separation. Based on the experiences Suiker Unie made over the years a couple of improvements were developed that have been realized when updating the system in 2015/16.

Analyzing final A-product with a size of approximately 600 µm was not possible with the original system. When updating the system the opportunity was taken to extend the measuring range to include A-product. The original light-microscope and automated XY-stage were extended with new components: an XY-stage controller, a high resolution camera and customized image analysis software, providing new methods of image grabbing and -processing. The new hardware and software configuration now allow measurements in the crystal size range of 10 µm to 2000 µm. Sample preparation procedure was modified to avoid false grain formation especially in suspensions of low purity. Samples are now directly fixated in sucrose-fructose solution. The updated system and procedure show less variance in Mean Aperture (MA) and coefficient of variation (CV) compared to the formerly used. Both seed crystals and the final A-product can now be analyzed, thus covering the complete sugar crystallization process.

Introduction

Crystallization is the most effective separation process in sugar refining, at the same time it is highly energy consuming. To get the most out of it in perspective of product quality and production costs, there is a strong demand for appropriate process-assessment, in particular the development of crystal size distribution and crystal morphology from seed-slurry up to final product.

In this paper the use and development of image analysis on massecuites at Suiker Unie is described and discussed.

[1] Suiker Unie, Noordzeedijk 113, 4671 TL Dinteloord, The Netherlands
[2] Cosun Food Technology Centre, Oostelijke Havendijk 15, 4704 RA Roosendaal, The Netherlands

Why image analysis?

Alternative methods

To analyze particle size distributions, there are a couple of methods (Table 1). Here a short description of the most relevant ones:

- Dry <u>sieving</u> is the standard method in the sugar industry worldwide, used by sugar producers and customers. It delivers mass-size-distributions in a particle size range from 50 µm up to 5 mm but doesn't say anything about crystal morphology. For dry bulk sugar no sample preparation is needed. But to analyse massecuites, one needs to separate crystal and syrup, wash and dry properly. Besides the fact that sample-preparation is very time consuming, there is a risk of structural deviation due to agglomeration during the sample preparation. Nevertheless, the method is cheap regarding capex and when carried out carefully it is useful for factory optimization.

- <u>Laser diffraction</u> determines particle size using light scattering. The method covers a wide size range between about 50 nm up to 3 mm. Laser diffraction is useful for all kinds of dispersions, which might justify a rather high price of the instrument. Especially for samples with high viscosity, proper dispersion is key to avoid structural deviation of the analytical result. However when using the method for massecuites, structural inaccuracy needs to be taken into account, f.i. due to optical effects caused by crystal edges.

- Surely the biggest advantage of <u>In-line image analysis (IA)</u> is getting the results directly without any delay and without spending any work on sampling and sample preparation. This provides a big disadvantage at the same time: with crystal contents above about 10 % particles start to overlap each other between light and lens. Today's algorithms are not able to detect and analyze single particles as long as they overlap.

Table 1: Alternative methods for crystal size measurements

Method	Sieving	Laser-diffraction	In-line-IAS	Off-line image analysis
Size range	50 µm ... 5 mm	0,05 µm ... 3 mm	4 µm ... 2000 µm	4 µm ... 2000 µm
Distribution	Mass	number → volume	number → mass	number → mass
Morphology	No	no	yes	yes
Preparation	no	stirring / ultrasound	no	wet dispersion
Use	bulk sugar	dispersions (carbonation sludge, crystal suspensions)	inline pan-boiling (crystal content limited)	• all crystal-suspensions - slurry - seed-crystal - final product • all purities • during whole batch

Also off-line IA provides a couple of disadvantages. The first is the most obvious: the existing systems use to be rather big and heavy. It takes two people and a big car-boot to transport it from factory to factory. Furthermore experience is essential to carry out sampling and sample preparation properly. A big advantage of off-line IA is the transparency of the analysis – what you see is what you get. Another big advantage is, due to the high quality of the microscopic picture, that also particle morphology can be determined.

The following arguments were relevant for Suiker Unie's final choice to go for off-line image analysis:

- easy sample preparation – compared to sieve analysis,
- what you see is what you get – compared to laser-diffraction,
- capable to cover complete range of crystallization products with one system – compared to in-line systems.

Applications

In the following chapter some typical applications of in-line image analysis at Suiker Unie are highlighted.

Validation of industrial crystal seed systems

As a regular service at start of the campaign the system is sent for a couple of days to the factories to fine-tune crystallizers batch control parameters. Special attention is paid on:

- Supersaturation at seeding,
- Size-mass-relation (d^3-rule) from seed to final product,
- Quality of seed material and final product.

Validation of seed-slurry/fondant

Whenever new seed-slurry needs to be tested, the IA is a useful tool. Not only due to the quantitative results such as MA, CV or conglomerate content. Also the photo-illustration – which comes as a by-product - adds value for reporting and expert discussion (Figure 1).

Conglomerate content in C-massecuite and influence on colour of C-melt

Already a couple of years old are the investigations that are carried out to investigate the influence of seed material, supersaturation and stirring on conglomerate forming (Wittenberg, 2001). The upper pictures in Figure 2 show C-product massecuites based on two different types of seed-material. Trial 2 came up with significantly less conglomerates than Trial 1.

The graphs in the lower picture show the color development of C-melt during these trials. Green dots show the C-melt-color produced with seed material 1, red dots give the color based on seed material 2.

The relation between conglomerate content and color is evident. Especially taking into account that wash-water dosing is reduced during trial-2 (purple line).

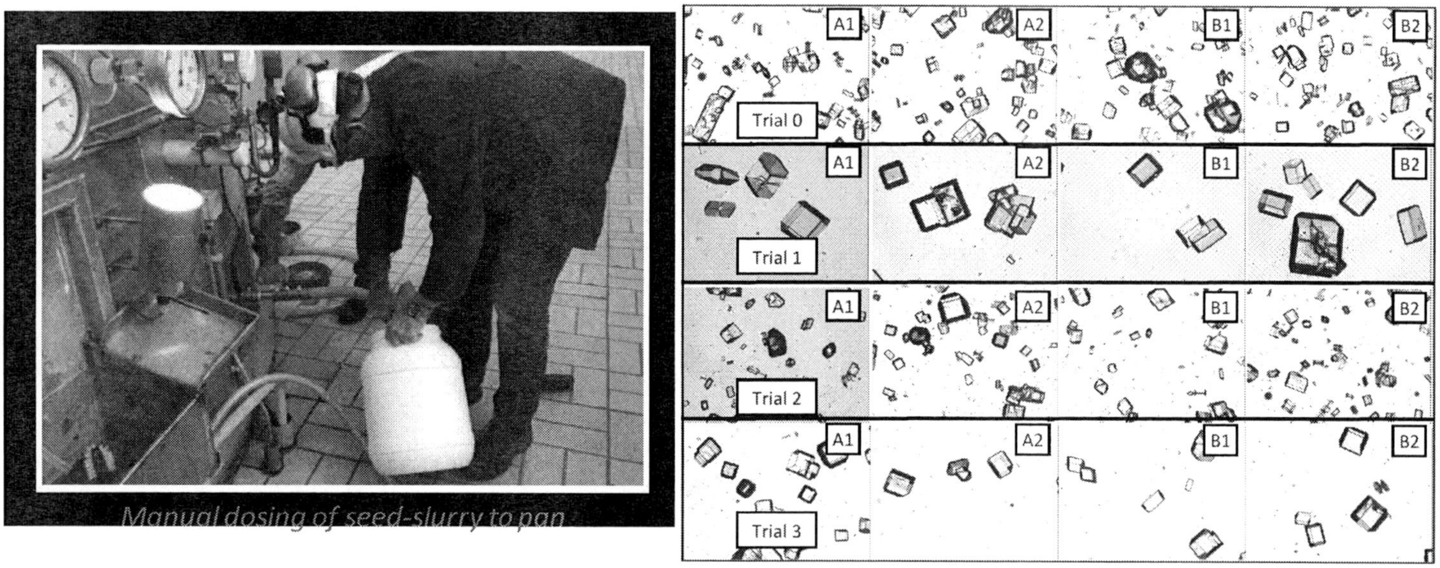

Figure 1: Validation of seed-slurry/fondant with off-line Image Analysis System

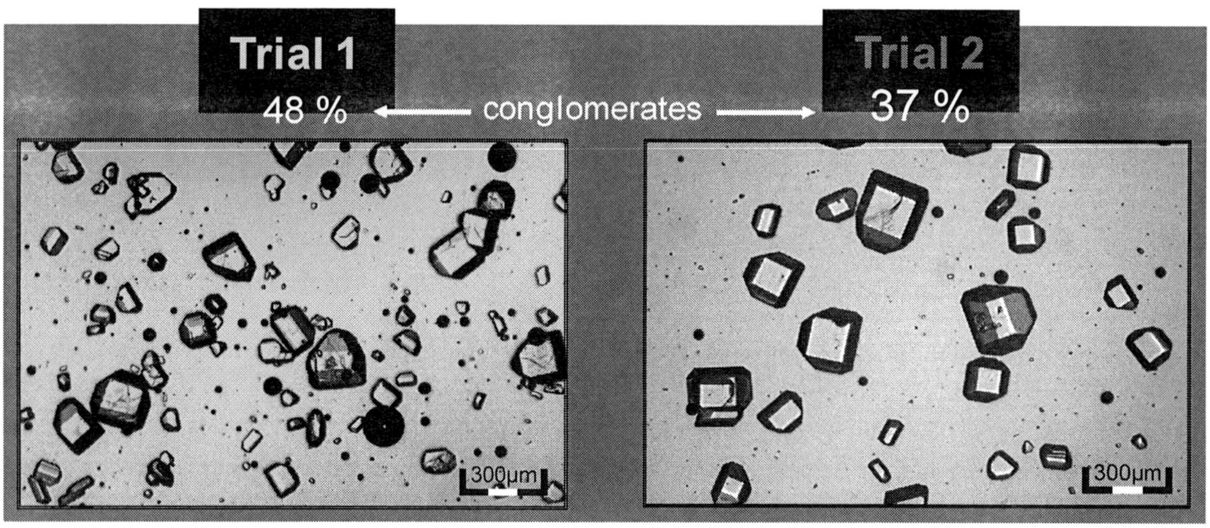

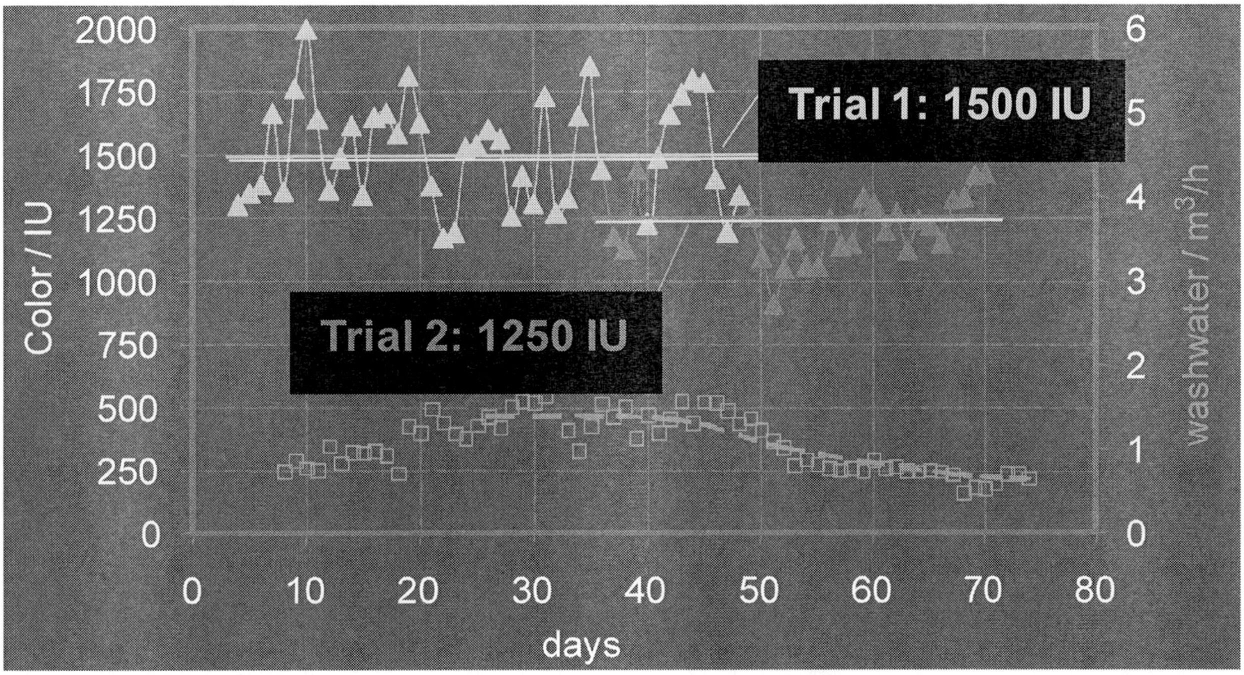

Figure 2: Conglomerates in C-massecuite and colour in C-melt

These examples prove that IA is a very valuable method for science-based optimization and development of industrial sugar crystallization.

Development of system and method

This chapter focusses on the recent system improvements developed and initiated by Suiker Unie and explains the working method of the improved system.

Limitations of existing image-analysis-system

In 2015 after 12 years of use, hard- and software of the existing IA-system were outdated. Support wasn't available anymore as the original supplier had disappeared. When initiating a project to update the system or buy a new one, an extended particle size range to cover the whole product range of our crystallization and a better mobility and were demanded.

System selection

After comparing a couple of technical solutions it was finally decided to keep the existing microscope and XY-stage (Figure 3). New are the camera with higher resolution, a 3D-stage controller and the notebook. Also new is the IA-system-software with an implemented sugar crystal script.

As the picture shows, the system is rather compact.

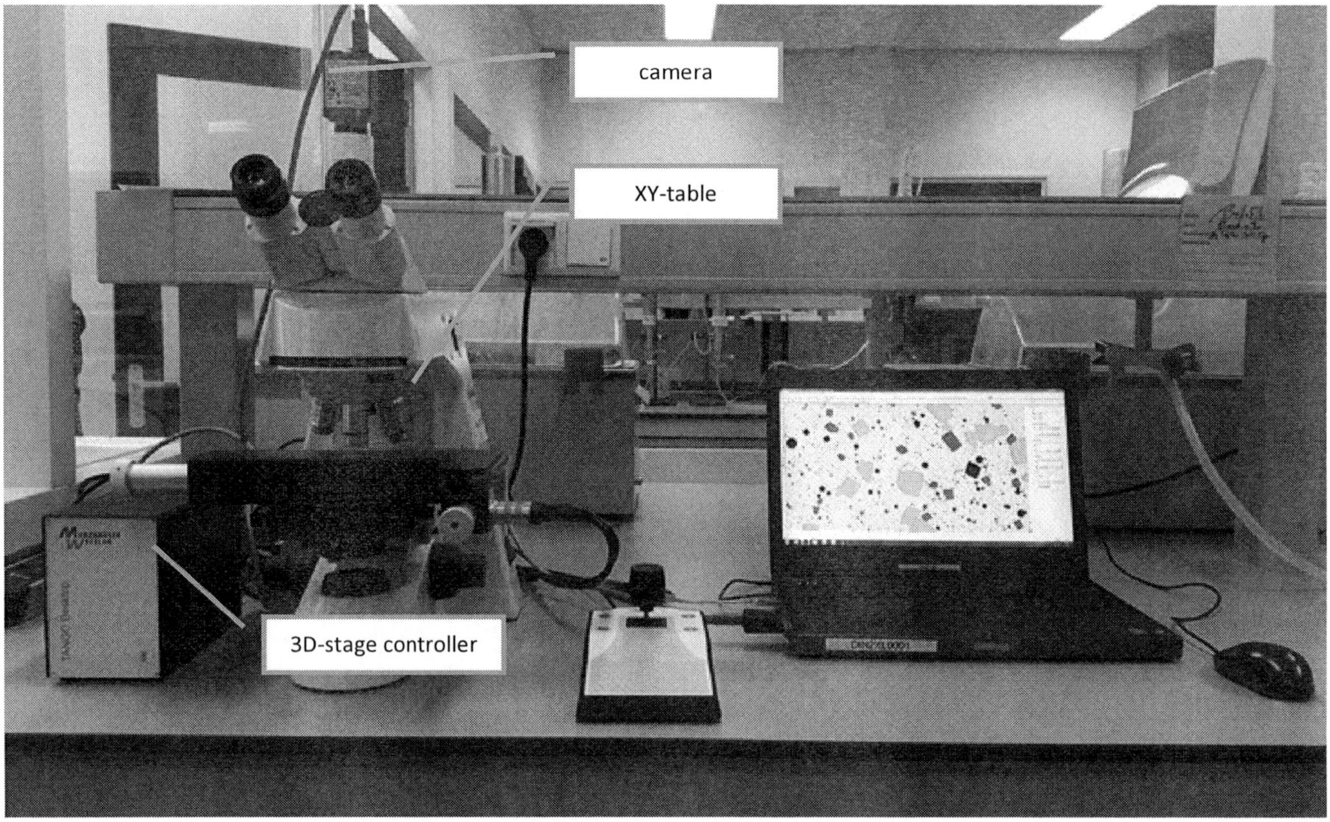

Figure 3: Updated IA-System 2016

Image analysis

The image analysis method and the algorithm used at Suiker Unie are based on the scientific work of Wagner (2003). Here just a short overview of the analytical procedure:

Image analysis (sequence)

1. Each massecuite sample is fixated and diluted; the crystal-suspension is distributed on a microscope slide.

2. Under the microscope a series of pictures is taken automatically. Whereas the old system needed to process picture by picture, the new one is able to combine all pictures to a large mosaic, which finally enlarges the particle size range up to 2000 µm.

3. A filter-algorithm rejects all non-crystal-objectives such as fibres and bubbles.

4. Finally all crystals are measured and classified in fines, single crystals, simple and complex conglomerates (Figure 4).

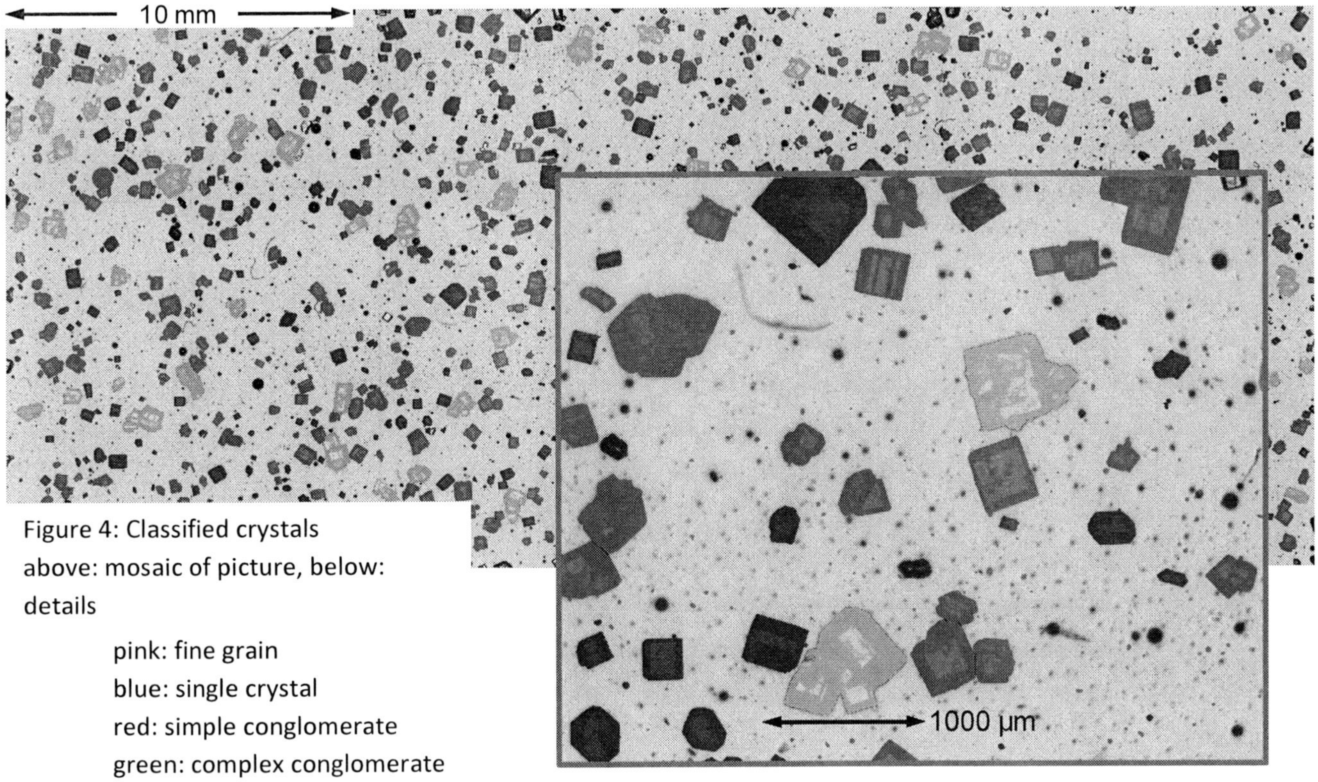

Figure 4: Classified crystals
above: mosaic of picture, below: details

pink: fine grain
blue: single crystal
red: simple conglomerate
green: complex conglomerate
grey / black: filtered out

Finally the whole set of digital information is stored and a quick sample report is generated gaining the most relevant information such as size distribution, distribution curves and some pictures for illustration.

System evaluation

Table 2 gives an overview of the main differences between the existing and the improved IA-system. Following aspects are explained in more detail:

- particle size range and analysis time,
- improved sample preparation,
- effect on particle size distribution.

Table 2: Key figures of initial and improved IA-system

	initial	improved
Particle size	Seed 1, Seed 2	Seed 1, Seed 2, A-product
Mobility	+ / -	✓

Sample preparation	20-25 min	15 min
Number of slides to be analysed per sample	1 – 3 slides depending on particle size	
Analysis time per slide	10 – 15 min	≈ 3 min

Particle size range and analysis time

As illustrated in Figure 5, composing the pictures to a large mosaic before image analysis, enhances the particle size range significantly. The improved system can process crystals up to 2 mm, covering the whole range of our industrial sugar production.

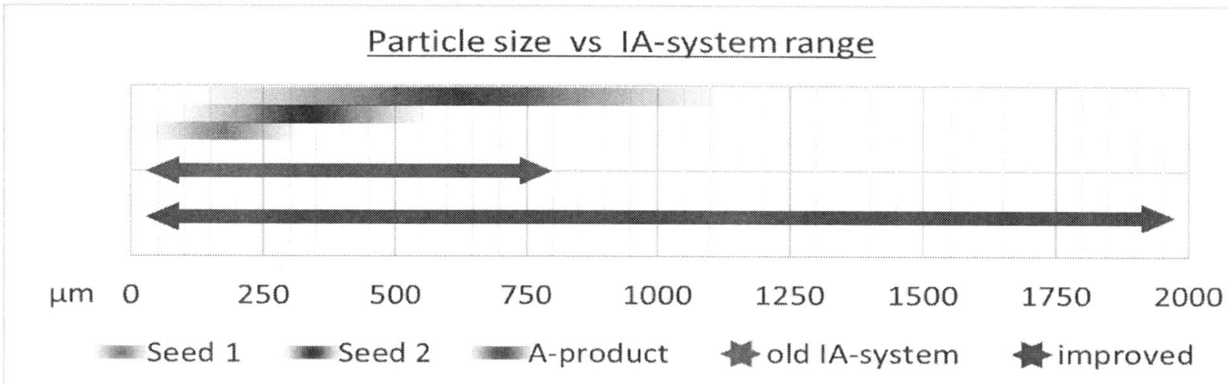

Figure 5: Particle size of crystallisation products and size range of Image Analysis

Due to modern hardware, the system is more mobile and user friendly. Improved sample preparation procedure saves up to 10 min per run. And thanks to today's PC- and controller-performance analysis time is reduced to about 1/3.

Sample preparation

Careful sample preparation is key. In the vacuum pan crystals should grow as fast as possible, but after sampling crystal growth needs to stop immediately. Initially fresh massecuite sample were first taken to the laboratory and there stabilized and dispersed in saturated sucrose-fructose-solution. If this took too long, crystals kept on growing or agglomerate- or false-grain formation occurred.

Therefore to stabilize the crystals immediately after sampling, sample dilution and stabilization is nowadays carried out directly at the pan: A certain number of droplets of the fresh massecuite sample are poured into a beaker with a certain amount of S/F-solution, which is then gently mixed.

With the new method reproducibility enhanced significantly and time for sample preparation is shortened with some 10 minutes.

Mean particle size

Based on the same massecuite samples the old and the new system are compared. On the y-as in Figure 6 the results of the old system are given, on the x-as the ones of the new system. The left plot shows the results with the old preparation method, the right plot the improved method.

Ideally all results should lay along the dotted line. This is the case for Seed 1 and 2, with a slide improvement by the new sample preparation method.

Regarding A-product it is obvious, that the old system gives too low results compared to the new one. With the new preparation method this improves but still in average a too small MA is determined. Here the effect of composing a large picture before image analysis becomes obvious. The total length of the outlines of the composed picture is much short than the sum of the outlines of all single picture. Hence the chance of crystal to be 'cutted' by outlines and therefore to be rejected from the analysis is much smaller.

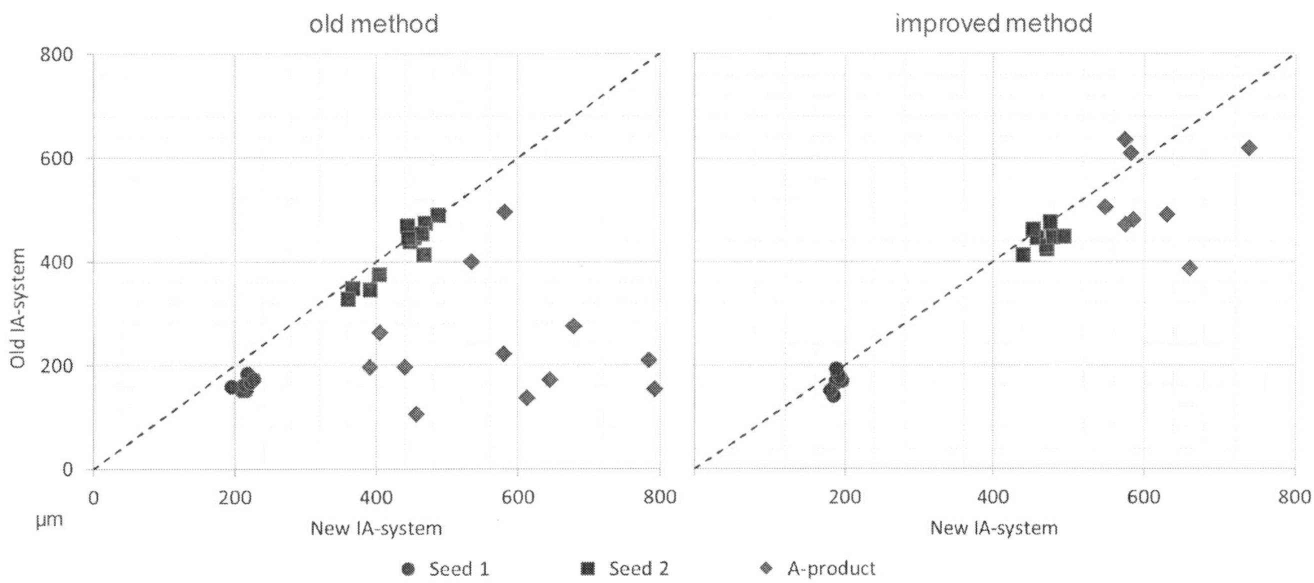

Figure 6: Comparison of mean particle size between old and improved method

Conclusion

Summing up the results of system development and regarding all experiences of practical use over the years, off-line image analysis is a valuable method for science-based optimization and development of industrial sugar crystallization.

The method gives insight in particle-size distribution (mass- and number) and crystal morphology. The range of a microscope based system covers the usual industrial sugar crystallization. The tool needs the hand of an expert-tool as careful sampling and sample preparation are essential.

The updated IA-system is available for sugar industry and –research. As development of analytical methods are pre-competitive, Suiker Unie is willing to cooperate in this field with other producers and research institutes.

Literature

G. Wagner (2003): Entwicklung eines automatisierten Bildanalyseverfahrens zur Beurteilung der Kristallqualität in Saccharosekristallsuspension, Berlin, Techn. Univ., Diss. 2003, dissertation.de – Verlag im Internet GmbH.

A. Wittenberg (2001): Beitrag zur Optimierung der technischen Saccharosekristallisation, Berlin, Techn. Univ., Diss. 2001, Bartens.

Appendix

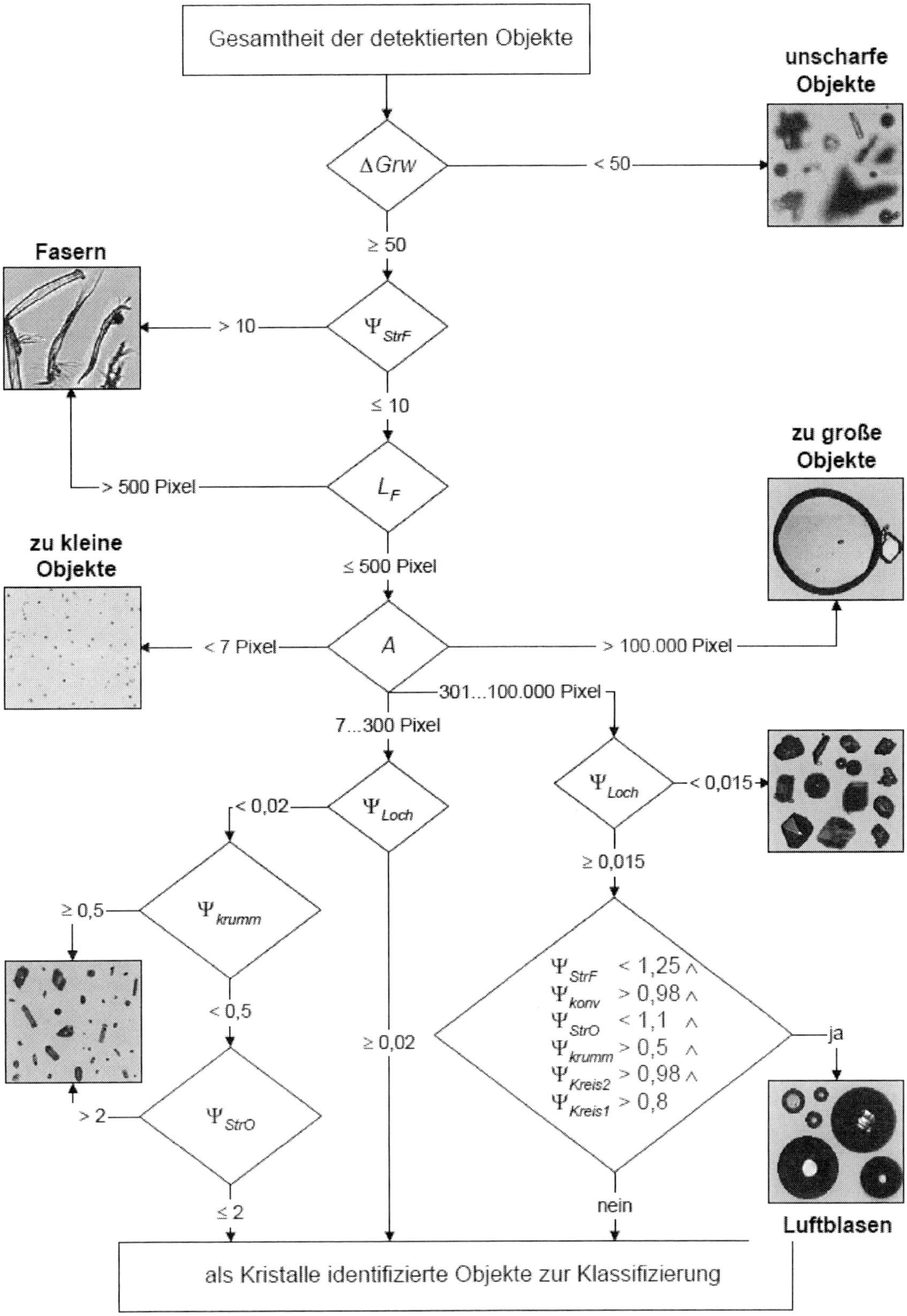

Figure 7:Algorithm for removal of non-crystal-particles (Wagner 2003)

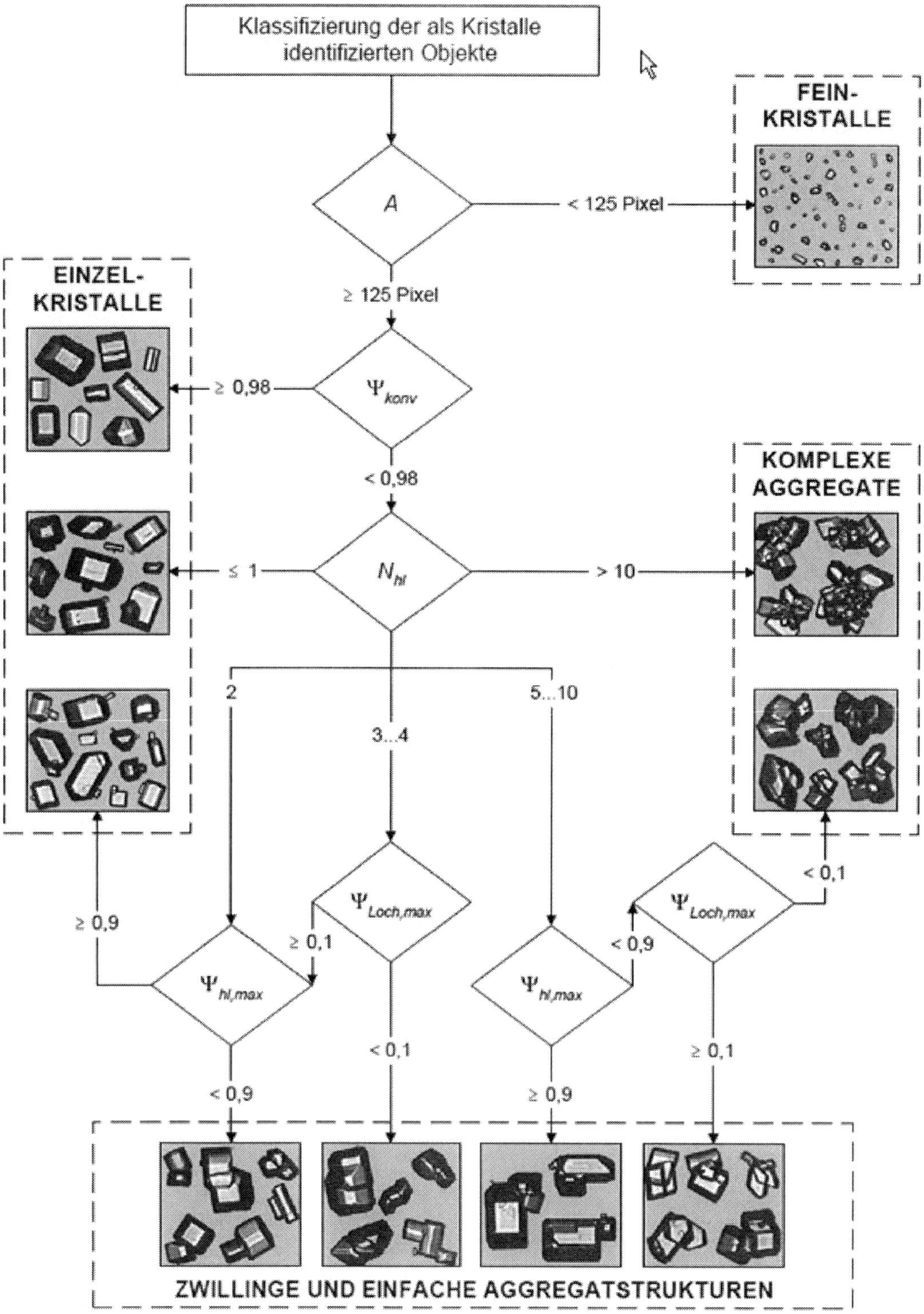

Figure 8: Algorithm classification of crystals (Wagner 2003)

S.I.T. Paper #1170

Improvement of Sugarcane Cultivation Technology for the Sustainable Development of the Sugar Manufacturing Industry
—(1) Analysis of Sugarcane Growth in North-eastern Thailand—

A Kamiya1, O Yatabe2,M.Mizu2,S Ito2, S Ando3,Y Kobori3,T Takei1,T Kawai2
W Ponragdee4 and A Sugimoto2
1)The kumphawapi Sugar Co. LTD, 2)Mitsui Sugar Company LTD, 3) Japan International Research Center for Agricultural Sciensies, 4) Department of Agriculture of Thailand

Abstract
The sustained development of the sugar manufacturing industry requires not only an improvement in manufacturing technology, but also a stable increase in raw material production even under unfavorable natural condition. In this study, we focused on Northeastern Thailand, where cane yield is low and unstable due to the harsh climate during the dry season. For improving cane yields with labor- and resource-saving agricultural way, we determined aspects of cane growth in this region. We created 144 plots both for plant cane and for ratoon cane, and studied yields and it's components in 2013 and 2014. The stalk yield was high in plant cane (97 t/ha, 105 t/ha) but low in ratoon cane (67 t/ha, 62 t/ha). There was no clear difference in the number of stalks. The stalk weight was heavy (1,594 kg, 1,626 kg) in plant cane and light (1,152 kg, 993 kg) in ratoons. It indicates that raising the yield of ratoon cane to the level of plant cane required increasing its stalk weight. Also, in both plant and ratoon cane, fields with high yields obviously had greater numbers of sugarcane stalks compared to fields with low yields. It indicates that improving the cane yield, it was also important to increase the number of stalks. Our study further revealed that the number of ratoon sugarcane stalks is closely related to the number of missing stalks when sugarcane is newly planted, and that the stalk weight is closely related to the depth at which axillary buds sprout in soil.

Introduction

Raw material production is one of the most important factors affecting the sugar manufacturing industry. With the world population exceeding 7 billion, the sustainable development of a sugar manufacturing industry requires the adoption of new technology to increase the raw material production, as well as advanced sugar manufacturing technology. From a global perspective, the standard approach to increasing raw material production is through the improvement of land productivity. This includes the development of technology to increase productivity in areas with low unit yield. Since the sugar manufacturing industry is expected to contribute to sustainable development and act as a key industry within the region, the adoption of new technology to increase the raw material production should also enhance the regional economy, improve the quality of life of producers, increase the profits of sugar manufacturing companies, and improve the natural environment as a whole. In other words, both labor productivity and land productivity should be enhanced by the adoption of cultivation technology that can help conserve the environment, resources, energy, and labor.

Thailand is the second largest sugar exporter in the world and the fourth largest producer of sugarcane, the raw material used to manufacture sugar. Any trend associated with the sugar manufacturing industry in Thailand will, therefore, have a huge impact on global food and energy production. Northeastern Thailand is the largest producing region in the country but is characterized by low unit yield and a low level of continuous ratooning. The low unit yield and low level of continuous ratooning can be attributed to the pronounced dry season, poor soil, and frequent outbreaks of white leaf disease. However, no precise information has yet been published on the key issues affecting sugarcane cultivation. Since such information is vital when developing technology to improve unit yield and the level of continuous ratooning, we conducted a survey on sugarcane cultivation in fields in northeastern Thailand in order to compare sugarcane growth in fields with high and low yields, and assess the difference between newly planted fields and ratooned fields. This report presents the beneficial information obtained from this survey regarding the potential approaches for improvement.

Materials and methods

Research sites were on the farms of the Kumphawapi Sugar Co. Ltd. and included large variation of cane yield but with similar basic agricultural management. We chose 24 fields from both plant cane and ratoons, and established 288 plots (144 plots each) for plant cane and ratoons. Each plot comprised one row (5 m long × 1.4 m wide): 3 areas were selected from each field and 2 plots from 1 area. The investigation plots were then classified into low-yielding plots (plant cane: below 50 t/ha, n=9; ratoons: below 40 t/ha, n=18); medium-yielding plots (plant cane: 50-80 t/ha, n=42; ratoons: 40-70 t/ha, n=74); and high-yielding plots (plant cane: above 80 t/ha, n=93 and ratoons: above 70 t/ha, n=52).

We studied: (1) Variations in cane yield and so-called contributors to yield (stalk weight, number of millable stalks, etc); (2) Relationships between cane yield and contributors to yield; and (3) Relationships among the contributors to yield. Relationships were evaluated by correlation coefficients. Differences between plant cane and ratoons, between high-yielding plots and low-yielding plots, and changes from plant cane to ratoons were determined. We measured: (1) cane yield; (2) number of millable stalks; (3) stalk weight; (4) number of missing plants - more than 50 cm or less than 100 cm of a blank row was ranked as a missing plant; (5) juice quality (Brix value, juice polarization, fibre content); (6) initial depth of re-sprouting stems (initiation depth).

Similar investigations were performed in 2013 and 2014 and similar results were obtained in both years. Here we mainly describe the results obtained in 2013.

Results

Differences in cane yield, number of millable stalks, and stalk weight between plant cane and ratoons are shown in Figure 1. Average cane yield in the 144 plots was higher in plant cane (96.7 t/ha) than in ratoons (67.2 t/ha) and the coefficient of variability (CV) in plant cane (35%) was lower than in ratoons (41%). The number of millable stalks was higher in plant cane compared to ratoons (plant cane: 60,700 /ha, CV 28%; ratoons: 58,300 /ha, CV 26%), but the difference was not significant. Stalk weight was heavier in plant cane (1594 g) than in ratoons (1152 g) and the CV value in plant cane (22%) was lower than in ratoons (31%).

The differences in contributors to yield were evaluated across the three yield groups. In both plant cane and ratoons, the number of millable stalks was lower in low-yielding plots than in high-yielding plots (Fig. 2). Differences in the number of millable stalks based on the three yield groups were: plant cane: low-yielding plots = 29,400/ha, medium-yielding plots = 51,540/ha and high-yielding plots = 67,940/ha; ratoons: low-yielding plots = 41,290/ha, medium-yielding plots = 53,890/ha, and high-yielding plots = 66,300/ha. The number of millable stalks in plant cane was lower than in ratoons of low- and medium-yielding plots. The coefficient of variance in plant cane and ratoons was higher in low- yielding plots compared to higher-yielding plots (plant cane: low-yielding plots = 29%, medium-yielding plots = 25% and high-yielding plots = 20%; ratoons: low-yielding plots = 30%, medium-yielding plots = 19% and high-yielding plots = 20%).

In plant cane and ratoons, stalk weight was lower in low-yielding plots and heavier in high-yielding plots, and ratoons showed lower stalk weights than plant cane (plant cane: low-yielding plots = 1308 g, medium-yielding plots = 1360 g, and high-yielding plots = 1729 g; ratoons: low-yielding plots = 802 g, medium-yielding plots =1040 g, and high-yielding plots = 1442 g). Stalk weights were lighter and the coefficient of variance was higher, especially in ratoons with low yields (plant cane: low-yielding plots = 24%, medium-yielding plots = 25%, and high-yielding plots = 17%, ratoon: low-yielding plots = 48%, medium-yielding plots = 20%, and high-yielding plots = 21%).

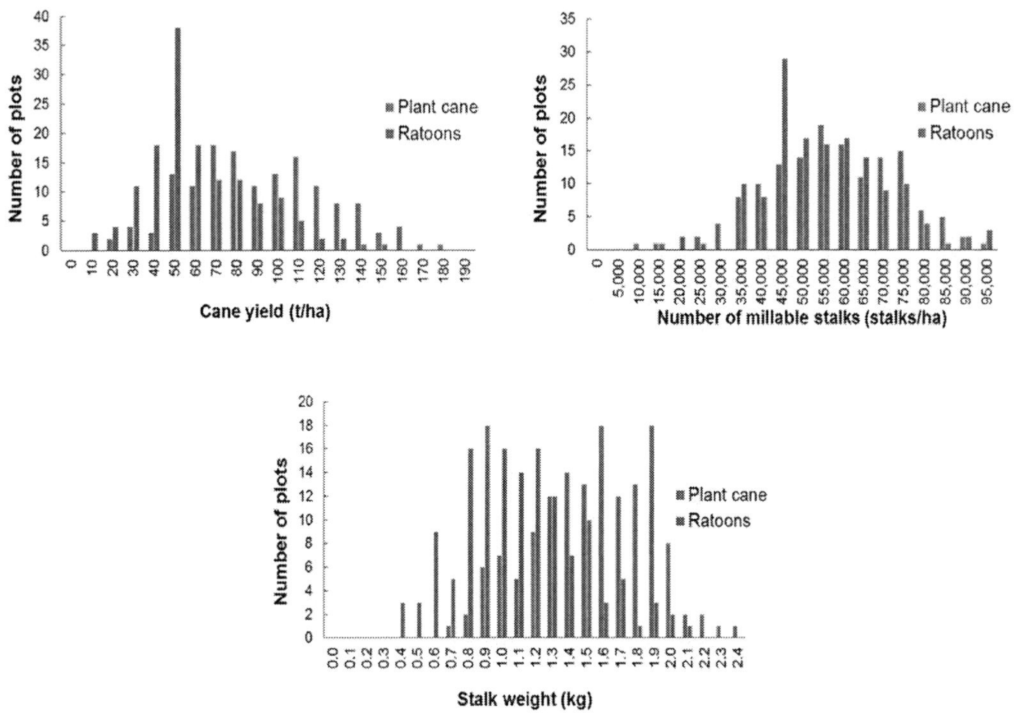

Fig. 1. Differences in cane yield, number of millable stalks and stalk weight between plant cane and ratoons in 2013.

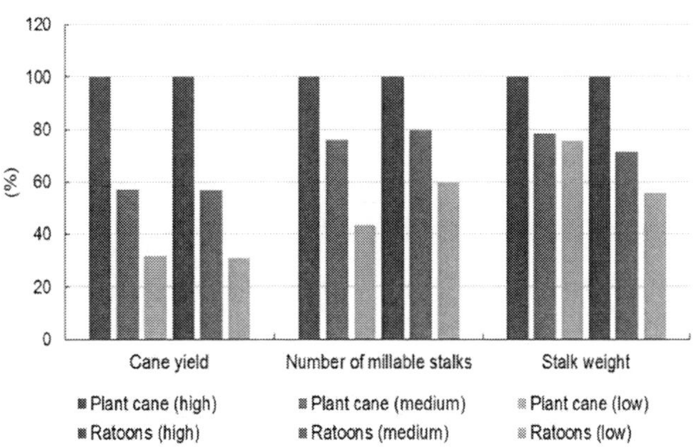

Fig. 2. Differences of contributors to yield among high-yielding plots, medium-yielding plots, and low-yielding plots. Vertical axis shows the percentage of each bar compared to the bar of high-yielding plots (100%).

Correlation coefficients among cane yield, number of millable stalks, and stalk weight are shown in Table 1. Cane yield was positively correlated with the number of millable stalks and stalk weight in all plots, in low-yielding plots, and in high-yielding plots of both plant cane and ratoons. In medium-yielding plots, cane yield was positively correlated with stalk weight, although it was not the case with the number of millable stalks. Correlations between the number of millable stalks and stalk weight were: in the low-yielding plots there was no significant correlation in plant cane and ratoons; in the medium- and the high-yielding plots there was a negative correlation both in plant cane and ratoons.

Table 1. Correlation coefficients among cane yield, number of millable stalks, and stalk weight of plant cane and ratoons in 2013.

	Number of millable stalks	Stalk weight		Number of millable stalks	Stalk weight
Plant cane (All plots)			Ratoons (All plots)		
Cane yield	0.89**	0.78*	Cane yield	0.84**	0.77**
Number of millable stalks		0.45**	Number of millable stalks		0.36**
Plant cane (Low-yield)			Ratoons (Low-yield)		
Cane yield	0.85**	0.69**	Cane yield	0.73**	0.56**
Number of millable stalks		0.23	Number of millable stalks		-0.12
Plant cane (Medium-yield)			Ratoons (Medium-yield)		
Cane yield	0.44	0.52*	Cane yield	0.09	0.44*
Number of millable stalks		-0.53*	Number of millable stalks		-0.81**
Plant cane (High-yield)			Ratoons (High-yield)		
Cane yield	0.73**	0.51**	Cane yield	0.55**	0.41**
Number of millable stalks		-0.21*	Number of millable stalks		-0.51**

Values where the correlation is significant are indicated with asterisks (* $P \leq 0.05$, ** $P \leq 0.01$)

Cane yield and contributors to yield were also compared between plant cane in 2013 and ratoons in 2014 in the same fields. There was a positive correlation in all parameters examined between plant cane in 2013 and ratoons in 2014: cane yield ($r = 0.32$**), number of millable stalks ($r = 0.18$*), and stalk weight ($r = 0.36$**). However, cane yields were significantly reduced when shifting from plant cane (96.7 t/ha) to ratoons (61.5 t/ha). The number of millable stalks reduced slightly when moving from plant cane (60,700/ha) to ratoons (58,300/ha). The stalk weight was significantly reduced with a shift from plant cane (1594 g) to ratoons (993 g).

Discussion

Our data show:

- Lower cane yield in ratoons than in plant cane was due to a marked reduction in stalk weight;

- The number of millable stalks in both plant cane and ratoons was higher and stalk weight was heavier in high-yielding plots compared to lower-yielding plots;
- The difference in the number of millable stalks in plant cane and the differences in number of stalks and stalk weight in ratoons were large;
- In all plots, cane yield showed positive correlations with number of millable stalks and stalk weight of plant cane and ratoons. The negative correlations were not observed between the number of millable stalk and stalk weight in low-yielding plots, although they were observed in medium-yielding and high-yielding plots.

These results also indicated that maintaining the number of millable stalks in plant cane to keep more stalks in ratoons and minimizing the reduction of stalk weight in ratoons should have priority to improve the yield in ratoons. As there was no negative correlation between number of millable stalks and stalk weight in low-yielding plots, it was considered to be possible to increase the number of millable stalks without reducing the stalk weight.

In our fields, the variability in the number of missing plants among fields was large, which was thought to be associated with number of millable stalks. Therefore, the associations of missing plants with cane yield and number of millable stalks were examined. Other characteristics thought to be associated with stalk weight were also investigated to determine what could contribute to improved cane yield in ratoons (Table 2).

There was no correlation between the number of millable stalks in ratoons and missing plants in plant cane (preceding year) at harvest time. However, the number of millable stalks of ratoon cane was negatively correlated with the missing plants after ratooning in March and in June. Cane yield of ratoon cane was also negatively correlated with the number of missing plants in March and June. In addition, the number of missing plants at harvest in the plant cane was positively correlated with the number of missing plants in March and June (after ratoon management practices). They, in turn, were strongly correlated with the number of millable stalks in the ratoons at the time of harvest. The number of millable stalks in ratoons was also positively correlated with the number of millable stalks in plant cane. The number of millable stalks in plant cane was negatively correlated with the missing plants in plant cane at harvest. These results suggest that the minimizing the number of missing plants in plant cane would be an effective way to increase the number of millable stalks in ratoons.

Table 2. Correlation coefficients among number of missing plants, cane yield, and various contributors to yield.

	Number of millable stalks in plant cane	Cane yield of plant cane	Number of missing plants in ratoons (March)	Number of missing plants in ratoons (June)	Number of millable stalks of ratoons	Cane yield of ratoons
Number of missing plants in plant cane	-0.36**	-0.24**	0.38**	0.36**	-0.15	-0.08
Number of millable stalks in plant cane		0.74**	-0.29**	-0.26**	0.18*	0.15
Cane yield of plant cane			-0.21*	-0.16	0.22*	0.32**
Number of missing plants in ratoons (March)				0.78**	-0.26**	-0.23*

Number of missing plants in ratoons (June)	-0.30**	-0.27**
Number of millable stalks in ratoons		0.85**

Values where the correlation is significant are indicated with asterisks (* P≤0.05, ** P≤0.01)

Changes in contributors to yield (stalk length, stalk diameter and number of stalks) during the growth period from June to November (time of harvesting) were also investigated (Fig. 3). Stalk length at harvesting time was shorter in ratoons than in plant cane because stalk extension was stagnant from July to September while the stalk extension in plant cane was active. There were no notable differences in the number of stalks and stalk diameter between plant cane and ratoons.

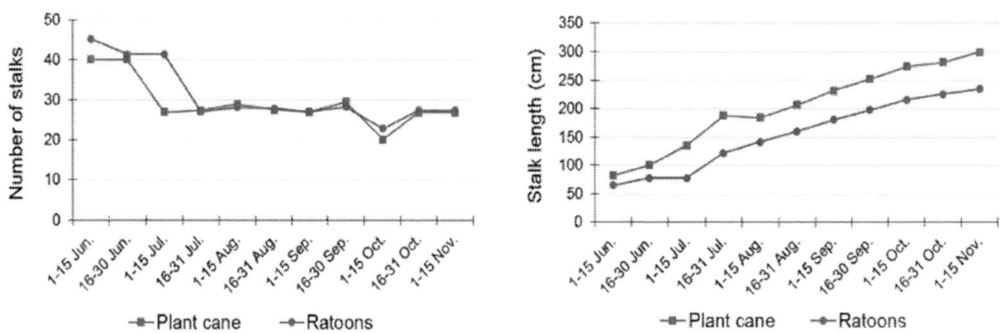

Fig. 3. Number of stalks and stalk length in 2013, June to November based on the survey in Kumphawapi farmer's fields; average of 80 fields.

As previously indicated, improvement ratoon yields was dependent on the improvement of stalk weight. During the investigation it was observed that the growth of sugarcane was better where large amounts of filter cake had been applied prior to planting. The applications of boiler ash and vinasse (from ethanol production) have also been reported as effective ways to improve cane yields (Prado *et al.* 2013). These reports support the importance of nutrient supply for stalk growth. In ratoon cultivation, if soil fertility is low and the nutrient supply is insufficient, sugarcane growth will stagnate and yield will be low. Thus, appropriate input of nutrients is needed in order to raise the cane yield in ratoons.

Table 3. Correlation coefficients between stalk weight and underground part characters (2014).

	Initiation depth	Number of underground nodes	Length of underground stems	Diameter of underground stems	Total number of roots
High cane yield					
Stalk weight	0.37**	0.25**	0.39**	0.39**	0.34**
Initiation depth		0.74**	0.96**	0.43**	0.61**
Number of underground nodes			0.73**	0.11	0.55**
Length of underground stems				0.46**	0.67**

Diameter of underground stems					0.36**
Low cane yield					
Stalk weight	0.24*	-0.10	0.24*	0.50**	0.21
Initiation depth		0.72**	0.96**	0.36**	0.76**
Number of underground nodes			0.72**	0.11	0.68**
Length of underground stems				0.42**	0.80**
Diameter of underground stems					0.42**

Values where the correlation is significant are indicated with asterisks (* $P \leq 0.05$, ** $P \leq 0.01$)

On the other hand it has also been reported that the depth of regrowth initiation (initiation depth) of ratoon stalks affects the growth of ratoon sugarcane (Miyahara and Kamiya 1984). We used our surveys of ratoon fields in 2013 and 2014 to investigate this concept in high- and low-yielding plots. Small areas were excavated to examine the underground plant parts and assess their relationship with stalk weight. The results were almost identical for both years. The results for 2014 are shown in Table 3.

A positive correlation was observed between stalk weight and initiation depth in good and poor growth fields. The initiation depth was also positively correlated with the number of underground nodes, the length of underground stems and the total number of roots protruding from the stems. Deeper initiation resulted in wider distribution of the roots and therefore greater access to soil moisture and nutrients. This finding suggests that one key point for improving ratoon yields is keeping the initiation position of stalk as deep as possible.

Conclusions

The project has resulted in the following information regarding improvements in cane yields in NE Thailand:

- For higher average yield in NE Thailand, improving yields of ratoon cane has priority.
- Increasing the number of millable stalks and stalk weight can improve cane yield of low-yielding fields of ratoons
-
- Maintaining the number of millable stalks in plant cane by reducing the number of missing plants, and maintaining stalk weight by preventing 'rising', the rising of the initiation positions in ratoon cane, are effective ways of improving low yield in ratoons.
- Minimizing the number of missing plants is important to ensure good stalk numbers in plant cane and maintain yield in ratoons. Improvements in germination and increases in initial growth rates are needed. It is also important to promote the use of good quality seed cane to minimize the onset of white leaf disease and smut.
- Retaining an appropriate depth of bud initiation after harvest of the plant cane will improve stalk weight in ratoons. The starting point to achieve this is to use relatively deep planting furrows at planting.

Refferences

Hanboonsong Y, Choosal C, Panylm S, Damak S. 2002. Transversial transmission of sugarcane white leaf phytoplasma in the insect vector *Matsumuratettix hiroglyphicus* (Matsumura). *Insect* Molecular Biology 11: 97-103.
Hemwong S, Toomsan B, Cadisch G, Limpinuntana V, Vityakon P, Patanothai A. 2009. Sugarcane residue management and grain legume crop effects on N dynamics, N losses and growth of sugarcane. *Nutrient Cycling in Agroecosystems* 83: 135-151.
Miyahara E, Kamiya J. 1984. Study on system of ratoon shoots in sugarcane. *Bulletin of the Okinawa Agricultural Experimental Station* 9: 56-66 (in Japanese with English summary).
Prado RM, Caione G, Campos CNS. 2013. Filter cake and vinasse as fertilizers contributing to conservation agriculture. *Applied and Environmental Soil Science* http://dx.doi.org/10.1155/2013/581984.
Wongwiwatchai C, Thippayarungs S, Sansayawichai T, Pintasen S, Matuso K. 2002. Prevalent agricultural practices and soil fertility in sugarcane cropping in Northeast Thailand. *JIRCAS Working Report* 30:103-109.

S.I.T. Paper #1171

AROMA ANALYSIS OF TAISUGAR DARK BROWN SUGAR AND MOLASSES

Tso-Lin Cheng Bo-Jhin Wu
Taiwan Sugar Corporation, Tainan, Taiwan

Abstract

This study combined the method of solid-phase microextraction (SPME) and the aroma analysis system of GCMS to identify aroma compounds in Taisugar dark brown sugar (TDBS) and Taisugar molasses (TM). The aroma strength of dark brown sugar and molasses are shown with scientific figures, and moreover, we observed the changes in aroma indexes when dark brown sugar and molasses were being stored.

We identified 13 aroma compounds: dihydro-2-methyl-3(2H)-furanone, 2,5-dimethyl-pyrazine, 2,6-dimethyl-pyrazine, 2-furanmethanol, 2-ethyl-6-methyl-pyrazine, trimethylpyrazine, 3-ethyl-2,5-dimethyl-pyrazine, 1-(6methyl-2-pyrazinyl)-1-ethanone, 5-methyl-1H-imidazole-4-methanol, 2-hydroxyl-3-methyl-2-cyclopenten, dihydro-3-hydroxy-4,4-dimethyl-2(3H)-furanone (or pantolactone), 1-(1H-pyrrol-2-yl-ethanone and 2,6-dimethoxy-phenol; we used dihydro-2-methyl-3-furanone□D2M3F□, 2-furanmethanol□2-FM□and 4-ethyl-2-methyl-pyrrole□2-AP□as aroma indexes and we also ran the quantitative analysis for these three compounds.

In our studies, we found that the concentration of 2-AP in TM was significantly higher than in TDBS. This can be used as a major aroma index to determine the "true molasses". The changes during storage in three aroma indexes of TDBS and TM are as follows: in both large and small packages, D2M3F has decreased over time. Within 5 months, it has reduced more than half. The retainability in the large package of TDBS was far less than the small package (as it relates to the packaging materials). 2-FM has also decreased within 5 months but not as significant as D2M3F. The concentration of 2-AP remained the same. In TM, the concentrations of D2M3F and 2FM have decreased significantly after being stored for 5 months, but the changes in 2-AP were irregular. In addition, we found that the concentration of three aroma indexes in TDBS and TM were different in each batch. It also indicated the flavor was different in each batch.

Keywords: Taisugar dark brown sugar (TDBS), Taisugar molasses (TM), Gas chromatography-mass spectrometer (GCMS), Solid-phase microextraction (SPME)

AROMA ANALYSIS OF TAISUGAR DARK BROWN SUGAR AND MOLASSES

Tso-Lin Cheng Bo-Jhin Wu
Taiwan Sugar Corporation , Tainan, Taiwan

Introduction

During the production process of dark brown sugar, the heat and dehydration occurs Maillard reaction, caramelization, oxidation and etc. Maillard reaction has features of complexity and diversity, it can produce hydroxylmethylfurfural (HMF) and several other carbonyl volatile compounds. Some caramel flavor compounds have been analyzed and proven, such as 4-hydroxy-2,5-dimethyl-3-furanone, acetic acid, formic acid, Pyrones and Diactyl compounds.

The regular process of Aroma Analysis is to extract or isolate volatile compounds first, then use GC-FID or GC-MS to run the test. In the meanwhile, we used different pre-treatment methods for indiviaul samples, including conventional methods such as liquid-liquid extraction, distillation, and Soxhlet extraction, or solid phase microextraction (SPME) which has been developing rapidly in recent years.

SPME is a solvent-free method for sample preparation. It was proposed by Pawlinszyn at Waterloo Univeristy, Canada in 1989. In 1994, Supelco, an American company, commercialized the method. As the instruments and equipments advanced, the method has been gradually expanded from the original use, environmental test, to food, natural products, pharmaceuticals, clinical chemistry, biochemistry, toxicology, and forensic medicine. This method uses special fused silica fiber to adsorb the test substance, then injects the fused silica fiber directly into the injector, heats and desorps the flavor compounds in the fused silica fiber so the subtance can go along with the carrier gas to the capillary column for the analysis. SPME has overcome the limitations of conventional sample pre-treatment methods, it combines sample collection, extraction, concentration and injection in one, boosts the speed of analysis.

SPME-GC/MS was used in this project, the aroma strength of Taisugar dark brown sugar (TDBS) and Taisugar molasses (TM) are shown with scientific figures, the changes in aroma indexes during the storage were also being observed.

Materials and Methods

We used SPME-GCMS for aroma analysis in the test. The one we used was SPME-Varian 3800GC-Saturn 2000 with the VF-Wax/MS column (FS 50 x 0.25 mm, DF = 0.2 μm). The operational condition for GC was to use ultra high purity Helium (purity >99.9995) with a flow rate of 1.0 mL /min as the carrier gas. We prepared and weighed 3.5 grams of the sample, put it into a special-use 20ml smaple glass vial, then sealed it with the teflon coated silicon spacer. We used SPME-fiber, DVB/CAR/PDMS(50/30 μm), to extract and absorb aroma compounds at a fixed temperature of 50°C for 30 minutes (see figure 1.). After that, we injected the fiber covered with aroma compounds into the injection port, heated and desorbed it at 270°C for 5 minutes. The inlet we used was splitless and the starting temperature for the column was 50°C. We kept it at 50°C for 5 minutes before raising the temperature to 240°C with a speed rate of 10°C /min. Then kept it at 240°C for another 6 minutes. The mass spectrometer was set to E I ion-source, at a temperature of 170 °C, and the mass-to-charge ratio of 40~550 au.

To quantify the concentration of three aroma indexes (dihydro-2-methyl-3(2H)-Furanone□D2M3F□, 2-Furanmethanol□2-FM□and 4-ethyl-2-methyl-pyrrole□2-AP□)in TDBS and TM, we stored several batches of TDBS and TM at room temperature for 5 months. Then tested and quantified the aroma indexes as above monthly to observe the changes of the concentration.

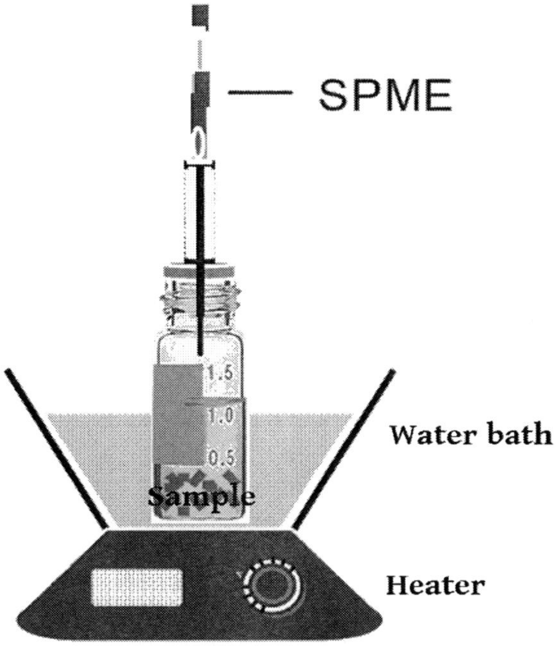

We prepared and weighed 3.5 grams of the sample, put it into a special-use 20ml smaple glass vial, then sealed it with the teflon coated silicon spacer. We used SPME-fiber, DVB/CAR/PDMS(50/30 μm), to extract and absorb aroma compounds at a fixed temperature of 50°C for 30 minutes.

Figure 1. SPME schematic

Results and Discussion

1.Estabalish the conditions for aroma analysis

1.1 Qualitative analysis of aroma compounds

We used SPME-GCMS to analysis the aroma compounds in TDBS and identified 13 aroma compounds (figure 2.), they are: dihydro-2-methyl-3(2H)-furanone, 2,5-dimethyl-pyrazine, 2,6-dimethyl-pyrazine, 2-furanmethanol, 2-ethyl-6-methyl-pyrazine, trimethyl pyrazine, 3-ethyl-2,5-dimethyl-pyrazine, 1-(6methyl-2-pyrazinyl)-1-ethanone, 5-methyl-1H-imidazole-4-methanol, 2-hydroxyl-3-methyl-2-cyclopenten, dihydro-3-hydroxy-4,4-dimethyl-2(3H)-furanone (or pantolactone), 1-(1H-pyrrol-2-yl-ethanone and 2,6-dimethoxy-phenol.

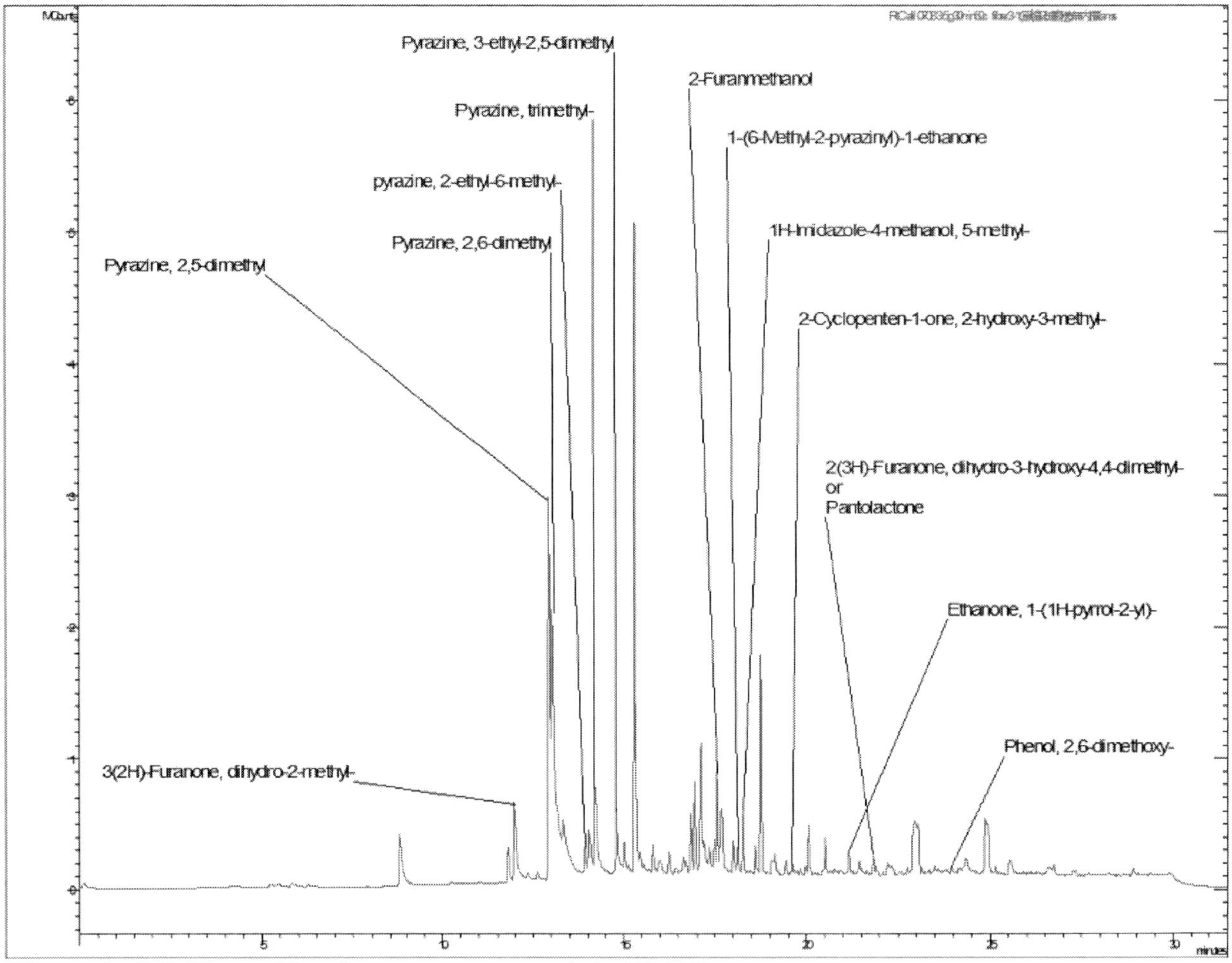

Figure 2. GCMS graphic of aroama compounds in TDBS.

1.2 Estabalish an internal standard and the regression curves of the aroma compounds' standard concentration

In order to recieve accurate data and reduce the interference from manual sampling error and different operating conditions of the equipement, it is necessary for us to estabalish the technic to add in and search a sutiable internal standard for aroma compounds in some special sugar (dark brown sugar and molasses) . In our research, we used a previous research, which is published by V. Ferreira in 2003, as our reference. We found out that when we used 2-octanol as an internal standard, its retention time did not overlap with the peak of the main compounds in special sugar (Figure 3.). We can use it as an internal standard for aroma analysis of special sugar .

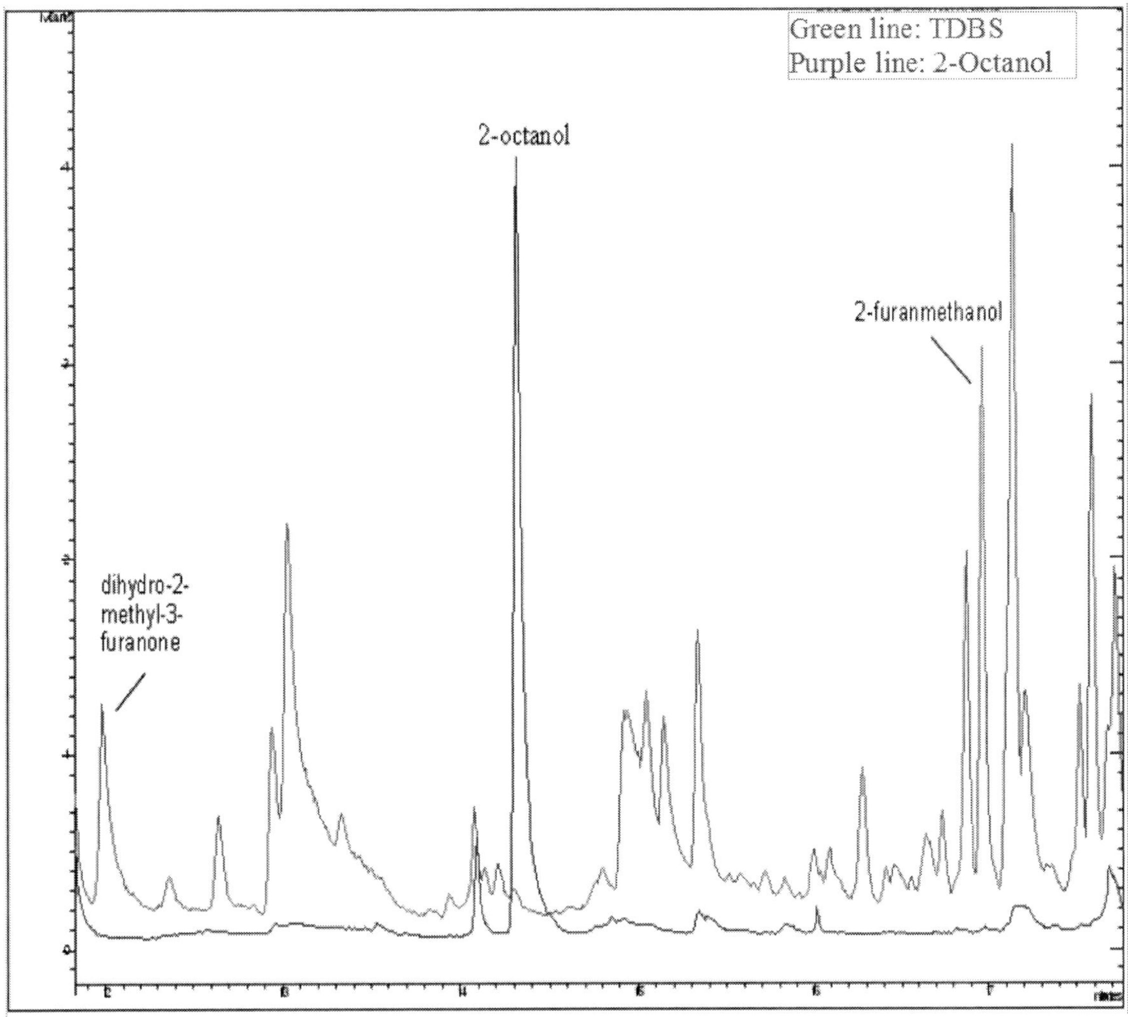

Figure 3. Analysis graphics of internal standard 2-octanol and TDBS

We diluted two aroma compound indexes in brown sugar as the standard solution, which are D2M3F (dihydro-2-methyl-3-furanone, purity 97%, MW:100.1, density 25□ :1.022 g/cm^3) and 2-FM (2-furanmethanol , purity:98%, MW:98.1, density 25□ :1.129g/cm^3), 500~10,000 times with methanol (LC grade, density:0.7918g/cm^3). The concentration of D2M3F is between 40 mg/kg~850 mg/kg, and 2-FM is between 45 mg/kg~950 mg/kg. We also diluted 2-Octanol 20,000 times with methanol to prepare it as internal standard solution. Then we used the standardized micro needle (0.5~10 μL) and the holder to draw internal standard solution and the standard solution of two indexes, 1.0 μL each, injected them into the sealed glass vials, repeated three times with SPME-GCMS. We divided peak height integrates of D2M3F and 2-FM (Hs) respectively by peak height integrates of internal standard 2-octanol (Hi), the data of Hs/Hi was used as standard curves for aroma indexes (see Figure 4.). The linear regression R^2 of D2M3Freached 0.9936, the other aroma index 2-FM reached 0.9844. These two indexes have excellent linear relationship within the concentration as mentioned above, and can be used as quantitative standard for aroma compounds in special sugar.

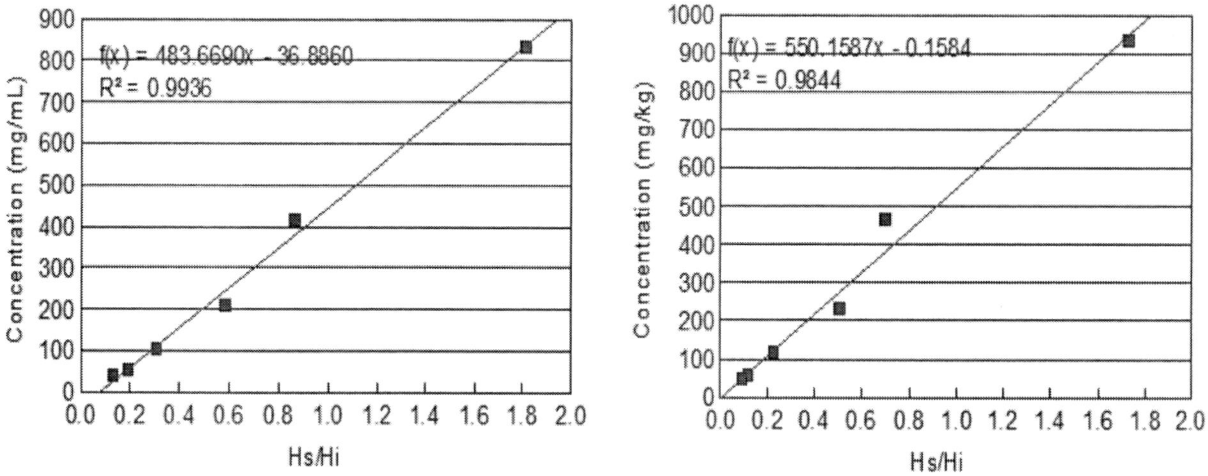

Figure 4. Peak height graphics of aroma indexes and internal standard solution
Left: dihydro-2-methyl-3-furanone(D2M3F)
Right: 2-furanmethanol(2-FM)
Hs/Hi =
Peak height of standard solution (index)/Peak height of internal standard solution

2. The concentration changes of aroma indexes in TDBS and TM during storage

The concentration changes of aroma indexes in TDBS and TM during 5-months storage is shown as figure 5-10. (01~05 represent different batches respectively). From the figure it is apparent that D2M3F decreased in both large(L) and small(S) pakages of TDBS. It decreased at least 50% within 5 months. The reductions of D2M3F in the TDBS(L) were far more than the TDBS(S). The amount of 2-FM in TDBS or TM went up and down, but overall it still went down during the 5-months storage, yet not as dramatically as D2M3F. The concentration of 2-FM showed no differences between TDBS(L) and TDBS(S). The concentration of 2-AP in TDBS went up or stayed at the same level.

The concentration of D2M3F in TM decreased with time. The concentration of 2-FM in TM is higher than in TDBS. 2-AP is the aroma index compound for true molasses, thus the concentration of 2-AP in molasses was way higher than in brown sugar. It increased significantly but it started to decrease in the fourth month.

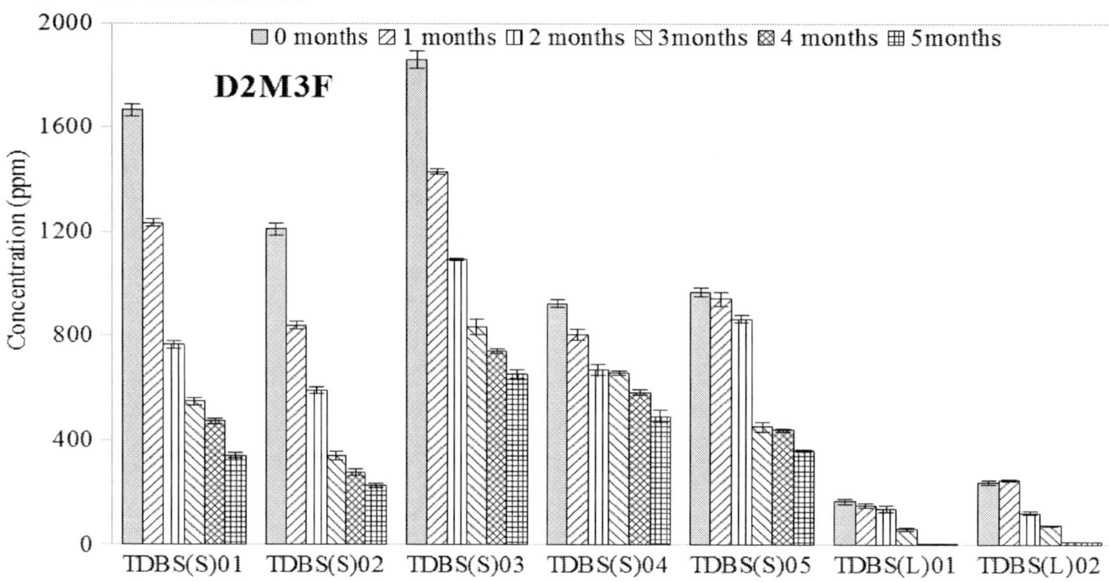

Figure 5. The concentration changes of D2M3F in TDBS during storage

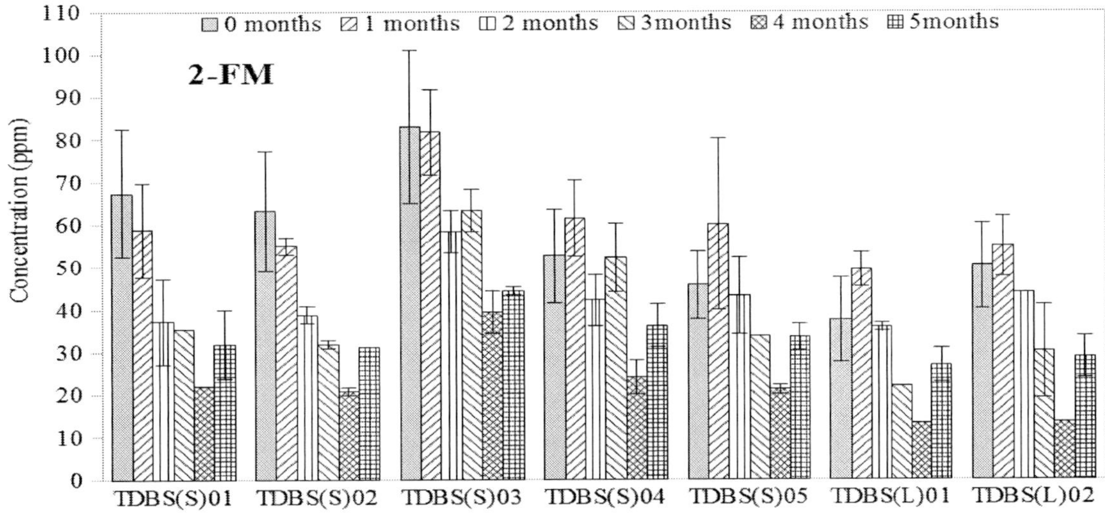

Figure 6. The concentration changes of 2-FM in TDBS during storage

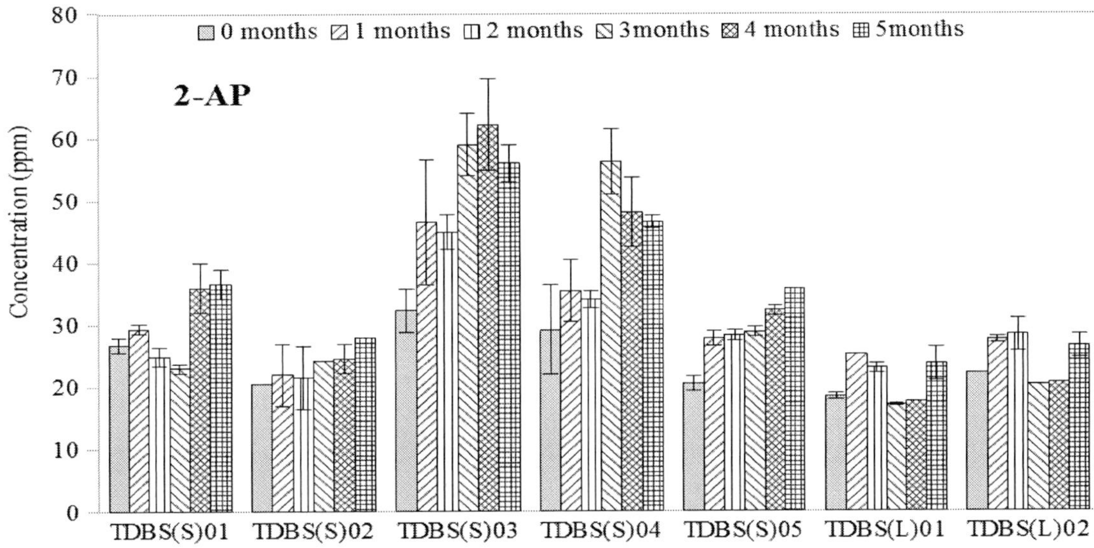

Figure 7. The concentration changes of 2-AP in TDBS during storage

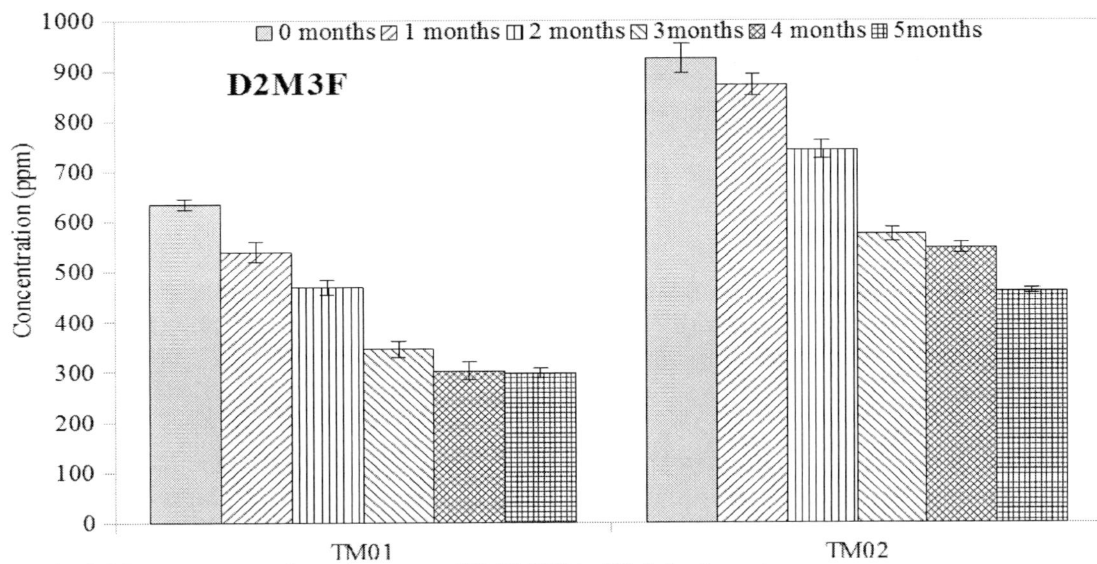

Fig.8 The concentration changes of D2M3F in TM during storage

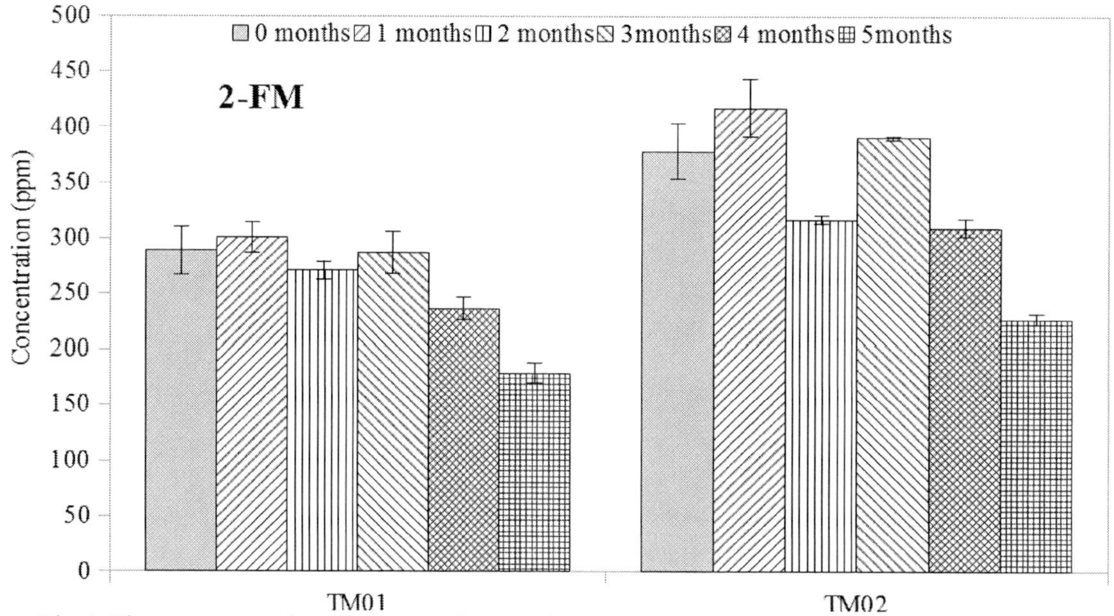

Fig.9 The concentration changes of 2-FM in TM during storage

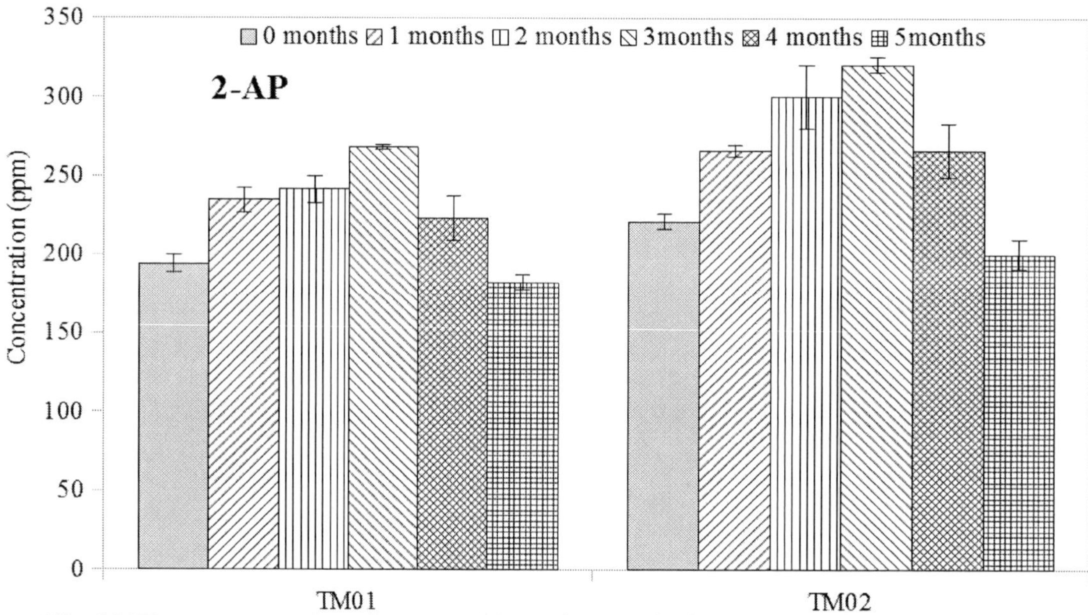

Fig.10 The concentration changes of 2-AP in TM during storage

Conclusion

1. Establish conditions for aroma analysis

Dihydro-2-methyl-3-furanone□D2M3F□(RT=11.9min), 2,5-dimethyl-pyrazine (RT=12.94 min) 2,6-dimethyl-pyrazine (RT=13.03 min) , and 2-furanmethano□2-FM□(RT=17.58 min), 4-ethyl-2-methyl-pyrrole□2-AP□(RT=21.52min) on GC graphics of TDBS showed larger peaks and associated with caramel aroma compounds (Nose, M., 1983□Tokitomo, Y., 1984). Partial peaks of 2,5-dimethyl-pyrazine and 2,6-dimethyl-pyrazine overlapped. This affected interpretation of the results, therefore we could not make an accurate quantitative analysis with these two compounds. Other componds only showed weak peaks and could not be quantitated, so they can't be used as indexes. For the aroma analysis for TDBS and TM, we only used D2M3F, 2-FM and 2-AP these three compunds as aroma indexes.

2. The concentration changes of aroma indexes in TDBS and TM during storage

We observed the concentration of aroma indexes in several batches of TDBS, and found out that the concentration of D2M3F showed significant changes in each batch (924~1861ppm in TDBS(S)). We speculated that it was caused by different factors during the manufacture of each batch (eg: the concentration temperature and time of heating, the remelting times). It also means that the flavor of each batch tasted slightly different from each other. During the storage, the concentration of D2M3F in TDBS(L) decreased dramastically, the small packages reserved more flavor compared with the large ones. We speculated that it was caused by the packing materials. The large package uses paper bag and PP film, it provides greater gas permeability. But the small package uses laminated film (PET/ PE/ CPP), which has low gas permeability. In addtion, they covered one side of the package with aluminum foil. Hence the small packages have better retainability of aroma than the large ones.

The concentraion of three indexes showed significant significant differences in each batch of molasses. It means in each manufacture batch of molasses, the concentration of aroma was not stable (the flavor was different in each batch). Comparing with drak brown sugar, 2-AP is much higher in molasses (we can use 2-AP as an aroma index for molasses and make a distinction between molasses and the "true" dark brown sugar.

Japanese scholar Tokitomo used GCMS identified several sugar aroma compounds from cane sugar molasses, including 3-hydroxy-4,5-dimethyl-2(5H)-furanone (sotolon), 4-Hydroxy-3-methoxy benzaldehyde (vanillin), 3-hydroxy-2-methyl-4-pyranone, 2-hydroxy-3-methyl-2-cyclopentenone, 4-nonanolide□4-vinylphenol and 2,6-dimethoxyphenol (Tokitomo, 1984). When we used it as our reference for the GCMS graphics of TDBS and TM, we did not find any compound mentioned in the reference in TDBS. But we identified three compounds in TM: 3-hydroxy-4,5-dimethyl-2(5H)-furanone (sotolon)□20.4min□, 2-hydroxy-3-methyl-2-cyclopentenone□20.01min□, 2,6-dimethoxyphenol□24.36min□. These three compunds were not the aroma index in our research. The reason why we did not use them as aroma indexes was because the peaks of some compounds were too weak (not major aroma). Also, the MS quality of these three compounds were all lower than 60% and can't be used as indicators, so we didn't not choose them for the quantitative analysis.

References

1.Ferreira, V., Jarauta, I., López, R., and Cacho, J. (2003), "Quantitative determination of sotolon, maltol and free furaneol in wine by solid-phase extraction and gas chromatography-ion-trap mass spectrometry", Journal of Chromatography, 1010(1)pp.95-103.

2.Nose, M., Matsui, M., Kobayashi, A., Yamanishi, T., and Takei, S. (1983), "Studies on the formation of sugary flavor sotolon in raw cane sugar", Journal of the Agricultural Chemical Society of Japan, 57(6)pp.557-561.

3.Pawliszyn J. (1997), "Solid Phase Microextraction: Theory and Practice", Wiley-VCH.

4.Tokitomo, Y., Kobayashi, A., Yamanishi.T. (1984), "Aroma Components of Fresh Sugar Cane Juice", Agricultural and Biological Chemistry, 48(11)pp.2869-2870.

S.I.T. Paper #1172

Treatment of the sweet water from filter cake in the sugar Refinery

Sheen Herng-Kuang[1,4], Lin Yong-Hua[2], Chen Wei-Dong[3]

1: Research engineer, Sugar Business Division of Taiwan Sugar Corporation
2: Former deputy chief executive, Sugar Business Division of Taiwan Sugar Corporation
3. Former director of the Xiaogang sugar refinery, Sugar Business Division of Taiwan Sugar Corporation
4: Correspondence to: Sheen Herng-Kuang, No. 54, Sheng-Chan Rd., Tainan, Taiwan, R.O.C.
E-mail: a60636@taisugar.com.tw

Abstract

The filtered washings was treated by the phosphate flotation system which was set up in Xiaogang sugar refinery in 2014. The most parts of equipment in the system were from old three-stage scum desweetening system, which was fitted into the Peikang sugar refinery. This sugar refinery has been closed by Taiwan Sugar Corporation since 2005. However, the defects were inevitable in the new system. Local improvement for some equipment and modification of pH control system of the filtered washings, had been able to grasp the phosphate flotation system. At pH 7.0, the removal percentage of color value and starch content in filtered washings after flotation were about 40% and 60% respectively. The amount of the scum after phosphate flotation was produced about 10% by volume (including bubble, solid and filtered washings). When all scum was thrown into sweet mud tank, and with mud were filtered together, it did not cause any problems in the sugar refinery.

Introduction

The supply and demand of sugar in Taiwan has been more than its produce since 1960. In order to compensate for the lack of raw sugar, Taiwan must import it from around the world. Xiaogang sugar refinery has imported raw sugar from many countries such as Brazil, Australia, Thailand, Guatemala, Salvador and India etc. respectively since 2006. Due to varies of qualities of raw sugar from the different exporting countries, some of which contain relatively high dextran and starch, while others contain high sediment and other impurities, led to increase difficulties of filtering in the carbonate process in this refinery. Generally, dextranase and amylase enzyme was added in the refined process to degrade the polysaccharides compound such as dextran and starch with huge molecules weight, to low molecular weight such as glucose or other compound, countermeasures taken by including Taiwan Sugar Corporation and the refineries all over the world. The researcher of Taiwan Sugar Corporation, Ting C.S. (1) observed that raw sugar with polarization content of 99.3 and the starch content up to 180ppm, also affected the efficiency of carbonation and filtration in the refining process. By adding amylolytic enzymes 2 ppm (for raw sugar) in the trial for five days, the efficiency of filtration can be improved. The effect of increasing the weight of dissolved sugar about 13%, but the consequences of this trial had led to an increase in the amount of reducing sugars, and have an influence on recovery of refining sugar. Usually, these enzymes are more suitable for low temperatures (60-70 □), to be able to exert its greatest effect. However, the significant operating temperature of the refinery was higher than limit of enzyme can bear, its activity was considerable influence and can not exert its due function [2]. The enzyme with high-temperature glycolytic enzymes if was used in process often lead to ineffective inert. Its residual should leave in the sugar product and final molasses, thereby affecting subsequent application of sugar processed foods and controversy.

In the 2011's research report explore that the color value of the affination sugar syrup which was dissolved in the filtered washings recovered from carbonation filter cake increased significantly even more than that of raw sugar in sugar refining manufacturing process [3]. In addition, dextran and starch in the syrup were higher than of the affination sugar in water. Obviously, the current operation is completely offset the function of the affination operation in the refinery. In the study report of 2013, the filtered washings was treated by several methods commonly used in sugar refining. It was found the phosphate flotation attached to the de-color resin unit can effectively remove the starch and colorant in filtered washings [4]. It reduced the re-introduction of impurities in the refining process. It is a considerable benefit to the sugar refinery. When raw sugar with large amount of starch and dextran was refinined, how to improve the quality of filtered washings, so that refining efficiency is not affected was the major work. PeiKang sugar refinery had three-stage scum desweetening system for syrup phosphate floatation system [5]. However, the refinery has been stopped refining by Taiwan Sugar coporation(TSC) since 2005, resulting in equipment idle. Xiaogang sugar refinery need to set up a phousphate flotation system of the filtered washings. Hence, the parts of the old equipment was moved to the Xiaogang Sugar Refinery and installed and modified into the current system to test the phosphate floatation of the filter washings process [6].

Materials and methods

1. Suitable pH range of filtered washings for phosphate floatation

Take six 1000-mL erlenmeyer flasks with screw caps, and put filtered washings, 500mL into each beaker, then each adding 0.25mL of 85% phosphoric acid respectively. finally, at the room temperature, adding 10% of the lime solution to adjust the pH to 4.5, 5.0, 5.5, 6.0, 6.5, and 7.0 respectively, and keep record whether precipitate occurrred in the solution. Place each treatment of these solution in microwave oven and heated solely, and measure the temperature after heating. Put 1 mL of 0.01% of the A110 floc agent into the hot solution immediately, then aerated by hand for 3min. Finally, place it into a 500-mL cylinder to observe the scum formation.

2. Flotation test of simulated sweet water of imported raw sugar for Xiaogang Sugar Refinery

Raw sugars from three different contries (Australia, Thailand and Brazil) were dissolved in water to meet 10Bx of concentration, were treated, as the foregoing method. After 30min, the floating and sinking amount of scum will be keeping record. The sweet water after flotation was filtered with Advantec # 2 filter paper, and the filtrate was analyzed for the character of sweet water.

Results and Discussion

1. Investigate the fluctuation of the concentration in filtered washings at the Xiaogang sugar refinery

The variations of the filtered washings at the Xiaogang sugar refinery were shown in Table 1, sampling period was once an hour, eight hour per day for three days, from December 19 to 21, 2013. The concentration and quantity of the filtered washings were ups and downs and relatively unstable. The fluctuation of the concentration which contains in sugar in the filtered washings daily from carbonation filter cake produced was quite large, which the highest and lowest was difference up to 20-30%,

Table 1. The changes of concentration of the filtered washings

Sampling time(hr)	December 19	December 20	December 21
		(^{0}Bx)	
09:00	13.12	15.12	14.58
10:00	11.68	15.09	13.06
11:00	12.5	12.66	13.5
12:00	12.03	15.11	13.55
13:00	11.45	15.51	11.85
14:00	11.73	14.26	13.54
average	12.09	14.63	13.35
Standard deviation	0.62	1.05	0.89
maximum	13.12	15.51	14.58
minimum	11.45	12.66	11.85
Standard error	5.15	7.15	6.66

2. Suitable range of pH for phosphate floatation for the filtered washings

The filtered washings was treated by different pH but same pre-treatment to let it layer three parts as the upper scum, clarified liquid and scum sunk in the bottom respectively. The amount of the scum at upper layer and layer that sunk in the bottom for phosphate floatation under different pH was shown in Table 2.

Under the range of pH to meet 4.5 to 6.0, no sediment and scum were observed when lime was added into the filtered washings. There were 45 mL of the scum and 5 mL of the sediment existing in filtered washings after 50 mL of the lime was added, then the pH of the filtered washings met 6.5. The pH of the filtered washings met 7.0 when added more lime into. A large amount of sediment and scum were produced in the operation of simulation phosphate flotation of the filtered washings and the amount of sediment was greater than that of upper scum.

Table 2. The change of amount of the scum formed by the filtered washings treated by phosphate floatation under different pH

Sample	amount of the sediment after neutralization (mL)	Operational temperature of the flotation (□)	Rise amount of the scum (mL)	sediment amount of the scum (mL)
pH 4.5	0	55	0	0
pH 5.0	0	55	0	0
pH 5.5	0	55	0	0
pH 6.0	0	55	0	5
pH 6.5	50	55	45	5
pH 7.0	100	72	40	50

The characters of the filtered washings after phosphate flotation at different pH were shown in table 3. The filtered washings with high color value and starch content was treated by phosphate floatation at pH 7.0. The removal rate of the color value and starch content were about 40%, and 60% respectively. The starch content in the filtered washings did not reduce, on the contrary, it increased at the range of pH 6.0 to 6.5. The phenomenon may be the scum formed what lime and phosphate reacted to fail to get the preferred particle agglomeration and precipitation in the filtered washings. However, the formation of suspended particles increased interference of the determination of the starch content, it was so much higher than that of original filtered washings. Obviously, the pH of the operating range of the phosphate flotation of the filtered washings was about 6.5 to 7.0 at the Xiaogang sugar refinery (Fig.1).

Table 3. The characters of the filtered washings after phosphate floatation under different pH

Sample	Operational temperature of the flotation (□)	Brix (Bx)	Color value (IU)	Starch content (ppm)
Original	42	9.76	2053	147
pH 6.0	43	9.8	1540	233
pH 6.5	44	9.77	1426	220
pH 7.0	44	9.70	1274	60

Fig. 1. The situation of the filtered washings after phosphate flotation at different pH

3. Study on the treatment of scum

The phosphate flotation system of the filtered washings was constructed in Xiaogang sugar refinery. It is most concerned scum about after the flotation, because the amount of the upper layer scum was about the volume of 10% (including bubble-containing solids and filtered washings) need to process. If available, dross could be thrown into the sweet mud tank and filtered together during the processing. Another way, is to incorporate into the waste water treatment system; therefore, the scum was filtered by Advantec 5C filter paper to simulate filtering process, collected the clear filtrates and analyzed their properties were shown in Table 4.

It was shown that the characters of the filtrate which scum was directly filtered by paper. It was same as that of the filtered washings treated by phosphate flotation. However, the impurities in the filtrate was significantly increased, it was washed out from the scum diluted by distilled water. The whole situation of scum filtered process seems to be also good and very smooth, yet this situation may be paper with inadequate loading. The residue scum after filtration was dried in the oven, then put in a furnace heated to a high temperature above 550 □, to get incineration ash residue of 12.5%.

Table 4. The characteristics of the filtrate of the scum after filtered washings treated by phosphate flotation

Sample	pH	Brix (Bx)	Reduced sugar (%)	Color value (IU)	Starch content (ppm)
The filtrate of scum	7.46	11.31	0.82	3392	20
The filtrate of diluted the scum (1:1)□	7.68	4.64	1.04	3572	0

1: Dilution ratio = Scum: Distilled water = 1: 1

4. Phosphate flotation of simulated sweet water in imported raw sugar for the Xiaogang sugar refinery

Xiaogang sugar refinery refined 300,000 tons/year of raw sugar from around the world. To test each raw sugar whether would affect the operation of the phosphate flotation system of the simulated sweet water, we collectted three raw sugars including Australia, Thailand and Brazil etc. from Xiaogang sugar refinery imported.

Table 5. The effect of simulated sweet water from Imported raw sugar on the amount of scum produced by phosphate flotation under different pH

Sample	operational temperature (□)			The sinking amount of the scum(30min) (mL)			The floating amount of the scum(30min) (mL)		
	Thai	Bra	Aus	Thai	Bra	Aus	Thai	Bra	Aus
pH 6.5	62	64	64	5	3	3	40	35	30
pH 7.0	64	64	64	5	3	10	45	30	60
pH 7.5	65	64	64	20	3	15	50	55	60
pH 8.0	64	64	64	10	3	5	50	55	55
pH 8.5	65	64	65	15	3	10	45	50	80
pH 9.0	65	64	65	15	5	5	45	55	55

1. Imported raw sugar: Australia (Aus), Brazil (Bra), and Thailand (Thai), respectively

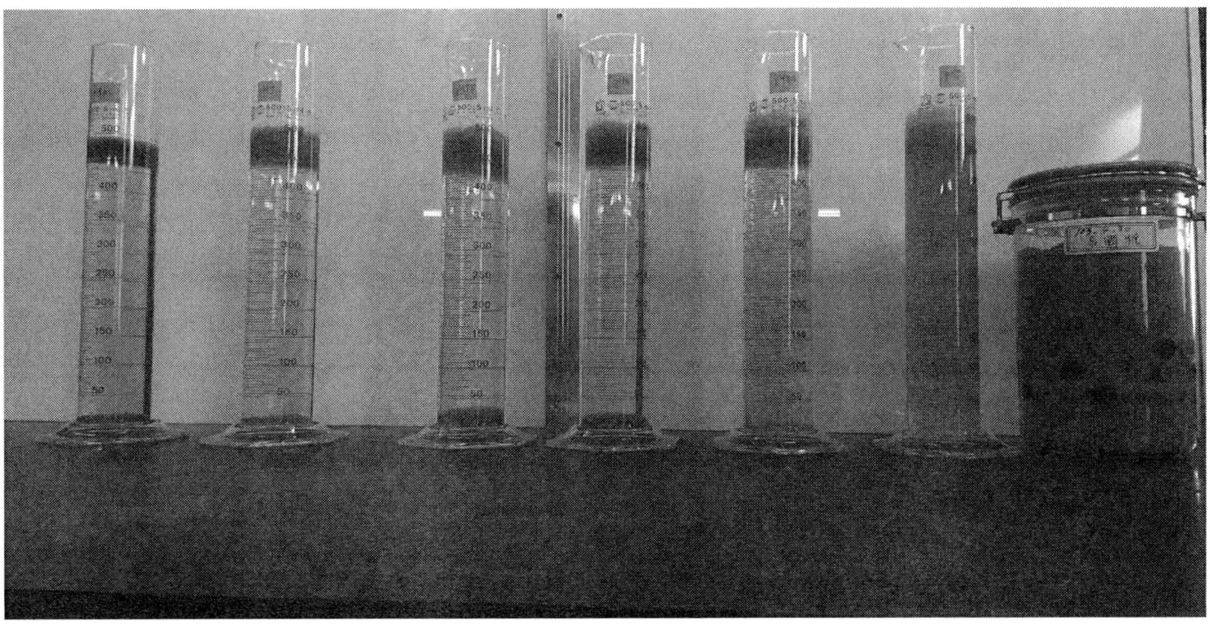

Fig. 2. The amount of the scum in simulated sweet water of the Thailand raw sugar after phosphate flotation

The amount of the scum in simulated sweet water with Thailand, Brazil and Australia's raw sugar at different pH of the phosphate flotation were shown in Table 5. Thailand raw sugar significantly increased the amount of subsidence, at 64-65 □, pH above 7.5(fig. 2). It showed that too much lime added into the simulated sweet water, may increased gravity and impurities of the scum, resulting in an magnification in the probability of the scum sank. The amount of the sediment in Brazil and Australia's raw sugar was not much, almost upper floated. The sweet water produced from that which refined Brazil raw sugar, was treated by the phosphate flotation, should be very suitable. And the character of those simulated sweet water were shown in table 6.

Thailand's raw sugar with starch content higher than previous that of Australian's raw sugar, but compared to the much lower than that of Brazil's raw sugar. The most starch content in sweet water on Brazil's and Thailand's raw sugar were removed after phosphate flotation process at different pH. It obviously showed raw sugar with high starch content was suitable for phosphate flotation process. These raw sugar as the increase of removal rate of color value with the rising pH. As to reducing sugar content, these raw sugar as the pH decreased with reversely rise, the phenomenon was worth to observe .

5. Establishment of pH Automatic Control System for Phosphate Floatation of filtered washings in Xiaogang Sugar Refinery

The engineering of filtered washings treated by phosphate flotation system in Xiaogang sugar refinery was constructed in mid-May 2014 and began trial operation on May 15, which the equipment was early devices of scum three-stage desweetening process. Its biggest problem could not immediately response to the changes of filtered washings. After some small modification to the device, the current phosphate flotation system with automatic pH control system had been able to control the pH within a narrow range to treat the filtered washings. The scum has been able to go up, as shown in Figure 3, and the filtered washings after flotation became quite clear, as well as the filtered washings with the very little scum residue inside (Figure 4).

Table 6. The impact of the properties of the simulated sweet water of the imported raw sugar treated by phosphate flotation under different pH

Samples	Reduced Sugar (%)			Ash content (%)			Starch content (ppm)			color value (IU)		
	Thai	Bra	Aus	Thai	Bra	Aus	Thai	Bra	Aus	Thai	Bra	Aus
original	1.06	0.30	0.34	0.23	0.17	0.32	105	222	40	3363	1846	2383
pH 6.5	1.04	0.39	0.43	0.40	0.28	0.50	36	67	0[1]	1423	813	1434
pH 7.0	1.12	0.52	0.50	0.29	0.25	0.38	23	37	0[1]	1506	797	1456
pH 7.5	1.13	0.63	0.46	0.25	0.23	0.34	23	67	41	1288	746	1338
pH 8.0	1.17	0.60	0.52	0.26	0.21	0.33	9	52	13	1204	709	1219
pH 8.5	1.19	0.64	0.57	0.25	0.20	0.34	0[1]	21	13	1188	651	1222
pH 9.0	1.19	0.68	0.62	0.25	0.20	0.34	23	52	41	1107	537	1129

1: The starch content was lower than regression value is negative, and was set to "0"

2: The imported raw sugar were diluted to 10Bx of the simulated sweet water

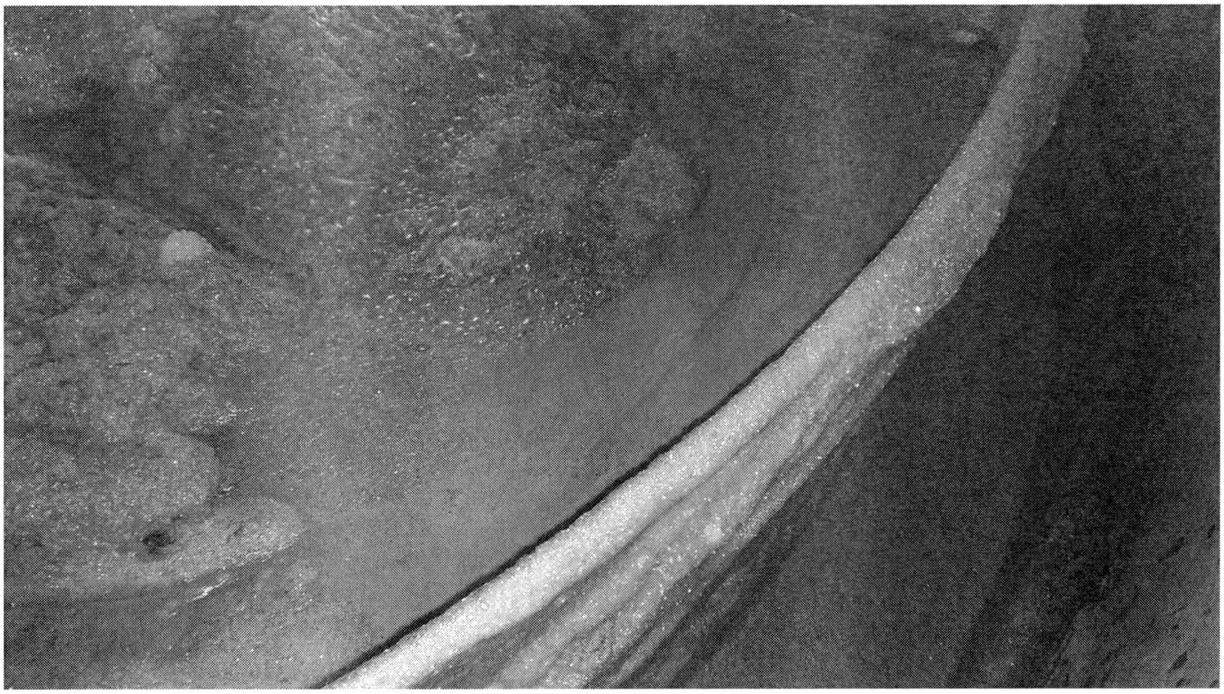

Figure 3. The operating situation of the filtered washings treated by phosphate flotation system in Xiaogang sugar refinery

Figure 4. The clear filtered washings after phosphate flotation

The characters of the filtered washings treated by the phosphate flotation system in the Xiaogang sugar refinery showed in table 7. The starch content in filtered washings before the flotation was an average of 104 ppm, and after the flotation was an average of 54ppm (removal rate of 47.9%), and then, after through a filter paper was an average of remaining 48ppm (removal rate of 54.8%). The color valve in filtered washings before the flotation was an average of 4092 IU, and after the flotation was an average of 2577 IU (removal rate 37.0%), and then through a filter paper was an average of 2322 IU (removal rate 43.3%). It was clear that some of the soluble starch and colorant of filtered washings which was from filtered cake, through the phosphate flotation system operation, could be removed, and would not return to the process of sugar refining to interfere with the normal procedure. The removal rate had met the target, 40-50% in the test.

Table 7. The characters of filtered washings before and after phosphate flotation

Samples	Sampling date[2]	pH	Brix (Bx)	Reduced sugar (%)	Ash content (%)	Starch content (ppm)	Color value (IU)
Filtered washings before phosphate floatation	2014/7/11	5.2	5.52	0.78	1.65	99	4110
	2014/10/7	7.0	11.01	0.95	0.79	41	3065
	2014/10/21	7.3	9.35	1.13	0.87	203	3692
	2014/11/10	7.6	8.00	1.21	1.03	93	5084
	2014/11/19	7.6	9.13	1.14	0.96	85	4508
average			8.60	1.04	1.06	104.2	4091.8
Filtered washings after phosphate floatation	2014/7/11	11.4	5.59	0.50	2.64	73	2190
	2014/10/7	10.9	11.09	0.80	1.11	28	1651
	2014/10/21	7.9	9.12	1.17	0.89	56	2740
	14/11/10	6.8	8.06	1.20	1.01	60	3280
	2014/11/19	7.1	9.11	1.16	0.96	55	3026
average			8.59	0.97	2.59	54	2577
Filtrate by paper from the filtered washings after phosphate flotation □	2014/7/11	11.4	5.60	0.48	2.52	23	1535
	2014/10/7	11.0	11.11	0.75	1.22	53	1090
	2014/10/21	7.9	9.31	1.14	0.87	85	2637
	2014/11/10	7.1	8.09	1.21	1.02	27	3350
	2014/11/19	7.3	9.15	1.13	0.96	55	2999
average			8.65	0.94	1.32	48	2322

1: # 2 filter paper
2: 7/11/2014 The sample is manually operated

The filtered washings treated by the phosphate flotation system was not smooth occasionally. It did not affect the sugar refining process, because of the filtered washings after flotation through a bag filter which removed the scum, that to be clean. In addition, the filtered washings after flotation was used for dissolving sugar into syrup, would be rechecked by carbonation and filtered process to remove the scum. In the Xiaogang sugar refinery, it was beneficial for the filtered washings treat by the phosphate flotation system.

Conclusions

The phosphate flotation system used to treat the filtered washings which was from filter cake in the Xiaogang sugar refinery, was consisted of the equipment from the old scum three-stage desweetening process and loaded in the Peikang sugar refinery. In the phosphate flotation systems, the pH of the operational range of the filtered washings was from 6.5 to 7.0. The color value and starch content in the filtered washings could be removed about 40% and 60%, respectively by this method. The scum of the upper layer of the phosphate flotation was produced approximately 10% of the amount of volume, all thrown into the sweet mud tank, and with mud were filtered together, did not cause any problems.

Cites

1. Ting C.S. 2001. Studies on Sugar Refining and Sugar Milling. Taiwan Sugar Research Institute research reports. Pp 149-151
2. https://zh.wikipedia.org/wiki/%CE%91-%E6%B7%80%E7%B2%89%E9%85%B6
3. Sheen H.K., Lin Y.H.. 2011. Studies on affecting factors of the filtration in Brazil's raw sugar. Sugar Business Division Taiwan Sugar Corporation Research reports, Kaohsiung, Taiwan
4. Sheen H.K., Lin Y.H. Cai W.C..2013. Improved assessment of the filtered washings in the sugar refinery. Sugar Business Division Taiwan Sugar Corporation Research reports, Kaohsiung, Taiwan.
5. Sheen H.K., Chen W.D. Lin Y.H. Yan Z.G.. 2014. Improved assessment of the filtered washings in the sugar refinery. Sugar Business Division Taiwan Sugar Corporation Research reports, Kaohsiung, Taiwan.
6. Ting C.S., J.S. Wu, F.M. Chen, W.D. Hsieh. 1997. The Development of a Three-stage scum desweetening system. Taiwan Sugar Research Institute Annual Report 1996-1997. p 28

Acknowledge

The authors thanked Cai Wencai (former deputy director), Yan Zhengwei (workshop director), Lin Shirong (section chief) and Chen Kaizhang (section chief) and others who were able to cooperate with the experiment during the trail experiment, successfully to completed the equipment of the filtered washings treated by phosphate flotation system , and to complete the test in Xiaogang sugar refinery.

S.I.T. Paper #1173

How Chemical Ripeners of Sugarcane Affect the Starch and Color Quality of Juices

Gillian Eggleston[1], Alexa Triplett[1], Eric Petrie[2], Kaitlyn Didier[1], Eldwin St. Cyr[1] and Gerald Myers[3]

[1]USDA-ARS-Southern Regional Research Center
1100 Robert E. Lee Boulevard
New Orleans, LA 70124, USA
gillian.eggleston@ars.usda.gov

[2]USDA-ARS-SRRC-Sugarcane Research Unit, 5883 USDA Road, Houma, LA, USA

[3]LSU Agricultural Center, School of Plant, Environment, and Soil Sciences, 104 MB Sturgis Hall, Baton Rouge, LA, USA

ABSTRACT

The application of chemical ripeners is an important component of sugarcane cultivation management to increase sucrose concentrations. Unfortunately, very little information is available concerning the effects of ripener on quality parameters which are critical to both factory and refinery processing, particularly starch and color. Thus, a large field study was conducted on the effect of two chemical ripeners with differing modes of action on sugarcane: Polado® (glyphosate) that suppresses the formation of new tissue at the apical meristem, and Moddus® (trinexapac-ethyl) that interferes with internode elongation. The ripeners were applied to nine commercial sugarcane varieties cultivated in south Louisana, USA, and hand-harvested 4, 5, 6, and 7 weeks after ripener application. Juice was extracted from whole-stalks with tops and leaves removed by roller milling. Results from the first year of the study showed that, overall, the effect of adding Polado and Moddus ripeners was to increase total starch (most starch in juice is insoluble) by 20.7 and 20.6%, respectively, although this was not statistically significant because of a strong varietal effect at the 0.01% probability level. Variation in the starch values was much higher than for Brix, pH, and color values, even in the untreated (control) juices; Polado caused worse starch variation than Moddus ripener. Neither ripener significantly affected the pH of the juice. The ripeners had less effect on color than starch, with no significant differences occurring among the treatments for ICUMSA color measured at pH 4.0, 7.0, 8.5, and 9.0 as well as the color indicator value I.V., although color variation increased with ripener application. A statistical analyses of the study is presented. Analysis of the second year of study is currently being completed to verify the first year trends.

INTRODUCTION

There is a lot of concern from Louisiana (LA) processors about the effect of chemical ripeners on starch and color in the sugarcane varieties delivered to and processed at factories. Chemical ripener is routinely applied to LA and other countries' sugarcane to accelerate maturity and increase sucrose storage within the stalk, leading to an increase in juice purity and sucrose content (van Heerden et al., 2014). The main advantage of chemical ripening is that it can suppress stalk and leaf growth much more rapidly and consistently than natural ripening processes such as decreasing daily incident sunlight, reduced seasonal temperatures or limiting soil moisture (Alexander, 1973; Legendre, 1975). The active compounds in two of the most well-known chemical ripeners have the following main effects on stalk (culm) and leaf growth (van Heerden et al., 2014):

Glyphosate (Polado®) – broad spectrum herbicide. At sub-lethal levels it is used as a ripener to suppress the formation of new tissues (apical meristem) by interfering with the synthesis of aromatic amino acids required for lignin biosynthesis. Fewer internodes are produced after application. Chlorosis of upper leaves occurs gradually. Auxiliary bud development (side-shooting) often occurs after application because of reduce apical dominance.

Trinexapac-ethyl (Moddus®) – which interferes with the hormonal regulation of stalk growth. Once absorbed through the leaves, the active compound transiently inhibits the conversion of an inactive precursor of the plant hormone gibberellic acid GA into one of its main bioactive forms. The action of this ripener leads to an inhibition of internode elongation.

As a general rule for chemical ripeners, enhancement of sucrose content in the sugarcane stalk is usually observed from 4 to 8 weeks after application (van Heerden et al., 2014). Unfortunately, very little information is available on how it affects quality parameters that are critical to both factory and refinery processing, particularly starch and color. Eggleston et al. (2007) reported that glyphosate (Polado®) action on juice starch, Brix, and pH varied with sugarcane variety and caused the production of fewer green leaves and more brown leaves. However, only two sugarcane varieties (LCP 85-384 and HoCP 96-540) were studied, the starch determined was mostly soluble starch, and no color was measured. It has since been unequivocally proved that most starch in extracted juice is insoluble (Eggleston et al., 2013). As quantities of starch delivered to and processed in LA factories have risen markedly in recent years because of one or a combination of: (i) increased production of green and combine-harvested sugarcane, (ii) varying environmental conditions, and (iii) the introduction of new sugarcane varieties with higher starch content (Eggleston et al, 2010), there is now a more urgent need to determine the effect of chemical ripeners on starch concentrations in delivered cane. At the same time the amount and control of color in sugarcane juices is proving a problem to LA sugar factories. This is because refiners are requesting higher purity (lower color) raw sugars. Natural colorants in sugarcane, e.g., phenolics and flavonoids, are small and monomeric, and can constitute up to 75% of the total color in raw sugar and which refiners have to remove (Smith et al, 1981; Paton and Smith, 1982; Eggleston et al, 2012).

This study was undertaken to investigate the effect of two chemical ripeners, Polado and Moddus on nine commercial sugarcane varieties, with special emphasis on how the ripeners affected starch and color quality attributes.

EXPERIMENTAL

Field experiments: Year 1

Raw juice was extracted from the second ratoon stalks of nine commercial LA sugarcane varieties (HoCP 00-950; L 03-371; HoCP 04-838; L 01-299; Ho 07-613; L 99-233; HoCP 96-540; L 99-226; L 01-283) grown at the Ardoyne Farm of the USDA-ARS-SRRC Sugarcane Research Unit in Shriever, Louisiana, USA. The maturity and other agronomy characteristics of these varieties are listed in Table 1.

Table 1. Characteristics of the nine commercial Louisiana sugarcane varieties in this study.

Varietal Characteristic	HoCP 96-540	L 99-226	L 99-233	HoCP 00-950	L 01-283	L 01-299	L 03-371	HoCP 04-838	Ho 07-613
Year of release:	2003	2006	2006	2007	2008	2009	2010	2011	2014
Maturity:	M*	M-E	M-E	E	E	M	M-E	M	E
Sugar per acre	VG†	VG	G	G	VG	VG	G	G	G
Sucrose Content	M-G	VG	M-G	VG	VG	G	G	M-G	VG
Tonnage	VG	G	VG	G	VG	VG	M	G	G

*E=early M=moderate, medium M-E=moderate to early maturing

† G=good P=poor VG=very good

Other general characteristics of these nine varieties are: HoCP 00-950 – a very high sucrose variety, but not popular because of low ratoon yields and short stalks; L 03-371 – very difficult to harvest due to brittle stalks; HoCP 04-838 – an expanding crop in Louisiana (LA) with 10% of the 2016 state acreage, and has very good cold tolerance but higher in fiber than other commercial varieties; L 01-299 – currently the most popular variety in LA with 30% of the 2016 acreage, has good ratoon ability; Ho 07-613 – only recently released; L 99-233 – not popular because of high smut susceptibility and high fiber; HoCP 96-540 – 30% of 2016 state acreage in LA; L 99-226 – high extraneous leaf material and declining acreage in LA with <10% acreage in 2016; L 01-283 – high sucrose concentration but also the probability of off-types resulting in very low cane weight; 10-11% of current LA acreage.

Research plots consisted of raised bed rows each measuring 1.8 m wide and were at least 9.14 m in length with four replicates. Sugarcane was planted with three whole-stalks of cane and a 10% overlap within the planting furrow. The cane was planted in September of 2012, which was 3 years prior to the ripener application. Plots were fertilized as normal for sugarcane at recommended rates (90 kg/ha N, 22 kg/ha K, 45 kg/ha P) and treated with pendimethalin plus metribuzin (2.242 plus 2.242 kg/ha) in March followed by pendimethalin plus metribuzin (2.242 plus 3.363 kg/ha) at layby time (mid-May). In 2015 the air temperatures were slightly above average as was the rainfall at 153 cm compared to the average rainfall of 140 cm.

On 1 Sept, 2015, two commercial ripeners (Polado® glyphosate; 5.3 oz/acre and Moddus®, trinexapacethyl; 11 oz/acre) were applied with a ground-operated backpack sprayer. The other treatment consisted of a non-treated control, and the experimental design was a randomized complete block with four replications per variety. Whole-stalk sugarcane stalks were harvested by hand, stripped of leaves and topped. Harvesting occurred on 29 Sept, 2 Oct, 13 Oct, and 20 Oct which are equivalent to week 4, 5, 6, and 7 after treatment, respectively. Each replication consisted of ~20 bundles of twenty whole-stalks. Juice was extracted through a Squier™ (Buffalo, NY, USA) three-roller mill at Ardoyne Farm (Houma, LA). First expressed juices were collected and combined; no imbibition water was added. The juice was filtered through a 0.6 mm mesh filter. All juices (~100 mL) were poured into two (120 mL) containers. One container of juice was immediately analyzed for Brix by refractometry, sucrose by polarimetry, and purity was calculated as the percentage ratio of pol to Brix. The second container of juice was frozen in dry ice, then transported to USDA-ARS-SRRC laboratories in New Orleans, LA and stored at -60 °C for further analyses.

Field experiments: Year 2

The first year of study conducted in 2015 was repeated in the 2016 season, with and without leaves and tops, to capture the starch and colorants in the leaves and top stalks. For without leaves, ten varieties were studied, which included eight of the nine varieties studied in the first year: HoCP 00-950; L 03-371; HoCP 04-838; L 01-299; Ho 07-613; HoCP 96-540; L 99-226; L 01-283; HoCP 09-804; Ho 09-840. Variety L 99-233 was removed from the study because of its unpopularity with growers and processors. Varieties HoCP 09-804 and Ho 09-840 were added. All the sugarcane was grown at the Ardoyne Farm of the USDA-ARS-SRRC Sugarcane Research Unit in Shriever, Louisiana, USA using the same cultivation conditions as in the Year 1 study. In 2016, the temperatures were consistently above average for all months, and rainfall was significantly ($P<0.05$) above average for all months at 238 cm compared to average 140 cm values. Due to limited availability of time and personnel, only five varieties were additionally studied with leaves and tops: HoCP 04-838; L 01-299; Ho 07-613; HoCP 96-540; L 01-283. Ripener treatments were the same as in Year 1 and applied on 23 Aug, 2016. Harvesting occurred on 20 Sept, 27 Sept, 4 Oct, and 12 Oct, which are equivalent to week 4, 5, 6, and 7 after treatment, respectively.

Measurement of total, insoluble, and soluble starch concentrations

Starch concentrations in the first three replicate juices were measured for total, soluble, and insoluble starch using the new microwave-assisted sonication/iodometric USDA method (Cole et al., 2014). Juices were assayed in duplicate; concentrations are quoted as average ppm on a per cent Brix basis.

Brix and pH

Juice Brix was measured in all four replicates using an Index Instruments (Kissimmee, FL) TCR 15-30 temperature controlled refractometer accurate to ± 0.01 Brix, and results expressed as an average of triplicates. Juice pH was measured in all four replicates on a Metrohm Brinkman 716 DMS Titrino (Riverview, FL, USA) with a Mettler Toledo (Columbus, OH, USA) xerolyte electrode.

Color and Indicator Values

Color of juices from all four replicate samples were measured as the absorbance at 420 nm and calculated according to the official ICUMSA method GS2/3-9 (1994). Samples were first diluted 50:50 in triethanolamine/hydrochloric acid buffer (pH 7.0) and filtered through a 0.45 µm filter. The color of some samples was also measured at pH 8.5, 4.0 and 9.0, by first adjusting the triethanolamine/hydrochloric acid buffer to this pH. The Indicator Value (I.V.) was calculated as the ratio of color at pH 9.0/pH 4.0.

Statistics

Results from the first year ripener study were statistically analyzed, with concentrations of quality parameters compared using one-way ANOVA, with means ranked using Tukey's HSD (honest significant difference) at the 5% probability level. The analyses were conducted using JMP v. 11.2.0 software (SAS Institute, Inc., Cary, NC). Comparisons were also made within each treatment type. All samples were pooled for each individual trial and compared to the control. Calculations of standard deviations and coefficient of variation values (CV) were calculated using Microsoft^{TM} Excel (ver. 2013), where CV = (standard deviation/mean) x 100.

RESULTS AND DISCUSSION

Brix and pH

In the first year of the study, both Polado and Moddus ripeners were applied to nine commercial sugarcane varieties at the USDA Ardoyne Farm near Houma, Louisiana, and juice was extracted from the whole-stalks only. As expected, the juice Brix significantly ($P<0.05$) increased with both Moddus and Polado ripener applications compared to the untreated (control) sugarcane, which indirectly indicated that sucrose had also increased (Table 2). Both Moddus (4.5%) and Polado (5.2%) ripeners also slightly improved precision for Brix values when compared to the control (5.3%). The mean juice pH results did not vary significantly with ripener treatment and pH precision data was also very similar among all the treatments, with Polado slightly better by CV0.1%. As the pH of fresh juice is mostly governed by natural acids in the plant tissues, including aconitic acid, these results suggest ripeners have no effect on acid levels.

Table 2. Overall Brix and pH results for the ripener treatment (first year study) of nine commercial sugarcane varieties. Results are the mean data for the four sampling dates (N=144). Precision results are presented as %CV (coefficient of variation also known as relative standard deviation RSD).

Ripener Treatment	Brix (%)	pH
	Mean Value	
Control (none)	18.13B	5.36A
Moddus	18.60A	5.37A
Polado	18.68A	5.38A
	Precision (% CV or RSD)	
Control (none)	5.3	1.5
Moddus	4.5	1.5
Polado	5.2	1.4

*Down columns only, differences in upper case letters were significant at the 5% probability level

Color

The color of the extracted juice was measured at pH 4.0, 7.0, 8.5, and 9.0 according to ICUMSA protocols, and the color Indicator Value I.V. was calculated (ratio of color at pH 9.0/color at pH 4.0). As expected, colors at pH 4.0 and 7.0 were lower than at pH 8.5 and 9.0

(Table 3), because color intensity depends on the solution pH due to changes in molecular structure, ionization, and association-dissociation equilibria (Dietz et al., 1959). The changes are not linear and color intensity changes most steeply in the pH range 6 to 8. This is one of the reasons that some U.S. sugarcane refiners measure color at pH 8.5 – a value at which the rate of color change has diminished (Godshall, 1997). The efficiency of industrial color removal by a select refinery process is dependent on the degree of ionization, polarity, and molecular weight of the colorants, and pH is known to strongly affect ionization (Riffer, 1988; Davis, 2001), which was another reason why color was measured at various pH values.

In the current study, as expected color values increased with pH (Table 3). Mean color values were lower than those typically found in the factory (Eggleston et al., 2012), because the leaves and tops had been removed from the whole-stalks. The addition of Polado and Moddus ripeners had very little effect on sample color values measured at pH 4.0, 7.0, 8.5, and 9.0 which were not significant, and tended to be slightly lower than the untreated (control) samples. Because of this, the I.V. values were not affected by the ripeners. Both ripeners had greater impact on color variation (precision) than mean color values (Table 3). The precision of the color measurements at different pHs, irrespective of treatment, followed the following general order: pH 4.0>pH 7.0>pH 8.5=pH 9.0. Thus, measurement of color at pH 4.0 and 7.0 was more variable compared to measurement at the higher pH values. This may be because the natural cane colorants are more sensitive to the higher pHs (Smith et al., 1981). Nevertheless, when compared to the control, moddus and especially glyphosate increased the variation in color measured at pH 4.0 and 7.0 (Table 3). In contrast, for colors measured at 8.5 and 9.0, Moddus decreased the color variation compared to the control, whereas Polado increased variation and, in turn, this strongly affected the variation of the I.V. values (Table 3). This strongly suggests that ripeners did have some effect on natural cane colorants present in the whole-stalks. It must be noted, however, that the variation in color values (Table 3) was much greater than the variation in Brix and pH values (Table 2)

Table 3. Overall mean color and I.V. results for the ripener treatment (first year study) on nine commercial sugarcane varieties. Results are the mean data for the four sampling dates (N=144). Precision results are presented as %CV (coefficient of variation also known as relative standard deviation RSD).

Ripener Treatment	Color pH 4.0	Color pH 7.0	Color pH 8.5	Color pH 9.0	I.V.
	Mean Color (ICU)				
Control (none)	5235A*	9057A	18596A	19211A	3.78A
Moddus	5115A	8910A	18210A	18880A	3.82A
Polado	5175A	8652A	18292A	18901A	3.81A
	Precision (% CV or RSD)				
Control (none)	14.6	11.5	8.5	8.3	9.9
Moddus	15.0	12.8	7.3	7.4	12.2
Polado	18.8	13.7	9.7	9.9	14.2

*Down columns only, the same upper case letters were not significantly different at the 5% probability level

Starch

It is well known that starch is an undesirable impurity that causes processing difficulties in the factory and refinery, and typically 250 ppm starch (using ICUMSA GS-17 method) in raw sugar is often quoted as the maximum amount compatible with satisfactory refining conditions in Louisiana. It is also now known that insoluble starch is very persistent across the whole factory, as well as in raw and even refined sugars (Eggleston et al., 2013), which has negative implications on amylase factory applications as well as syrup/liquor viscosity and sugar recovery (Eggleston et al, 2015). Fortunately, a new USDA Starch Research method is now available to measure both insoluble and soluble starch (Cole et al, 2014), which was used in this study to measure total, insoluble, and soluble starch (Table 4).

Most starch in the juices was insoluble because the juice had not been subjected to any heat. The small amounts of soluble starch that were detected (Table 4) may be due to natural amylase (diastase) enzymes solubilizing some starch in the extracted juice before the samples were frozen. Overall, results showed that the effect of adding Polado and Moddus was to markedly increase total starch by **20.7** and **20.6%**, respectively, compared to the control, although this was not significant at the 5% probability level because of high variation among the field replicate samples and varieties. Furthermore, there was a strong varietal effect ($P<0.0001$) for total starch concentrations as illustrated in Fig. 1. Such an increase in starch concentrations by ripeners could markedly impact sugarcane processing (Eggleston et al., 2015). Compared to the mean Brix, pH, and color values the variation in the starch values was much higher (compare precision results in Table 4 to Tables 2 and 3), even than in the controls. This strongly suggests that the natural variation of starch in the whole-stalks is higher than for Brix, pH, and color. From an engineering viewpoint, it is always easier to process incoming cane with low variation in quality and processing parameters. It is also shown in Table 4, that Moddus reduced starch variation compared to the untreated control, whereas Polado increased it. Thus glyphosate not only increased total starch values but also increased its variation as well.

Table 4. Overall total, insoluble, and soluble starch results for the ripener treatment (first year study) of nine commercial sugarcane varieties. Results are the mean data for the four sampling dates (N=108). Precision results are presented as %CV (coefficient of variation also known as relative standard deviation RSD).

Ripener Treatment	Total Starch* ppm/Brix	Insoluble starch* ppm/Brix	Soluble Starch* ppm/Brix
		Mean Value	
Control (none)	916.5A‡	899.4A	21.0A
Moddus	1105.9A	1080.0A	30.7A
Polado	1105.0A	1092.0A	44.2A
		Precision (% CV or RSD)	
Control (none)	30.6	29.8	151.6*
Moddus	24.7	25.7	143.0*
Polado	35.7	34.2	160.1*

‡Down columns only, the same upper case letters were not significantly different at the 5% probability level

*The very high %CV values for soluble starch were a function of the very low mean values

Total starch values for each variety at every sampling week are also shown in Fig. 1, as well as the effect of Moddus (M) and Polado (P) ripeners. Variety L 99-233 contained ($P<0.05$)

dramatically more starch than the other varieties, and both ripeners caused the greatest increase (at least 102%) in total starch in this variety. Fortunately, this variety has become unpopular

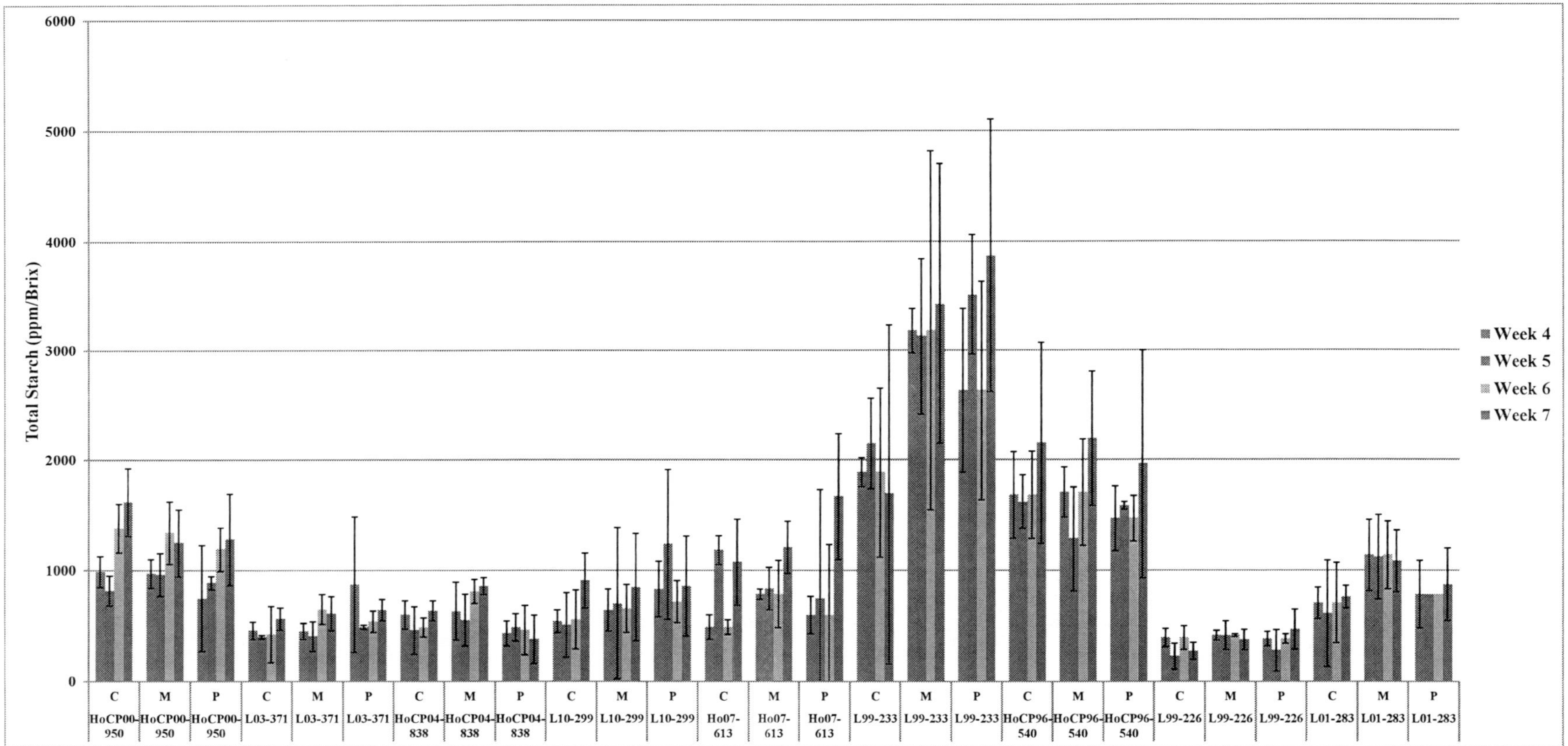

Fig. 1. Effect of ripeners on the total starch concentration in the juice extracted from the stalks of nine commercial varieties in 2015 (first year of ripener study). Samples were collected 4, 5, 6, and 7 weeks after ripener applicaton. C=control (no treatment), M=Moddus ripener treatment, and P=Polado ripener treatment. Lines on each column indicate the range of standard deviation.

with sugarcane growers and processors anyway because of its high fiber values which negatively impacted the tandem mill rollers. It is also fortunate that many other varieties contained ($P<0.05$) relatively lower starch concentrations compared to L 99-233 (Fig. 1), thus offering growers and processors much needed flexibility to control starch in their cane management systems.

As illustrated in Fig. 2, by week 4 after ripener application, for most of the varieties the total starch in the control (average across all varieties was 831.2 ppm/Brix) juices was already lower than in the Moddus (944.4 ppm/Brix) and Polado (969.6 ppm/Brix) treated juices. By week 7, total starch in the Polado treated juices was slightly higher than for Moddus treated juices, and both ripeners had considerably higher starch than the untreated (control) juices, although this was not significant because of the high variation in starch concentrations (Fig. 2).

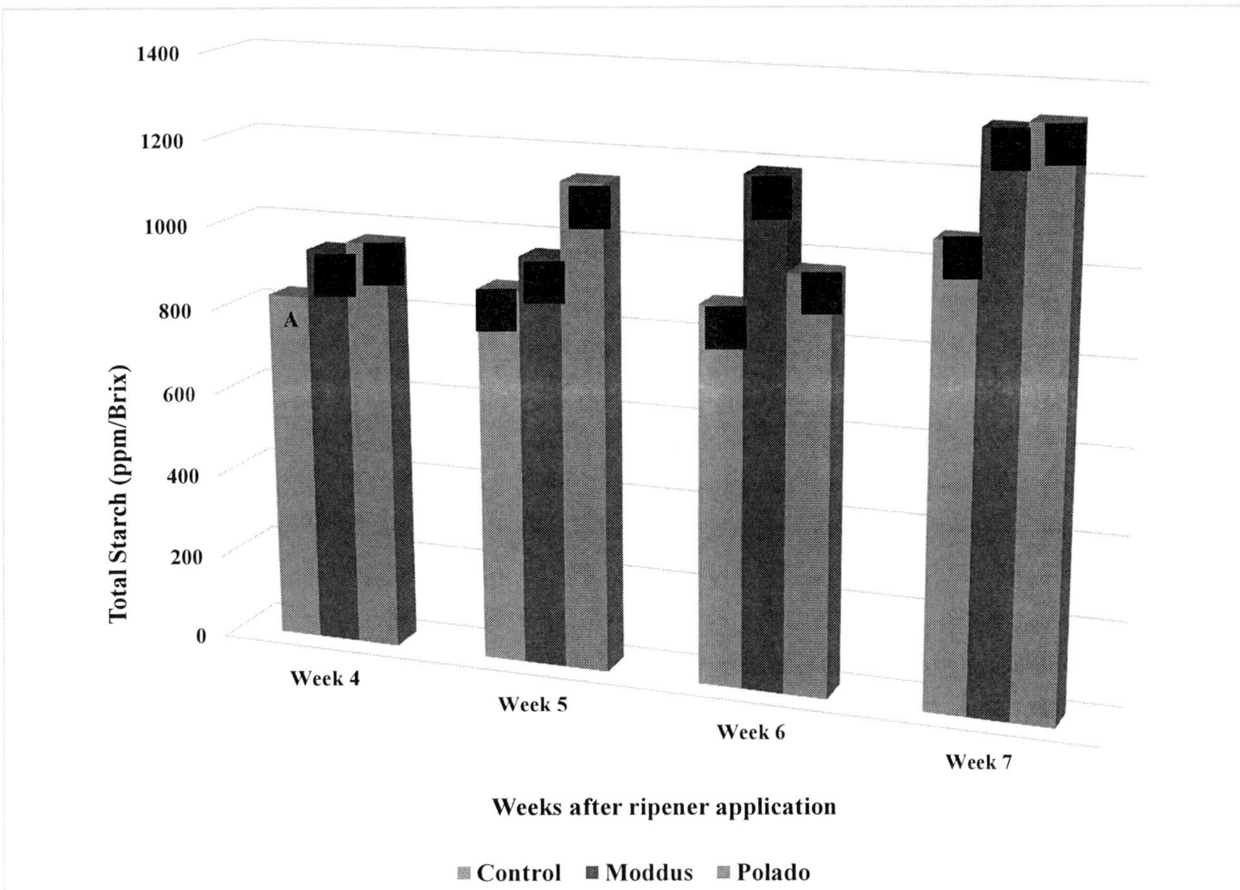

Fig. 2. Overall total starch contents in juices extracted from sugarcane stalks in the weeks following the application of chemical ripener treatments. Results are the mean data of all nine commercial varieties studies for the four sampling dates. Across columns for each week only, the same upper case letters were not significantly different at the 5% probability level.

Second Year Study: 2016

In the second year study (2016) both ripeners were similarly applied again to ten pre-commercial and commercial sugarcane varieties at Ardoyne Farm of USDA near Houma. Color measured at pH 4.0, 7.0, 8.5, and 9.0, and the color I.V., total, insoluble and soluble starch were measured in the juice extracted from stalks. The effect of leaves was also studied in five of the ten varieties. Samples are still being analyzed. A full statistical analyses of the two year study will be undertaken including G x E (genotype by environment) interactions.

CONCLUSIONS

The pre-harvest application of chemical ripeners to increase sucrose content and storage in sugarcane stalks is an established technology in many sugarcane-producing areas of the world. Unfortunately, very little information is available on how it affects quality parameters that are critical to both factory and refinery processing, particularly starch and color. A large field study was conducted on the effect of two chemical ripeners (Polado and Moddus) with differing modes of action on nine commercial sugarcane varieties cultivated in south Louisana, USA. Juice was extracted from whole-stalks with tops and leaves removed by roller milling. Results from the first year of the study showed that, overall, the effect of adding Polado and Moddus ripeners was to dramatically increase total starch (most starch in juice is insoluble) by 20.7 and 20.6%, respectively, although this was not significant because of a strong varietal effect ($P<0.0001$). The variation in the starch values was much higher than for Brix, pH, and color values, even in the untreated (control) juices. Polado caused worse starch variation than Moddus ripener. The mean juice pH results did not vary with ripener treatment and pH precision data was also very similar among all the treatments. As the pH of fresh juice is mostly governed by natural acids in the plant tissues, including aconitic acid, these results suggest ripeners have no effect on acid levels. The ripeners had less effect on color than starch with very little differences occurring for ICUMSA color measured at pH 4.0, 7.0, 8.5, and 9.0 although color variation increased, particularly in sugarcane juices applied with Polado. A second year of study is currently being completed to establish if the trends found in this first year of study are consistent, and to evaluate the impact of environment and the addition of leaves and tops (trash).

ACKNOWLEDGEMENTS

The authors thank the staff at the USDA-ARS-Sugarcane Unit Research Unit at Ardoyne Farm, Shriever, LA, USA for their help in harvesting, milling, and on-site juice analyses. Thanks to Herman Waguespack, Jr. of the American Sugarcane League for useful discussions, and the American Sugar Cane League for partial funding of this research. Mention of trade names or commercial products in this article is solely for the purpose of providing specific information and does not imply recommendation or endorsement by the U.S. Department of Agriculture. USDA is an equal opportunity provider and employer.

REFERENCES

Alexander AG (1973). In *Sugarcane Physiology. A Comprehensive Study of the Saccharum Source-to-Sink System.* Elsevier, Amsterdam.

Cole M, Eggleston G, Gilbert A, Chung YJ (2014). Development of a research method to measure insoluble and soluble starch in sugarcane factory and refinery products. *Proc. 2014 Sugar Industry Technol. Mtg*, Canada, **Vol. LXXIII**, 212-229.

Davis SB. (2001). The chemistry of colour removal: A processing perspective. *Proc. SASTA*, **75**, 328-331.

Deitz, VR, Carpenter FG, Rieger CJ, Rootare HM. (1959). The influence on pH on some properties of char liquors. *Bone Char Research Project Tech. Report,* No. 53.

Eggleston G, Antoine, A, Montes B, Stewart D (2010). Seasonal variations in optimization of intermediate temperature stable α-amylase in raw sugar manufacture. *Internat. Sugar J.,* **112**, 472-480.

Eggleston G, Birkett H, Gay J, Legendre B, Jackson W, Schudmak C, Monge A, Andrzejewski B, Viator R, Charlet T (2012). How harvesting cane with different levels of trash affects production and processing. Part II. Pilot plant processing to sugar. *Internat. Sugar J.,* **114**, 169-178.

Eggleston G, Cole C, Gilbert A, Rose I, Andrzejewski B, St. Cyr E' Stewart D (2013). A new look at starch in the sugarcane factory and refinery: The presence of soluble and insoluble starch. *Proc. 2013 Sugar Industry Technol. Mtg*, China, **Vol. LXXII**, No. 1066, 326-339.

Eggleston G, Sugawara A, Toyamasu T, Cole M, Montes B, Triplett A (2015). Re-examining the applications of amylase in the sugar industry: Conquering the control of insoluble and soluble starch. *Proc. 2015 Sugar Industry Technol. Mtg*, Japan, **LXXIV**, 36-59.

Eggleston G, Viator R, Grisham M (2007). Glyphosate ripener effects on the processing quality of different sugarcane tissues. *Proc. ISSCT.*, **26**, 1460-1467.

Godshall MA. (1997). Color analysis. *Proc. 1997 Sugar Industry Technol. Mtg*, Canada, **LVI**, 211-231.

Legendre BL (1975). Ripening of sugarcane: Effects of sunlight, temperature, and rainfall. *Crop. Sci.,* **15**, 349.352.

Paton NH and Smith P. (1982). A HPLC study of the changes in colorant composition following factory decolorisation of raw liquors with bone char, resin and granular carbon. *Proc. Sugar Proc. Res. Conf.*, 1-23.

Riffer R. (1988). The nature of colourants in sugarcane and cane sugar manufacture. In: *Chemistry and Processing of Sugar beet and Sugarcane*. Eds. M.A. Clarke and M.A. Godshall, Elsevier Science, Amsterdam, 186-207.

Smith P, Paton NH, Delaney HR, Ames R (1981). Colour studies in milling. *Proc. Aust. Soc. Sugar Cane Technols.,* 71-80.

Van Heerden PDR, Eggleston G, Donaldson RA (2014). Ripening and postharvest deterioration. In *Physiology, Biochemistry, and Functional Biology of Sugarcane*, Moore PH and Botha FC (Eds). Wiley-Blackwell, Chapter 4, 55-84.

S.I.T. Paper #1174

Development of a Factory/Refinery Method to Measure Total, Soluble, and Insoluble Starch in Sugar Products

Marsha R. Cole[1,2] and Gillian Eggleston[1*]

[1,*] USDA-ARS Southern Regional Research Center, 1100 Robert E. Lee Boulevard, New Orleans, LA, 70124, USA; gillian.eggleston@ars.usda.gov; +1 (504) 286-4446

[2] Fellow, Oak Ridge Institute for Science Education Research Participation Program at USDA-ARS

ABSTRACT

An easy, rapid, and inexpensive method was developed to measure total, soluble, and insoluble starch in products at the factory and refinery, using microwave-assisted neutralization chemistry. The method was optimized using the previously developed USDA Starch Research method as a reference. Optimal acid and base (alkali) combinations and concentrations plus microwave time and power were determined to completely solubilize an insoluble corn starch reference. The final method solubilizes <4,000 ppm insoluble starch in 2 min, has acceptable precision (7% CV, coefficient of variation), accuracy (≥94%), uses a corn starch reference, and incorporates a color blank to remove contribution from natural colorants found in industrial samples. The method was validated using both simulated (5%) and factory samples and found applicable to sugarcane crusher juice (8% CV), clarified juice (12% CV), massecuite (2% CV), molasses (4% CV), syrup (2% CV), and raw sugar (8% CV) with 95.4% satisfactory performance z-scores. The USDA Starch Factory/Refinery and Research methods were very similar with an overall mean difference of ~6.5%.

INTRODUCTION

Accurate starch analysis has been an ongoing concern for the international sugar industry where starch solubilization procedures, detection methods, starch standards, and whether to correct the results for color have been questioned (Eggleston, 2014; Godshall, 1997, 2005; Harrison, 2012; Thielecke, 2002, 2004). In general, industrial starch methods used in the sugar industry measure only total starch, require that the sugarcane sample be diluted, precipitated with alcohol, or boiled for <15 min with or without acidified calcium chloride solution, before adding an iodometric solution and measuring the blue color formed with a spectrophotometer (Cole et al., 2017a). Aside from the consensus that the use of potato starch or potato amylopectin as a reference for sugarcane starch is unsuitable and that corn starch is a better reference (Cole et al., 2014; Godshall et al., 1991), correcting starch results for sample color is a major limitation also affecting starch analysis in the sugar industry (Cole et al., 2017b). This is because the absorbance of the solubilized sugarcane starch-triiodide (I_3^-) complex must be greater than the absorbance of the actual color of the product, which varies in color intensity, types of colorants, and cannot be standardized. In spite of this, majority of the industrial starch methods available in the sugar industry have primarily focused on solubilizing all of the sugarcane starch, rather than correcting starch results for sample color, since sugarcane starch solubilization is challenging (Cole et al., 2013; Eggleston, 2017; Godshall et al., 2004).

It is now known that not all of the insoluble starch is solubilized during sugar manufacture and therefore existing methods can often "underestimate the total amount of starch in the sugar product" which has major processing consequences (Eggleston et al., 2017). Thus, it is very important that the industry can measure soluble and insoluble starch as well. Among analytical starch methods, only the USDA Starch Research method, which is based on microwave-assisted probe ultrasonication (Cole et al., 2014), is able to efficiently solubilize the insoluble granular sugarcane starch. The USDA Starch Research method quantifies soluble, insoluble, and total starch concentrations, uses a corn starch standard, incorporates a color blank, is unaffected by other impurities present in the sugar product, and is applicable to all sugarcane factory and refinery products (Cole et al., 2014). Although the USDA Starch Research method is very accurate and precise, it is not considered an industrial tool because of the relatively expensive ultrasonication probe (~$7,000 USD) which is also not readily available worldwide. Therefore, it was imperative that an industrial factory/refinery starch method be developed with the same benefits as the USDA Starch Research method but without the probe ultrasonicator. For the USDA Starch Factory/Refinery method to be successful, it needed to be i) inexpensive, ii) easy, iii) rapid, iv) compatible with existing analytical equipment and reagents at the factory and refinery, v) uses easily accessible materials, and vi) has comparable high accuracy and precision to the USDA Starch Research method. In this paper, we introduce a newly developed USDA Starch Factory/Refinery (Industrial) method that is based on microwave-assisted neutralization chemistry that has been validated on juices, syrups, molasses, massecuites, and raw sugar following ICUMSA procedures.

EXPERIMENTAL

Chemicals and sugar samples

Corn starch, sucrose (99.95%), other chemicals were analytical grade, and the Corning 6595 96-well microtiter plates were purchased from Sigma Aldrich Company (St. Louis, MO, USA). Raw sugars were kindly provided by Imperial Sugars (Port Wentworth, GA, USA). Sugarcane crusher and clarified juice, final evaporated syrup, molasses, and massecuite were kindly provided by Alma Plantation (Lakeland, LA, USA).

Brix (% Dissolved Refractometric Solids)

The soluble solids content of the simulated juices were determined using degrees Brix in triplicate with a temperature-controlled refractometer (model TCR 15-30; Index Instruments, FL) to an accuracy of ± 0.001 Brix.

Solubilizing Swollen Starch with Acid after Microwaving

Corn starch (2,000 ppm, 15 Brix sucrose) solution was microwaved for 1 min at 100% power to create swollen starch. To eight 40 µL serial dilutions (1:1) of hydrochloric (HCl), acetic, sulfuric, and citric acids ranging from 0.2-60% w/v or v/v in a 96-well microtiter plate, 40 µL of the swollen corn starch sample was added. Changes in starch solubilization was then measured in 30 s intervals for 30 mins, in which 20 µL of 0.25M HCl and 100 µL of iodometric reagent (1 mM KIO_3 and 5 mM KI) was added to the respective wells before measuring at 600 nm using a Tecan Sunrise microplate reader (Life Sciences, San Jose, CA, USA) equipped with Magellan software.

Microwaving Insoluble Starch with Acid

Ten mL of a well-stirred insoluble corn starch solution (2,000 ppm, 15 Brix sucrose) was added to four 125 mL volumetric flasks. Each flask was treated with 0.05-1M HCl, acetic, sulfuric, or phosphoric acid, weighed, covered with a watchglass, and microwaved at full power in a commercial microwave oven (Kenmore, Sears-Roebuck and Company, IL, USA) with a maximum output of 1,100 W at 2.45 GHZ together for 1 min. Then, the samples were removed from the microwave and placed in an ice bath to cool to room temperature. Any water that was lost during microwaving was added back by weight with water. Both filtered and unfiltered, treated sample (800 μL) were placed in a cuvette and mixed with 200 μL of 0.25 M HCl and 1 mL of iodometric reagents (1 mM KIO_3 and 5 mM KI) to a cuvette. The absorbance for each cuvette was measured at 600 nm using a UV-Vis spectrophotometer (UV mini-1240; Shimadzu, Houston, TX, USA). The absorbance of the filtered and unfiltered samples were compared to ensure complete solubilization and was also compared to the same sample after being analyzed with the USDA Starch Research method for accuracy.

Optimizing Microwave Power, Time, and Number of Flasks

The effects of microwave power, time, and number of flasks were completed using the same conditions above except the microwave power was varied between 20-100%, microwave time ranged from 20 s to 1 min, and the number of flasks from 1 to 4. The position of the flasks on the microwaving plate was also assessed. The absorbance of the filtered and unfiltered, treated sample (800 μL) were placed in a cuvette and mixed with 200 μL of 0.25 M HCl and 1 mL of iodometric reagents to a cuvette. The absorbance for each cuvette was measured at 600 nm using a UV-vis spectrophotometer. The absorbance of the filtered and unfiltered samples were compared to ensure complete solubilization and was also compared to the same sample after being analyzed with the USDA Starch Research method for accuracy.

Effect of Acid and Base on Starch Solubilization

Starch solubilization of 2,000 ppm insoluble corn starch in 15 Brix sucrose by use of both acid and base (neutralization) was also investigated. Known concentration of base, i.e., sodium hydroxide (NaOH), urea, and ammonia, were added to equal known concentrations of HCl, acetic, sulfuric, and phosphoric acids to determine if maximum solubilization, accuracy, and precision could be obtained. Acid and base concentrations ranged from 0.01-5M, depending on the acid studied. All of the experiments were conducted in four flasks in which microwaving (1 min, 100% power) either occurred (i) before the addition of both acid and base or (ii) after microwaving with base and before adding acid. The filtered and unfiltered samples were also analyzed as previously mentioned at 600 nm.

Final USDA Starch Factory/Refinery Method to Measure Total, Soluble, and Insoluble Starch in Sugar Products

Ten mL of a well-stirred, ~15 Brix simulated product (i.e., 1,333-26,667 ppm/Brix insoluble corn starch) or factory product were added to four 125 mL volumetric flasks. Concentrated sodium hydroxide (36 μL NaOH) was added to each sample, solutions were gently mixed, and then weighed in the flask. The weight of the samples and flasks were recorded before covering each flask with a watch glass and microwaving them for 1 min at 100% power. After microwaving all of the flasks in a diamond configuration, the sample were gently mixed again, cooled in an ice bath for 1 min to room temperature, and then reconstituted with water to the original weight. A second round of microwaving (1 min, 100% power) was performed similarly but after adding 58 µL of concentrated hydrochloric acid (HCl). After cooling the sample and reconstituting the water lost by weight, the treated samples were filtered (0.45 µm Whatman PVDF syringe filter). Three cuvettes per sample flask was prepared for analysis by adding 800 µL treated, filtered sample, 200 µL of 0.25 M HCl, and 1 mL iodometric reagents (1 mM KIO_3 and 5 mM KI). The absorbance for each cuvette was measured at 600 nm using a UV-vis spectrophotometer (UV mini-1240; Shimadzu, Houston, TX, USA). If needed, some treated samples were diluted with water to be within the absorbance range of 0.2-1.5 absorbance units (a.u.) at 600 nm.

To determine the concentration of soluble starch in the factory sample, the 15 Brix, untreated sample was filtered, added to a cuvette, and treated with the iodometric reagents mentioned above before measuring the absorbance at 600 nm.

Each treated and untreated sample was corrected for the iodine color and sample color with an iodometric blank (I_b) and color blank (C_b), respectively, before determining starch concentration. The I_b consists of 800 µL water, 200 µL of 0.25 M HCl, and 1,000 µL iodometric reagents, whereas the sample C_b consists of 800 µL untreated or treated sample at the same dilution as the sample, 200 µL of 0.25 M HCl, and 1 mL of water. Both the absorbance of the I_b and C_b were measured at 600 nm and subtracted from all sample absorbance measurements. After extrapolating the final ppm concentrations from the corn starch calibration curve, the concentration of insoluble starch was calculated by subtracting the soluble starch concentration (untreated sample) from the total starch concentration (treated sample), divided by the initial sample Brix, and reported as ppm/Brix (Fig. 1).

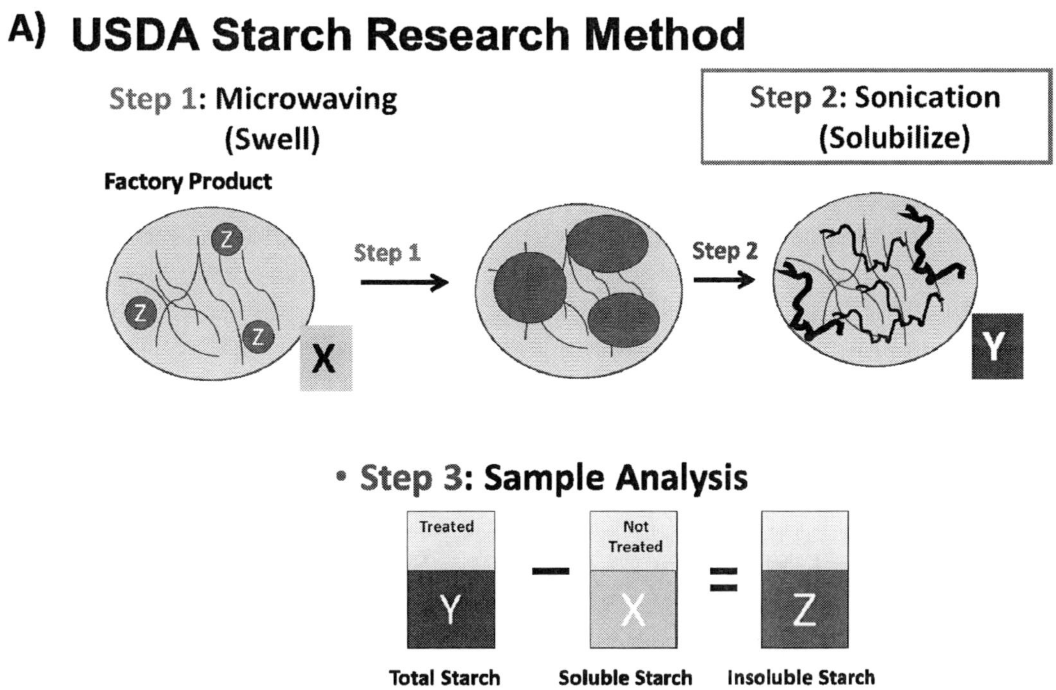

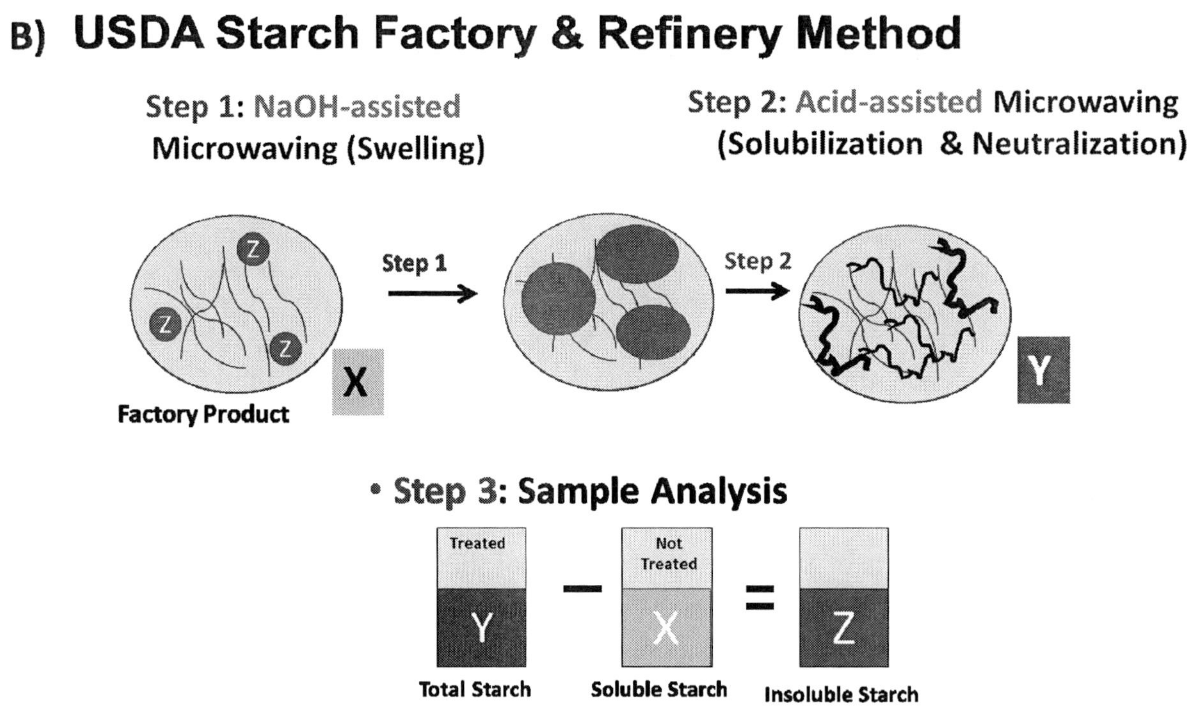

Fig. 1. Principles summarizing the A) USDA Starch Research method and B) the new USDA Starch Factory/Refinery method.

The calibration curve was made using 2,000 ppm insoluble corn starch dispersed in a 15 Brix sucrose solution. Starch solubilization was performed as mentioned above. Ten serial dilutions (1:1) to a concentration range (0–500 ppm) were made in water before reacting with iodometric reagents and measuring the absorbance at 600 nm using the procedure described as outlined above.

USDA Starch Research Method

The USDA Starch Research method, which measures total, soluble, and insoluble starch concentrations in a variety of sugar products and by-products, was used without modification (Cole et al., 2014). Briefly, 10 mL of a ~15 Brix sugar sample was microwaved for 1 min (100% power) and sonicated for 5 min at 60% amplitude. After microwaving and sonicating the sample, any weight that was lost from heat was reconstituted with water. The treated sample was filtered through a 0.45 m Whatman PVDF syringe filter. Then, 800 μL of sample was mixed with 200 μL of 0.25 M HCl and 1 mL iodometric reagents (1 mM KIO_3 and 5 mM KI) before measuring the absorbance at 600 nm on a spectrophotometer.

Accuracy, Precision, and Validation of the USDA Starch Factory/Refinery Method on Simulated and Factory Samples

The USDA Starch Research method was validated following ICUMSA protocols, where every sample in this study was measured in eight separate, independent executions and with triplicate analyses per run. The final USDA Starch Factory/Refinery Starch method was studied on simulated samples containing 1,333 – 26,667 ppm/Brix insoluble corn starch and factory raw sugars and processing by-products, i.e., crusher juice, clarified juice, final evaporated syrup, massecuites, and molasses. The mean of the data set (X_{mean}) and standard deviation (SD) were calculated, as well as the relative standard deviation (RSD) or coefficient of variation (% CV) (Thompson, 1988):

$$RSD=\left(\frac{SD}{X_{mean}}\right)x\ 100$$

Accuracy (%) was determined by comparing the starch concentration (ppm) obtained experimentally to the known amount added to the sucrose in the simulated samples and to the starch concentration determined using the USDA Starch Research method. Thus, the mean of the total starch concentration for X sample (X_{mean}) was divided by the theoretical starch concentration either added or determined by the USDA Starch Research method ($X_{theoretical}$) and multiplied by100:

$$\%\ Recovery=\frac{X_{mean}}{X_{theoretical}}\times 100$$

Differences between the methods were determined by plotting the USDA Starch Factory/Refinery method by the USDA Starch Research method and evaluating the slope. Total equatability between the two methods will yield a slope of 1, slope values greater than or less than 1 indicate differences in method equatability.

Statistical Analysis

All method performance, comparison, and accuracy/precision statistics were determined using a modified version of the Method Comparison JMP add-in in JMP version 11.2 statistical software (SAS Institute, Cary, NC, USA). Pearson correlation (R) coefficients were used to assess the relationship and significance between the two methods and their starch results. Performance z scores were calculated in Microsoft Excel^{TM} version 2007 using the following formula (Thompson et al.,2006):

$$z=\frac{x-X}{SDPA}$$

Where x is the data value, X is the assigned value, and SDPA is the standard deviation for proficiency assessment. $|z|$ scores <2.00 are satisfactory, $|z|$ scores between 2.00–3.00 are questionable, and $|z|$ scores >3.00 are unsatisfactory (Thompson et al. 2006).

Results and Discussion

Initial Method Development to Totally Solubilize Starch

It is well known that sugarcane (grass) starch acts more like corn (grass) than potato (tuberous root) starch because it is considerably more difficult to solubilize when boiled in water (Cole et al., 2014). All of the existing sugar industry starch methods use potato starch and boiling to solubilize starch in sugar products, although it has been unequivocally proven that not all of the industrial starch methods efficiently solubilize insoluble starch (Cole et al., 2017a). This is because starch solubilization is affected by many properties unique to the botanical source where traditional boiling can not always easily and efficiently solubilize the granule (Jambraka et al., 2010; Xue et al., 2008). Hence, the persistence of insoluble starch across factory and refinery processing into final raw sugars and refined sugars (Cole et al, 2014 ; Eggleston 2017).

During the development of the USDA Starch Research method, it was found that boiling and microwaving only solubilized <50% and <65% insoluble corn starch, respectively, depending on initial starch concentration and heating time, but never reached 100% solubilization (Cole et al., 2014). Thus, an ultraprobe sonication step was needed to rupture the swollen granules and release the soluble starch for analysis (Fig. 1).Therefore, in the development of the industrial method we had to find another way to rupture and release the soluble starch.

Post-microwaved acid solubilization of gelatinized starch

Initial solubilization studies to replace the ultrasonication probe included the action of acids, of varying strength, on microwaved corn starch samples to determine if acid alone could rupture the swollen starch granules. This is because early studies by Nägeli (1874) and Lintner (1886) reported that strong acids like hydrochloric (HCl) and sulfuric acids can be used to produce "hot water soluble starch," which was the template used for the commercial production of acid-hydrolyzed starch in the late 19th century (Whistler, 2008). This approach has even been applied to starch granules in dilute aqueous or acidified alcoholic solutions, but without heat (Wang & Copeland, 2015).

Various acids (0.2-60% w/v or v/v HCl, acetic, sulfuric, phosphoric, and citric acids) were initially applied to swollen starch, i.e., 1,000 ppm insoluble after microwaving for 1 min at 100% power, in 15 Brix sucrose. Starch solubilization for the samples was monitored in 30 sec increments over 30 min. The starch results for each sample continued to change over the 30 min (Fig. 2) resulting in variable and inconclusive results that required more optimization (Fig. 2). Fig. 2 shows that the weaker acids, i.e., citric and acetic acids, resulted in more starch-I_3^- color indicating that some long-chain soluble starch was released from the swollen starch; whereas, the strong sulfuric and HCl acids either formed short oligosaccharides that did not produce a color with iodine or were overall ineffective in solubilizing the starch. Because it was very difficult to determine whether the starch granules had been solubilized or if the acids were effective, microscopy had to be used. The samples were also very difficult to filter and some concentrations still had precipitate and starch granules present, which suggested that using acid on already swollen starch was not effective, thus it was no longer studied. Whistler (2008) similarly reported that acids did not always solubilize starch and that it depended on botanical source and even the physico-chemical structure of the starch granules obtained within their cultivars. This agrees with Cole et al. (2017a) who also found a difference ($P<0.05$) in the solubilization of potato and corn starch in simulated sugars by current industrial starch methods, which some used acetic acid to help with solubilization, that was also concentration dependent.

We next investigated if the acids could solubilize the insoluble starch if added during the microwaving step. Phosphoric acid was added in place of citric acid, because it is a stronger hydrolyzing agent than citric acid.

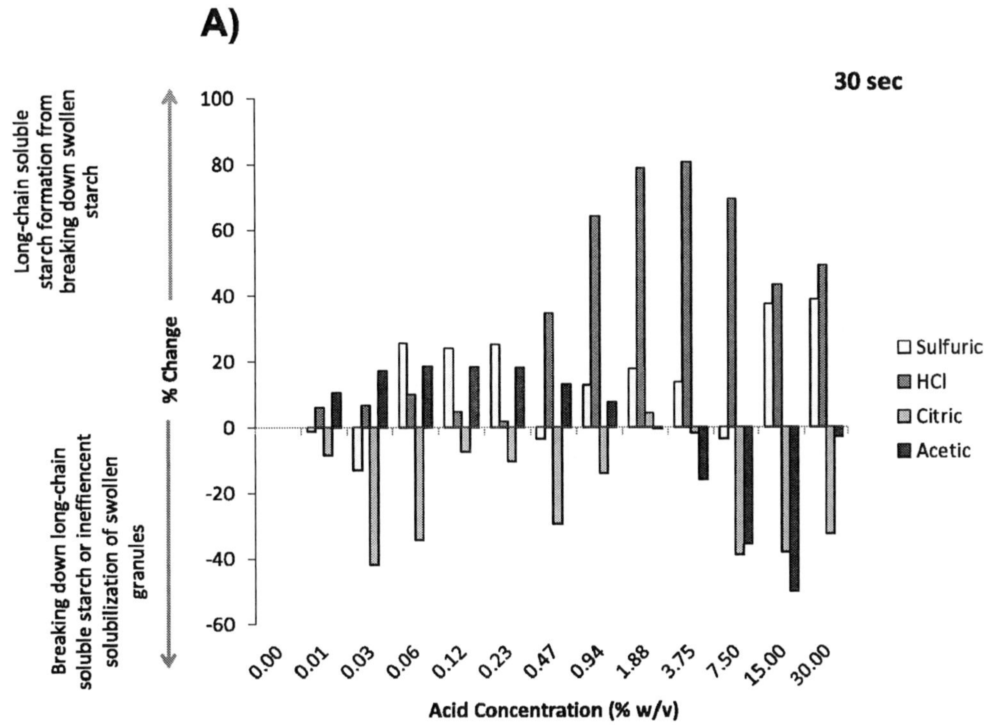
A)
30 sec
Long-chain soluble starch formation from breaking down swollen starch
% Change
Breaking down long-chain soluble starch or ineffiencent solubilization of swollen granules
100
80
60
40
20
0
-20
-40
-60
0.00
0.01
0.03
0.06
0.12
0.23
0.47
0.94
1.88
3.75
7.50
15.00
30.00
Acid Concentration (% w/v)
Sulfuric
HCl
Citric
Acetic

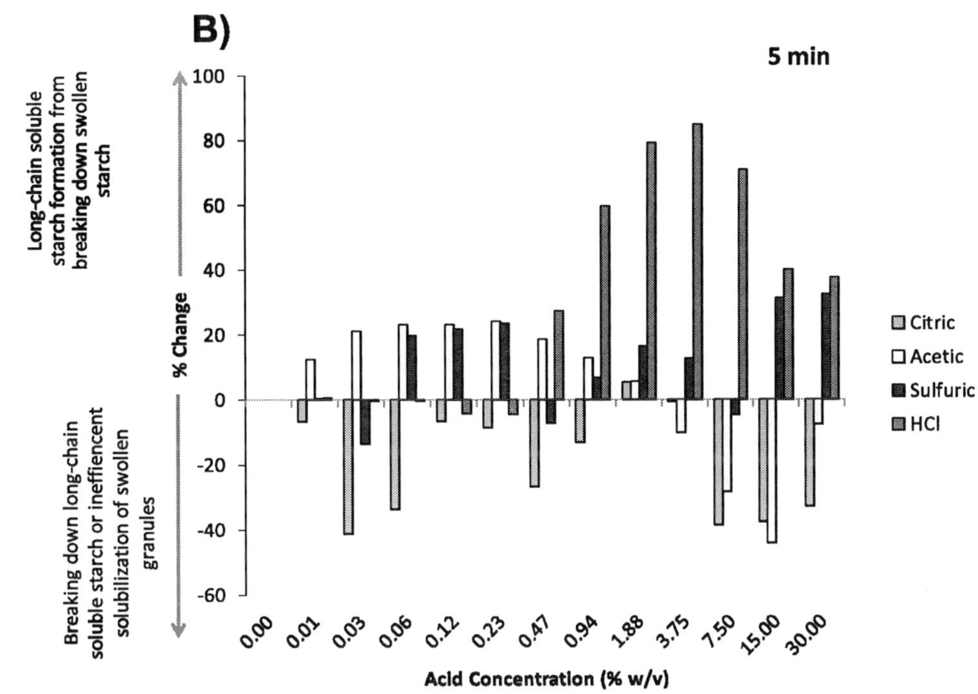
B)
5 min
Long-chain soluble starch formation from breaking down swollen starch
% Change
Breaking down long-chain soluble starch or ineffiencent solubilization of swollen granules
100
80
60
40
20
0
-20
-40
-60
0.00
0.01
0.03
0.06
0.12
0.23
0.47
0.94
1.88
3.75
7.50
15.00
30.00
Acid Concentration (% w/v)
Citric
Acetic
Sulfuric
HCl

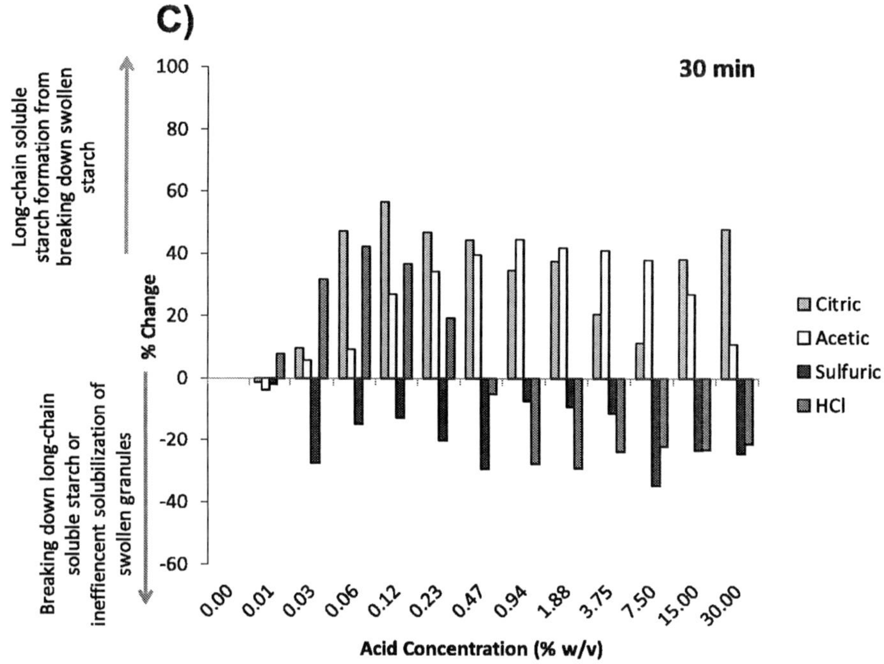

Fig. 2. Changes in the amount of solubilized starch produced by acid-assisted solubilization of swollen starch after using 0.02-30% w/v or v/v citric, acetic, sulfuric, and HCl acids for A) 30 sec, B) 5 min, and C) 30 min. Four flasks were simultaneously heated. Positive values indicate that swollen starch was converted to long soluble starch chains. Negative values show that long chain soluble starches were broken into smaller fragments that were undetectable with iodometric reagents or acids were ineffective.

Acid-assisted microwave solubilization

Because the concentration of acids used in the first experiments with swollen starch were very high, i.e., <60% or between 0.02–11 M, the concentration range studied was reduced to 1M maximum, for safety, and four flasks were heated simultaneously for efficiency. Molarity was also used since all of the acids studied were now liquids (citric acid is a solid and was no longer considered) and it was easier to compare.

Upon microwaving the corn starch for 1 min at 100% and with <1 M HCl, it was quickly determined that 1M HCl was still too concentrated and therefore also ruled out. The 1M starch solution changed from colorless to a bright orange color, an indicator of acid hydrolyzed glucose, and had begun smoking after 1 min of microwaving. However, the solution was clear, had no residual starch granules, and filtered very easily. This indicated that all of the insoluble starch had become solubilized despite the high HCl concentration. At lower HCl concentrations (<0.1 M), the solutions were almost colorless (which indicated that little sugar degradation products had occurred), translucent, and very easy to filter. The solution filterability (filtered vs not filtered) of the HCl solutions were >90% (based on starch-I_3^- absorbance). In comparison to the probe sonicator, solubilizing insoluble starch with HCl in the microwave gave accuracies that decreased with increasing HCl concentration (Fig. 3). Since the 0.1M HCl treatment yielded absorbances >100%, it was apparent that extra fragmentation of the solubilized starch had occurred (Fig. 3), and the optimal HCl concentration needed ranged between 0.05-0.1 M.

Heating the insoluble starch with <1M acetic acid resulted in colorless solutions but none were clear. Also, the microwaved samples had a bad odor and were very difficult to filter, which indicated that a large % of the starch granules were still swollen. Compared to the probe sonicator, acetic acid gave yields between 90-140% (Fig. 3), and would need >1M acetic acid to solubilize the insoluble starch. However, after adding up to 5M acetic acid, accuracy decreased (data not shown). Therefore, acetic acid was also ruled out as an effective solubilizing agent.

When the starch samples were heated with sulfuric acid, the samples were also clear and easy to filter similar to HCl; at <0.1M HCl the filtered vs not filtered samples were 99.4% equal. Similarly to acetic acid, sulfuric acid solubilized the starch sample giving yields >100%. This indicated that the optimal sulfuric acid concentration was found to also range between 0.05-0.1 M when compared to the probe sonicator (Fig. 3). Although sulfuric acid efficiently solubilized the insoluble corn starch in the microwave, the samples with >0.05M sulfuric acid turned dark brown and red indicating that erythrodextrin sugar degradation products, i.e., oligosaccharide that were formed under extreme hydrolysis conditions and produce a red color with iodine (Duncan, 1898), had formed. It was also very difficult to control the reproducibility of the sulfuric acid treated samples and prevent the color from forming.

Very similar results were obtained for phosphoric acid with 93.8% difference between filtered and not filtered samples and 91.7% maximum accuracy when compared to the probe sonicator (Fig. 3). The samples that were treated with phosphoric acid were clear, but were green. Therefore, among the acids studied, HCl showed the most promise and was used for further optimization studies. Further interpretation of the data in Fig. 3, indicated that 0.08 M HCl may be the optimal concentration needed.

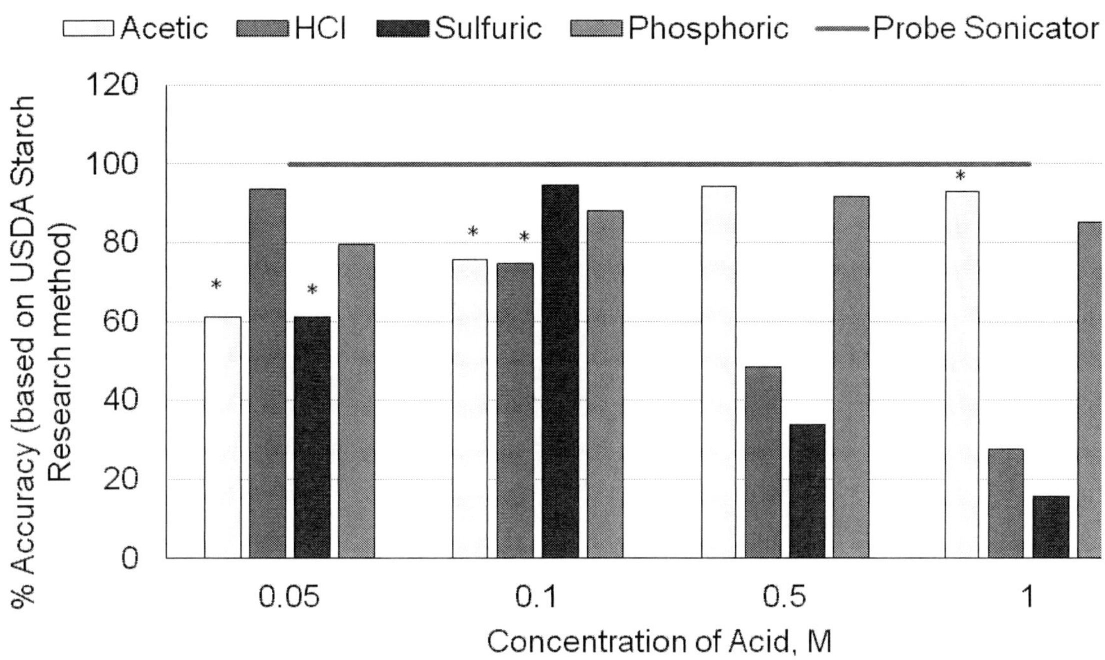

Fig. 3. Percent accuracy of the acid-assisted microwave solubilization approach on 2,000 ppm insoluble starch (four flasks simultaneously heated) in comparison to the USDA Starch Research method. The bars with asterisks (*) gave starch-I_3 values that were greater than 100%.

Effect of microwaving power, heating time, and number of flasks

After finding 0.08M HCl as the most useful acid to solubilize starch, the microwave power, heating time, and number of flasks were analyzed. To control the heating power administered to each starch sample in the microwave oven and to fully obtain complete solubilization, we varied the microwaving power (20-100%), heating time (20-60s), and number of flasks that were simultaneously heated from 1 to 4 during the microwaving step.

Decreasing the microwave power while heating for only 1 min did not improve starch solubilization with increasing HCl concentration (Fig. 4); as more HCl was used, accuracies decreased significantly (Fig.4). Concentrations <0.1M HCl and >80% power, gave the highest accuracy (Fig. 4). When 100% microwaving power and 0.1M HCl was used, starch-I_3 values greater than 100% were obtained which indicated that ~0.1M was the best to solubilize starch among the concentrations studied. After a few more studies, it was also confirmed that 0.08 M HCl was the optimal concentration to dissolve starch.

In an effort to ensure complete solubilization and control the heating power of the sample in the presence of 0.08M HCl, since HCl can act as a solubilization catalyst and continue fragmenting solubilized starch if it is not stopped, the number of flasks microwaved and the time that the flasks were microwaved was varied from 20 sec to 1 min. It was found that starch solubilization for only one flask either did not occur (i.e., <40 sec) or resulted in the formation of unwanted degradation products (i.e., 50 sec and 1 min) and, therefore, microwaving time needed to be reduced (data not shown). Adding more flasks reduced the amount of water evaporated per sample and minimized the formation of unwanted sugar degradation, but critically decreased starch solubilization when less power or heating time was used (Figs. 4 and 5). Thus, it was important to balance the boiling time in the microwave with microwave power and the number of flasks heated. Cole et al. (2014) found that at least 30 sec was needed for one flask of a 10 mL sucrose solution to boil in the microwave and thus, it was important that the samples in this industrial method were boiled but with control and the same efficiency. It was also important to ensure that the flasks were positioned in the microwave evenly following the positions illustrated in Fig. 5A. Microwave power is cumulative and should evenly distributes its energy among all items being heated. By increasing and decreasing the number of flasks in the microwave that are being heated, the % power can be better controlled per sample. For example, if one flask is microwaved for 1 min it will receive 100% of the power; whereas, if four flasks are microwaved for 1 min at 100% power, each flask will receive 25 % power. This approach was far more effective than reducing the microwave power for one sample, because at 100% power, more heating power was distributed to all of the samples whereas, reducing the power caused less heat to be distributed to each sample. Fig. 5B shows that microwaving one flask at 100%, whether on the rim or in the center of the microwave, caused ~75% of the sample to be hydrolyzed to glucose or short starch chains that could not be detected with iodine (starch chains with >6 glucose molecules are needed to bind effectively to iodine, Baldwin et al., 1944). However, more flasks helped distribute the heat better in the microwave and allowed for better controlled boiling, yielding ~98.8% accuracy (Fig. 5B). The optimal heating time for 0.08M to solubilize starch by the number of flasks studied is listed in Table 1. The minimum equivalent times to heat four flasks in microwaves with varying power capacity are listed in Table 2.

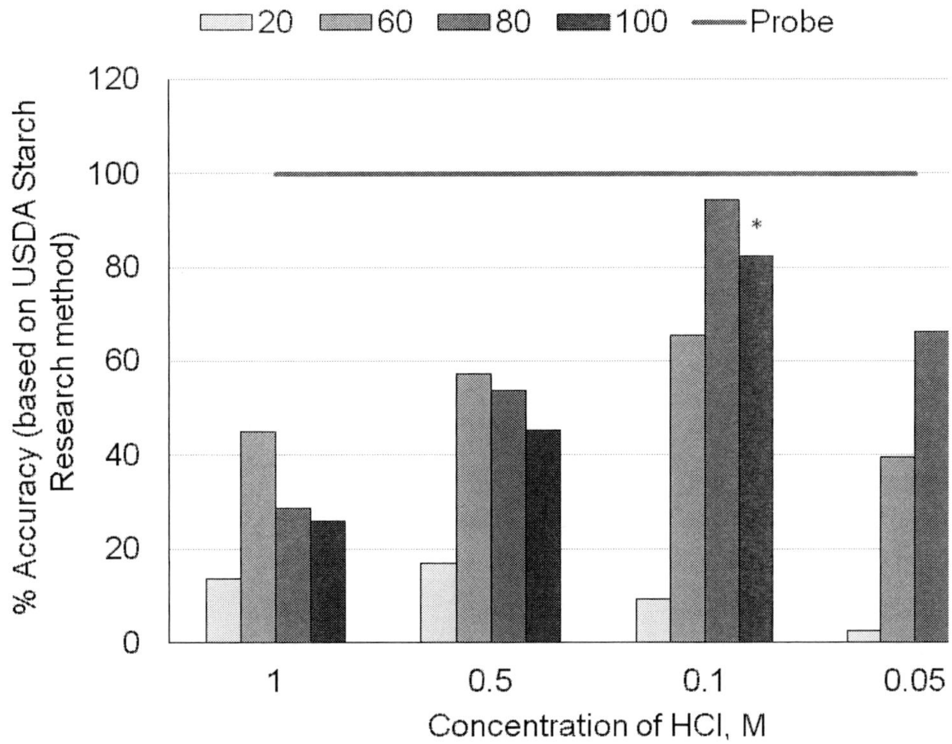

Fig. 4. Percent accuracy of the acid-assisted microwave solubilization approach on 2,000 ppm insoluble starch in comparison to the USDA Starch Research method with changes in microwave power (20-100%). Four flasks were simultaneously heated. The bars with asterisks (*) gave starch-I_3 values that were greater than 100%.

With four flasks evenly distributed in the microwave, we were able to microwave the samples long enough so that they would boil and solubilize the insoluble starch. Considering that HCl was also included during microwaving, the proper balance between heating time and power was needed to minimize further sugar degradation or evaporation of too much water. Although ~98.8% accuracy was very impressive results obtained with 0.08M HCl, one disadvantage to the use of HCl, or the other acids as solubilizing agents, is that the hydrolytic activity is uncontrollable and can further break down the starch with longer exposure time and with large irreproducibility. Therefore, the presence of an equally strong base was needed to neutralize the acid hydrolysis of the starch, stop the over fragmentation of the solubilized chains, and maintain the high ~98.8% accuracy.

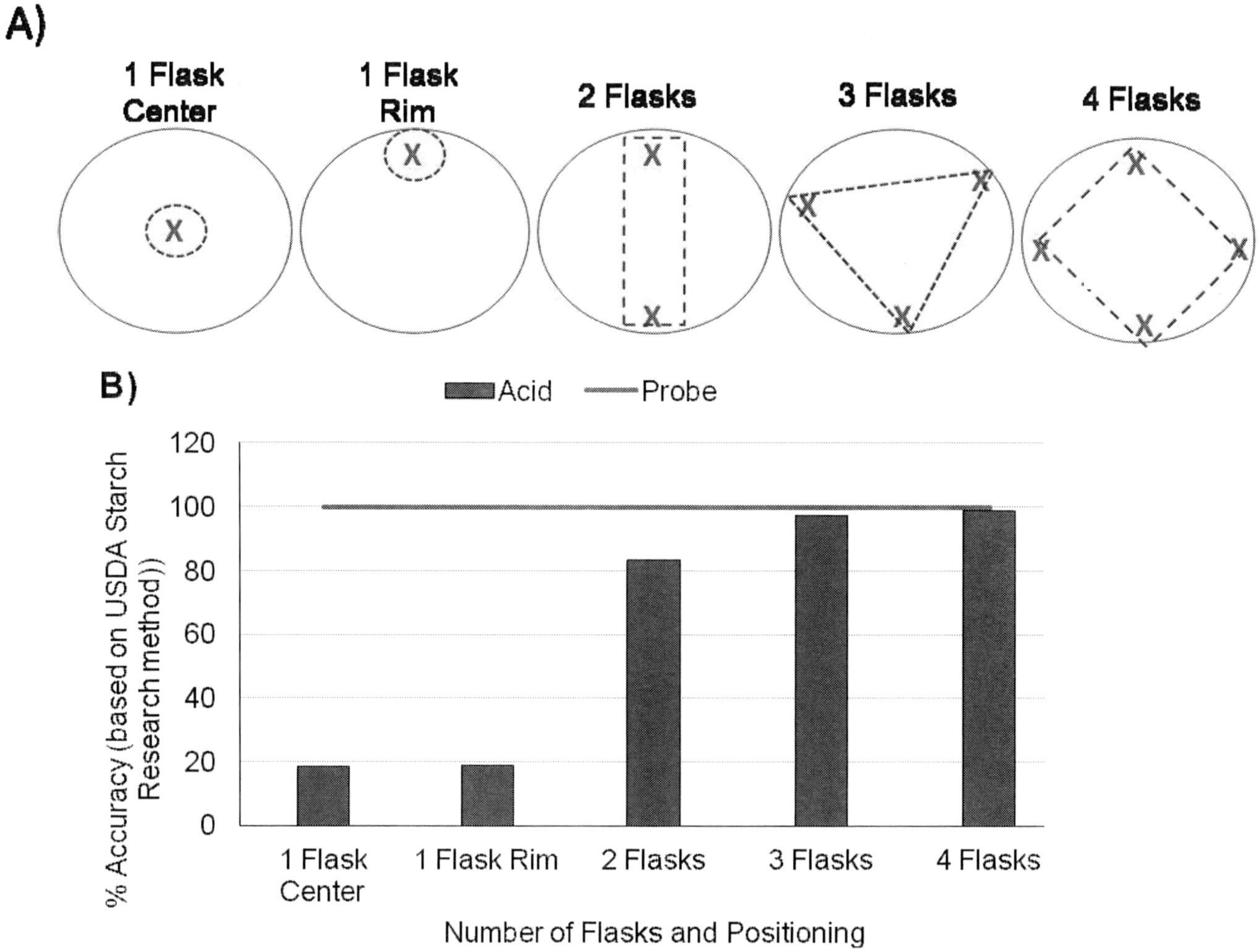

Fig. 5. A) Arrangement of multiple flasks in the microwave to control heat distributed to the HCl acid-assisted microwave solubilization of 2,000 ppm insoluble starch at 100% power and 1 min. B) Accuracy (%) of 0.08M HCl solubilization of insoluble corn starch by the number of flasks that were simultaneously heated and in comparison to the USDA Starch Research method (probe).

Table 1. Minimum heating time needed to solubilize 2,000 ppm insoluble corn starch with 0.08M HCl at 100% power.

Number of Flasks	*Microwave Heating Time (sec) at 100% power*
1 (center)	N.d.*
1 (rim)	20 to 30
2	30 to 40
3	50 to 60
4	60

*N.d.= could not be determined.

Table 2. Equivalent heating time for microwaves with different power capacity.

Microwave Power (W)	*Microwave Heating Time (sec) at 100% power*
1500	44
1300	50
1200	55
1100*	60
1000	66
900	73
850	77
800	82
750	88

*Microwave power used in the USDA Starch Research and Factory/Refinery methods.

Acid-alkali neutralization-assisted microwave solubilization

Initially we investigated the use of urea, ammonia, and sodium hydroxide (NaOH) to control the acid reactions, however, preliminary results and the technical handling indicated that NaOH was much better. Also, fundamental chemistry supports that combined use of NaOH and HCl will instantly neutralize the other's activity and produce the non-reactive compound sodium chloride. We found that this step was essential to ensuring reproducible results and safety.

Sodium hydroxide is a very well-known gelatinizing agent for starch, which depends on the botanical source, and can completely disrupt and dissolve granules. Maher (1983) reported that the alkali:starch ratio is not equal for all starch types and that even different varieties of corn starch (high amylose vs dent corn) also responded to NaOH differently. However, it was determined that NaOH did cause the starch to swell and lose it crystallinity, which is one indicator of starch solubilization. In this study, we expanded on this approach and microwaved 2,000 ppm insoluble corn starch with equal concentration of NaOH and HCl, i.e., 0.01-0.25M, with the intent to control reproducibility and maintain high accuracy. As expected, microwaving the corn starch suspension with both NaOH and HCl initially added together, did not yield any solubilized starch because the NaOH and HCl neutralized each other immediately and formed NaCl. Separately microwaving the NaOH first (100%, 1 min) before adding the HCl, however, did considerably improve starch solubilization and reproducibility (Fig. 6). It should be noted that microwaving NaOH without HCl caused color to form in the sample but had improved filterability and accuracy (data not shown). Adding the HCl stopped color formation and even reduced the intensity of any color that did form when 0.08M NaOH and insoluble starch were microwaved together.

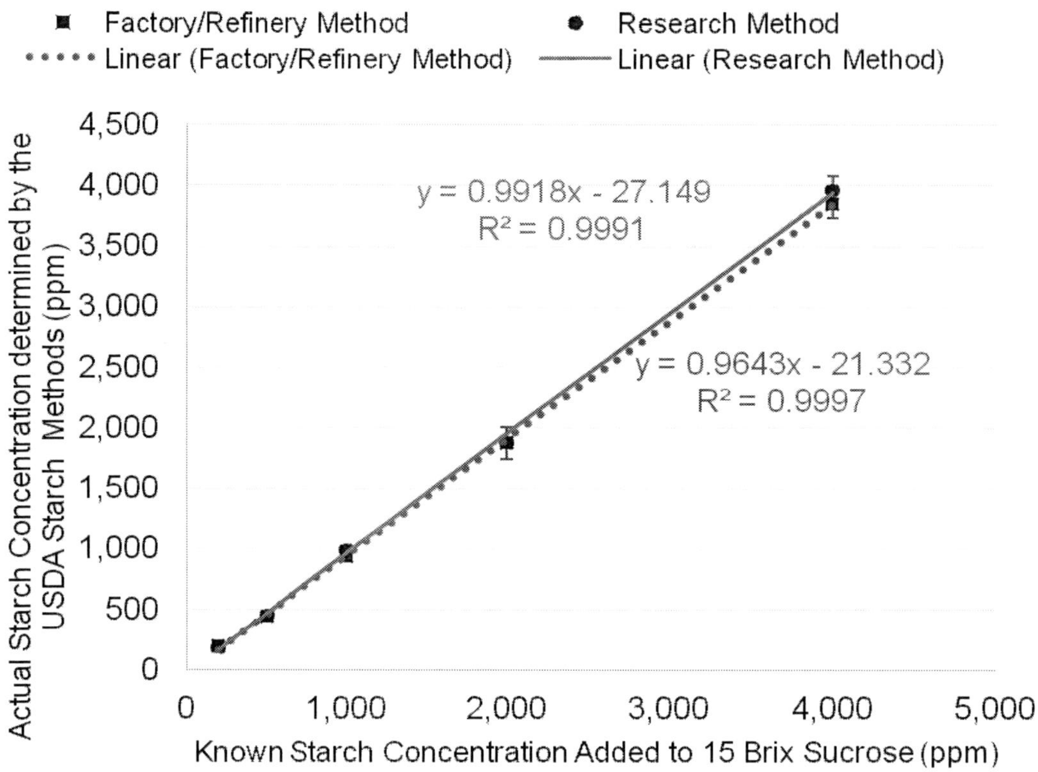

Fig. 6. Accuracy and precision of the USDA starch methods on sucrose samples with known concentrations of corn starch added.

Microwaving the samples with equal concentrations of NaOH and HCl yielded solutions that had high filterability but precision was not acceptable. During some experiments, the samples would be clear and colorless, but other times the samples would be cloudy. By keeping the concentrated HCl acid cold in an ice bath, only neutralizing the reaction after the microwaved sample had cooled to room temperature, and adding a second round of microwaving after neutralizing the NaOH with HCl, the reproducibility of this part of the method was improved. Thus, the final conditions of the method are: 0.08M NaOH, 0.08M HCl, two rounds of microwaving, four flasks, 100% power, and 1 min. The accuracy, precision, and performance of the final USDA Starch Factory/Refinery method on sucrose samples with known starch concentration is shown in Fig. 6. Although high linearity and a very close equatability between the two USDA starch methods was obtained, the accuracy deviated at >4,000 ppm insoluble corn starch concentration. This result could challenge its use on some high starch crusher and mixed juices, however, the factory juice sample can be diluted to ensure the results are within the solubilization range of the method and calibration curve.

Overall Accuracy, Precision, and Validation of the USDA Starch Factory/Refinery Method on Simulated Sugars, Factory Raw Sugars, and other Processing By-products.

The USDA Starch Factory/Refinery Method was validated according to ICUMSA guidelines. The accuracy of the method was determined by using the previously developed USDA Starch Research Method (microwaving + probe sonication) as a standard reference (Cole et al., 2014).

Starch solubilization was validated on both simulated sugars with known amounts of corn starch and factory products with unknown amounts of sugar cane starch. Several method validation parameters were determined such as linearity, accuracy, and precision (repeatability). Since no certified reference material was available for the analytes and matrices of the factory raw sugar and processing by-products, the accuracy of the USDA Starch Factory/Refinery method was determined using USDA Starch Research Method as an alternative to trueness. Precision (repeatability measured by the coefficient of variance, % CV) results were obtained by the same operator on eight separate days in triplicate, i.e., twenty four replications. Further training and validation at sugar factories were initially studied however, it was found that each factory did not have the equipment listed in the method which halted the validation study. A second training and validation study will be pursued during the 2017 processing season.

The accuracies of the simulated sugars (200-4,000 ppm corn starch) to the known amount of insoluble starch added to sucrose were between 88.1-97.7% and 91.8-99.8% when the USDA Starch Factory/Refinery and Research methods were used, respectively (Fig. 6). Using the USDA Starch Research method as a reference method to analyze the starch in the simulated sugars, the average method accuracy for the simulated sugars ranged 94.6-100.0% (Fig. 6). Results confirmed that this method was very repeatable on simulated sugars and had precision values of <6% CV (Fig. 6).

Fig. 7 illustrates the accuracy and precision for the USDA Starch Factory/Refinery and Research Starch methods on factory raw sugars and processing by-products. Accuracies of the factory products were 98.6, 88.7, 86.1, 94.1, 96.3, and 95.3 % for massecuite, A molasses, clarified juice, crusher juice, FES, and 8 raw sugars, respectively (Table 2). The low accuracy (88.7%) obtained for the A molasses was attributed to its high color and low total starch concentration that interfered with starch-I_3^- absorbance. The C_b for the A molasses was 64% of the total absorbance at 600 nm. The accuracy obtained for clarified juice (86.1%) was because this industrial starch method yielded 1.5-2.75x higher starch-I_3^- absorbances which indicated that the already solubilized starch in this juice may have become fragmented forming more starch-I_3^- complexes that can be detected at 600 nm. Precision for the methods were also very good (Table 2). The high overall imprecision values obtained with the raw sugars and simulated sugars listed in Table 2 were due to the high number and diversity of samples studied since these results are an average of five and eight samples, respectively. Expanded precision values (data not shown) for most of the raw sugars ranged between 2.6-8.4 %CV, with the exception of one with a %CV of 26.8. This high %CV was attributed to the low starch concentration, i.e., 1.7 ppm, and standard deviation; the low starch content in the raw sugars and difficult repetitive sampling is a contributing factor to lower precision observed in raw sugars since the mean starch concentration and standard deviation can both be very small numbers. When the mean concentration value is a small number, than the CV values can become quite large simply because a small (but acceptable) standard deviation is being compared with another small number (Coleman & Vanatta, 2012). Also, since this low starch value is below the USDA Starch Factory/Refinery method's limits of detection (LOD) and quantification (LOQ), i.e., 3.4 and 10.2 ppm, respectively, it is not reliable contributing to the high imprecision. The average z scores for the USDA Starch Factory/Refinery method are 95.4%.

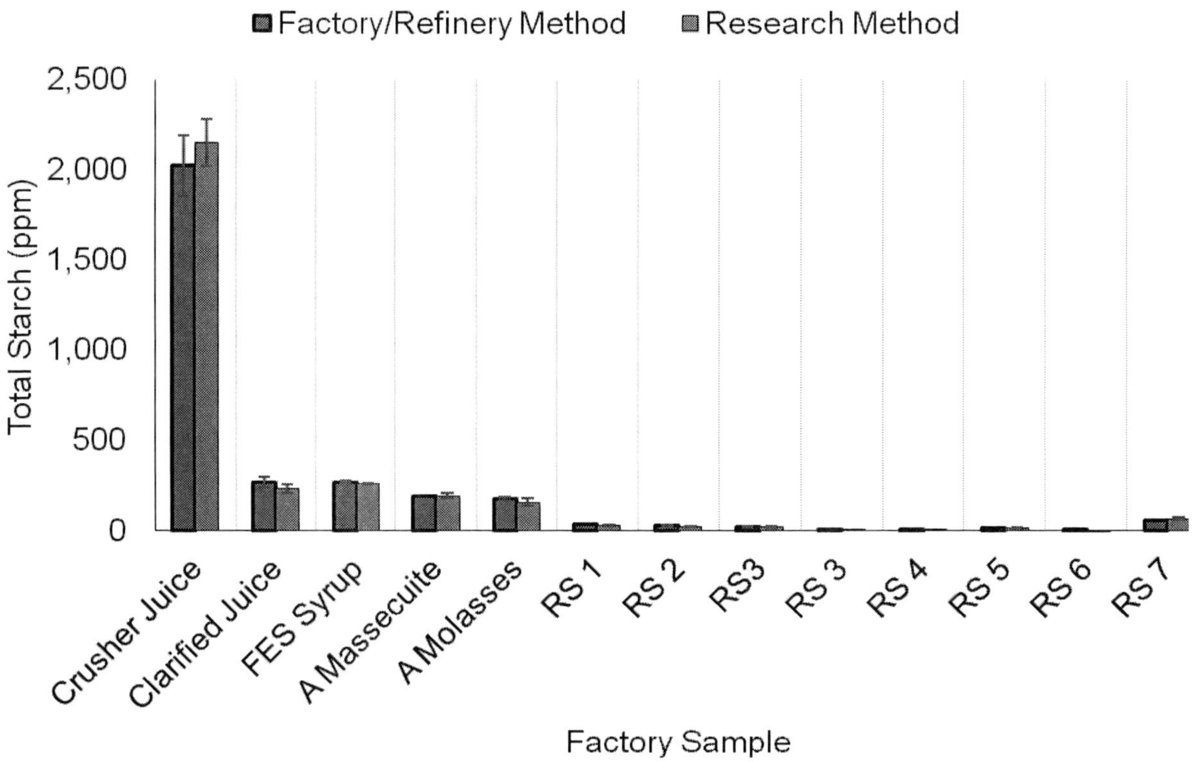

Fig. 7. Accuracy and precision of the USDA Starch Factory/Refinery methods on factory raw sugars and processing by-products. RS = raw sugar.

Having such a low LOQ value will be extremely helpful in quantifying starch in refined sugars, because of the negligible background color of the solution. Since factory products from crusher juice to raw sugar are colored, the LOD and LOQ will be higher to compensate for the variable color contributed by these samples. This is because it is important that the absorbance starch-I_3^- color be higher than the background of the natural factory raw sugar and processing by-product color. Considering the different ICUMSA color at pH 7 of the factory raw sugars studied, the LOD and LOQ will increase to 27 and 34 ppm, respectively, for a high colored raw sugar (ICUMSA color = 5,400 IU) and 21 and 28 ppm, respectively, for a low colored raw sugar (ICUMSA color = 822 IU) although the quantity and type of colorant in the factory sample will vary among samples. Most importantly, this method will detect high starch raw sugars that may violate the existing 250 ppm starch limits (ICUMSA GS1-17 starch method) set by U.S. carbonatation refineries, with minimal interference of raw sugar background color.

Table. 2. Total starch concentration, repeatability, and percent recovery of factory products and by-products.

Sample	**Factory/Refinery Method**	**Research Method**	**Comparison between USDA Starch Factory/Refinery & Research Methods**

	Total Starch (ppm)	% CV	Total Starch (ppm)	% CV	(% Accuracy)
Simulated Sugars (5 sugars)	1,463.7	91.0	1,500.2	91.3	97.6
Crusher Juice	2,027.0	8.3	2,154.7	5.9	94.1
Clarified Juice	268.2	12.2	235.5	10.9	86.1
FES Syrup	270.0	2.4	260.5	1.7	86.1
Massecuite	189.0	0.9	191.7	7.8	98.6
A Molasses	176.3	6.3	158.4	11.7	88.7
Raw Sugar (8 sugars)	20.8	81.9	19.4	109.9	92.5

Comparison of the USDA Starch Methods

High linearity (R^2=0.999) between the USDA Starch Factory/Refinery and Research methods was obtained using all eighteen samples (Fig. 8). Regression results indicated that the USDA Starch Research Method may always give greater starch concentrations than the USDA Starch Factory/Refinery Method, but statistical analyses confirmed that this difference was not statistically significant. Statistics also revealed that precision among the factory sample types were statistically different ($P<0.05$) suggesting that the method's precision varies according to sample type analyzed, but was still remarkably low. However, the precision between the USDA Starch Factory/Refinery and Research methods indicated that both methods were equally precise at the 95% confidence interval for all sample types studied. Overall, the USDA Starch Factory/Refinery and Research methods were very similar with an overall mean difference between ~6.5%. The final comparison of the two USDA starch methods is summarized in Fig. 9.

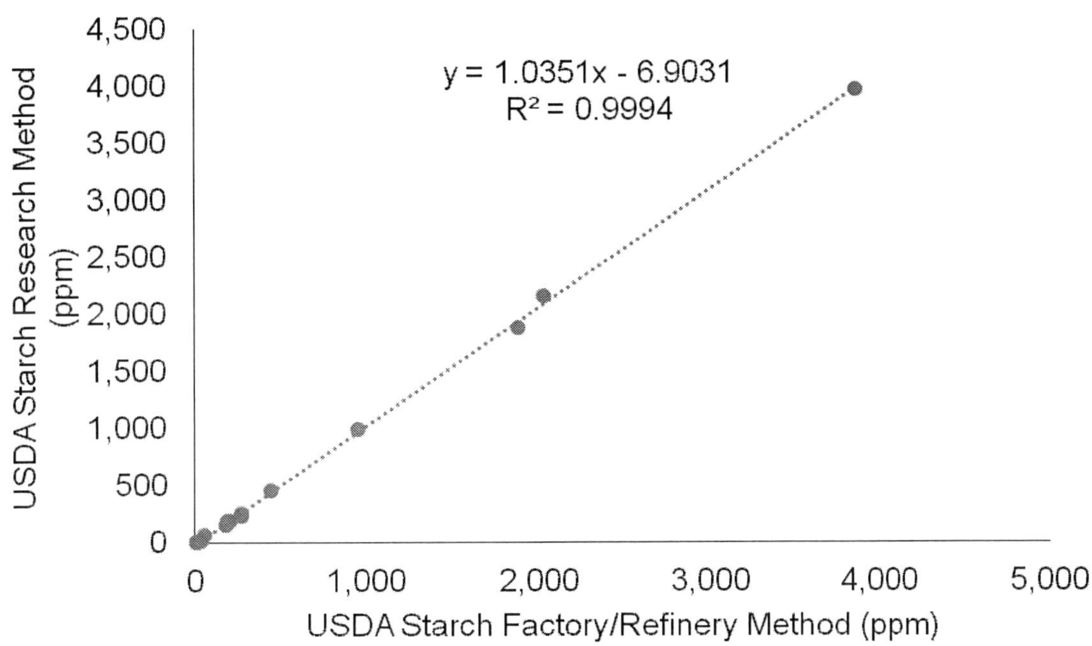

Fig. 8. Linearity between the USDA Factory and Research Starch methods.

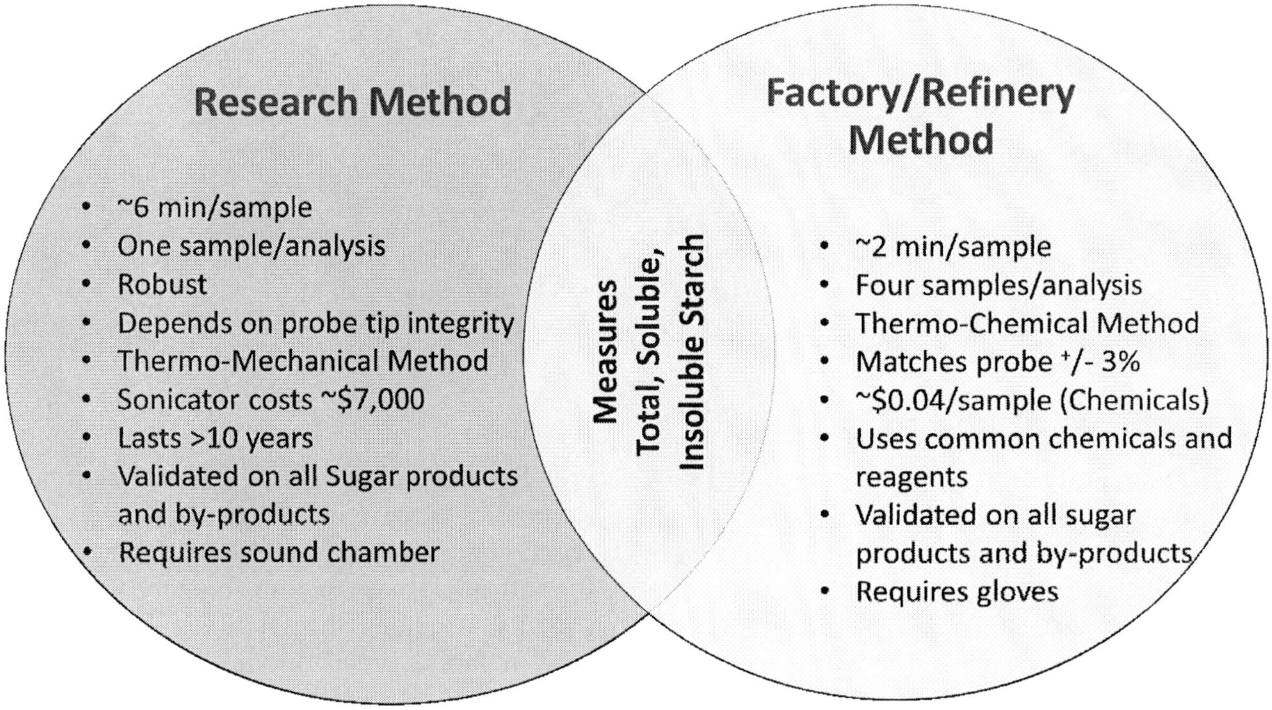

Fig. 9. Comparison of the USDA Starch Research and Factory/Refinery methods.

CONCLUSIONS

- A rapid and precise industrial starch method based on microwave-assisted neutralization chemistry has been developed to measure the total amount of starch and percent contribution from soluble and insoluble starch in factory raw sugars and processing by-products.
- The method is sensitive, highly selective to starch and, compared to current sugar industry starch methods, it is more accurate, flexible and can rapidly analyze all samples with broad starch concentration ranges (i.e., juices, syrups, molasses, massecuites and raw sugars) and wide Brix values.
- The USDA Starch Factory/Refinery method uses two rounds of microwaving with strong acid and base to replace the expensive ultrasonication probe step used in the USDA Starch Research method, with similar accuracy and precision values.
- This method also uses known starch-iodine complex absorbance to quantify starch on a spectrophotometer at 600 nm and with a corn starch standard curve.
- Overall accuracies ranged between 86-98% and had precision values (% CV) between 2-12%, depending on the sample type.
- Future research is now needed to further validate and deploy this method at international refineries.

ACKNOWLEDGEMENTS

The authors gratefully thank Ms. Deriesha Gaines for her technical support and helping validate the USDA Starch Factory/Refinery method. This research was supported in part by an appointment to the Agricultural Research Service (ARS) Research Participation Program administered by the Oak Ridge Institute for Science and Education (ORISE) through an interagency agreement between the U.S. Department of Energy (DOE) and the U.S. Department of Agriculture (USDA). ORISE is managed by ORAU under DOE contract number DE-SC0014664. All opinions expressed in this paper are the author's and do not necessarily reflect the policies and views of USDA, ARS, DOE, or ORAU/ORISE. Mention of trade names, commercial products, or methods in this article is solely for the purpose of providing specific information and does not imply recommendation or endorsement by the U.S. Department of Agriculture (USDA). USDA is an equal opportunity provider and employer.

REFERENCES

Baldwin, R. R., Bear, R. S., & Rundle, R. E. (1944). The relation of starch-iodine absorption spectra to the structure of starch and starch components. *J. Amer. Chem. Soc.*, *66*(1), 111–115.

Cole, M., Eggleston, G., & Triplett, A. (2017a). Analytical evaluation of current starch methods used in the international sugar industry: Part I. Food Chemistry, 228, 226–235.

Cole, M., Eggleston, G., & Triplett, A. (2017b). Comparison of international method for the determination of total starch in raw Sugars: Part II. *Food Chemistry, in review.*

Cole, M., Eggleston, G., Gilbert, A., & Chung, Y. J. (2014). Development of a research method to measure insoluble and soluble starch in factory and refinery products. *Proc. Sugar Ind. Technol.*, *73*, 212–229.

Cole, M., Eggleston, G., Gilbert, A., Rose, I., Andrzejewski, B., St. Cyr, E., & Stewart, D. (2013). The presence and implication of soluble, swollen, and insoluble starch at the sugarcane factory and refinery. *Internat. Sugar J.*, *115*(1380), 844–851.

Coleman, D., & Vanatta, L. (2012). Statistics in Analytical Chemistry: Part 48—Relative Standard Deviations (RSDs). *Amer. Lab.*, 30–33.

Duncan, W. (1898). The Production of a Brilliant Beer. *J. Fed. Inst. Brew.*, *4*(4), 477–508.

Eggleston, G. (2014). *Cane Sugar Processing* (Referee's Report No. General Subject 7). Sao Paulo, Brazil: ICUMSA., 94–112

Eggleston, G. (2017). *Cane Sugar Processing* (Referee's Report No. General Subject 7). Warsaw Poland: ICUMSA, 1–28

Godshall, M. A. (1997). Color Analysis – Symposium on Sugar Color. *Proc. Sugar Ind. Technol.,* 56, 211–231.

Godshall, M. A. (2005). Understanding and Controlling Color Developments in Mills and Refineries. *Proc. Sugar Ind. Technol.*, 64, 167–174.

Godshall, M. A., Clarke, M. A., & Dooley, C. D. (1991). Starch: Process problems and analytical developments. *Proc.1990 Sugar Proc. Res. Conf.* 244–264.

Godshall, M. A., Triche, R., & Moore, S. J. (2004). A rapid starch test for use in cane mills. *Proc. 2003 Sugar Proc. Res. Conf.,* 428–448.

Harrison, A. (2012). *Cane Sugar Processing* (Referee's Report No. General Subject 7). Bartens, Germany: ICUMSA., 25–35.

Jambraka, A. R., Hercega, Z., Šubaric, D., Babić, J., Brnčić, M., Brnčić, S. R., … Gelo, J. (2010). Ultrasound effect on physical properties of corn starch. *Carbohydr. Polym.*, 79, 91–100.

Lintner, C. J. (1886). Studien uber diastase. *J. Prakt. Chem.*, *34*, 378–386.

Maher, G. G. (1983). Alkali gelatinization of starches. *Starch - Stärke*, 35(7), 226–234.

Nägeli, W. (1874). Beiträge zur näheren Kenntniss der Stärkegruppe. *Just. Lieb. Annal. Chem.*, 173(2), 218–227.

Thielecke, K. (2002). *Cane Sugar Processing* (Referee's Report No. General Subject 2). Berlin, Germany: ICUMSA., 169–189.

Thielecke, K. (2004). *Cane Sugar Processing* (Referee's Report No. General Subject 2). Berlin, Germany: ICUMSA., 141–189.

Thompson, M. (1988). Variation of precision with concentration in an analytical system. *Analyst (London)*, *113*, 1579–1587.

Thompson, M., Ellison, S. L. R., & Wood, R. (2006). The international harmonized protocol for the proficiency testing of analytical chemistry laboratories. *Pure Appl. Chem.,* 78, 145–196.

Wang, S., & Copeland, L. (2015). Effect of Acid Hydrolysis on Starch Structure and Functionality: A Review. *Crit. Rev. Food Sci. Nut.*, 55(8), 1081–1097.

Whistler, R. L. (2008). In J. R. Daniel, R. L. Whistler, H. Röper, & B. Elvers (Eds.), *Ullmann's Encyclopedia of Industrial Chemistry* (7th ed., pp. 11–20). Weinheim: Wiley-VCH.

Xue, C., Sakai, N., & Fukuoka, M. (8). Use of microwave heating to control the degree of starch gelatinization in noodles. *J. Food Eng.*, 87(3), 357–362.

S.I.T. Paper #1175

PRE-CARBONATATION WITH PURE CO_2 – NEW DEVELOPMENTS

José Manuel Chorão

RAR – Refinarias de Açúcar Reunidas, Portugal (jmchorao@gmail.com)

Abstract

The addition of pure CO_2 to limed liquor in a pre-carbonatation step proved to be a successful means to overcome the difficulties raised by the low CO_2 content of flue gases from boilers burning natural gas. The pre-carbonatation system has been in continuous operation for three years, as described in a previous S.I.T. meeting (New York, 2016).

In the present work, we describe the modifications made to the system in order to make it simpler, easier to operate and less energy consuming.

The new pre-carbonatation system has been in operation since November 2016 with excellent results.

Introduction

RAR refinery is a carbonatation cane sugar refinery, using a typical equipment setup with calandria type saturators. Melt liquor and lime are mixed and sent to three saturators in parallel, working at 80 °C, a pH of 9.4 to 9.7, and 50 minutes residence time. The final pH of 8.1 to 8.3 is achieved in a fourth saturator, with 18 minutes residence time. Carbon dioxide for the carbonatation reaction is added to limed liquor in two steps: *i*) in a pre-carbonatation step, where pure CO_2 is mixed with limed liquor by means of a static mixer, before the A saturators; *ii*) injection of boilers' flue gas (after scrubbing and neutralization) in the saturators through saw-tooth type distributors. Pure CO_2 injected in the pre-carbonatation step represents 30% to 40% of stoichiometric. Details are given by Chorão *et al.* [1]. A simplified flowsheet of the current carbonatation setup is presented in figure 1.

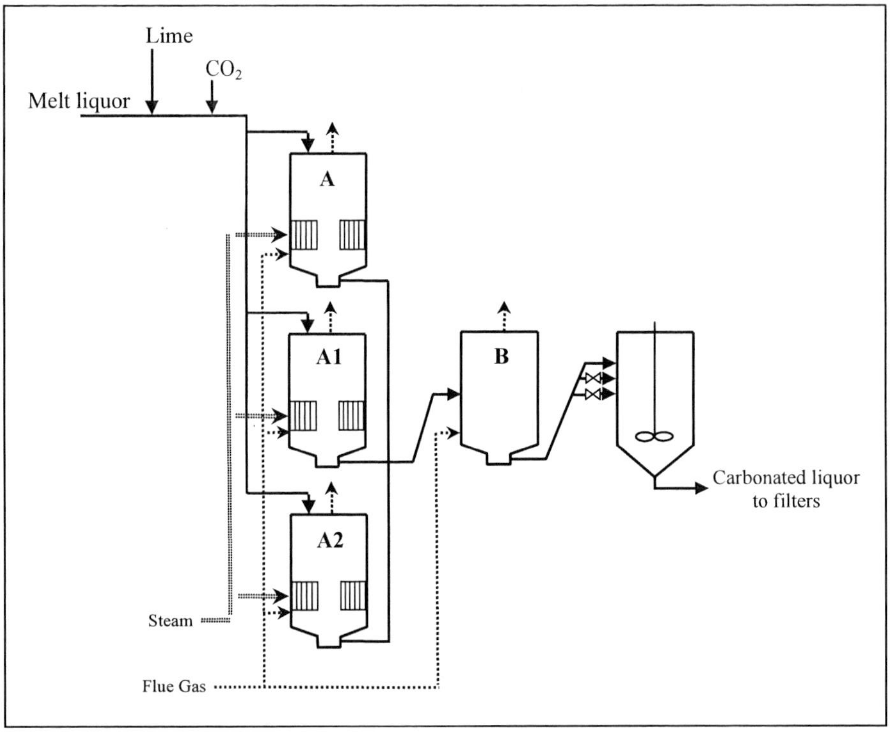

Figure 1. Flowsheet of Carbonatation.

Modifications to the installation

The mixing of lime and liquor before entering the A saturators was accomplished in a tank equipped with an agitator. Because of the detrimental effect of high pH on the liquor, the tank was by-passed and the lime is injected in the liquor pipe. The recommended setup is to have an in-line mixer [2], but given the length of the pipe from the injection point of lime to the injection point of carbon dioxide (32 meters length, with 3 curves) no mixing device was installed. No changes in the operation of saturators and filters were noticed after this modification, indicating that the mixing of lime and liquor is efficient.

In the previous pre-carbonatation step, pure CO_2 was mixed with limed liquor by means of a static mixer and a recirculation pump (figure 2). The distance from the static mixer to the saturators was 24 meters, and the last 15 meters where suffering scaling and had to be periodically cleaned. This problem was an indication that there was enough time for the absortion of CO_2 in the limed liquid, the reaction of CO_2 with calcium hydroxide, and the formation of calcium carbonate. This suggested that the distance from the injection point of pure CO_2 to the saturators could be shortened, and that an ejector in the pipe could replace the work of the static mixer and circulation pump. The new CO_2 injection setup is diagramed in figure 3.

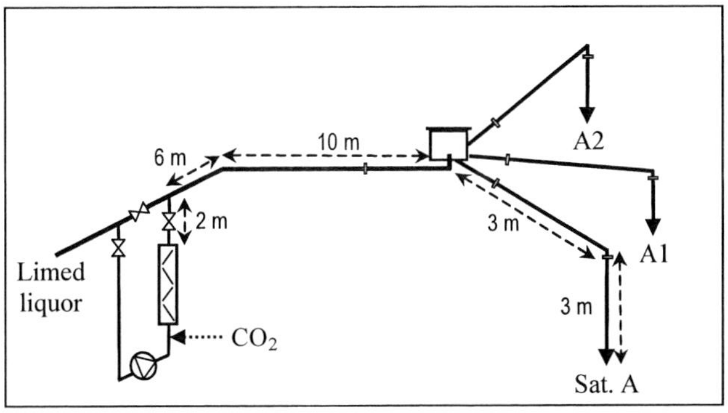

Figure 2. Previous CO_2 injection setup.

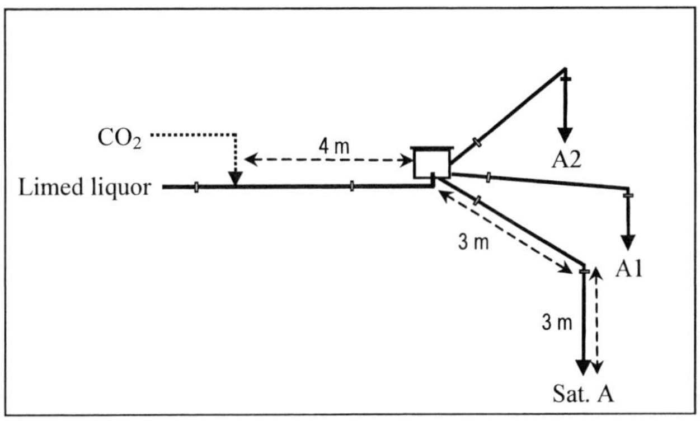

Figure 3. New pipe arrangement and location of CO_2 ejector.

The injection point in the new pipe arrangement is the same as in the experimental setup, in 2014 [1]. It was also noticed in the experimental setup that incrustation was less severe than in the final installation, in 2015 (figure 2).

New pre-carbonatation setup

The length of pipe from the mixing point of CO_2 to the saturators must be regarded as a reactor. In the traditional carbonatation process the mass transfer of carbon dioxide from gas bubbles to the liquid phase is the limiting step [2]. When using pure CO_2 this limitation become negligible, and the molecules of carbon dioxide dissolved in the liquid phase undergo a fast reaction with the hydroxyl ions present in high concentration ($pH_{20°C}\approx 11$) to form calcium carbonate.

After one month of operation with the new setup the pipes and ejector where inspected. Figure 4 shows the pipe where carbon dioxide is injected. Some incrustation can be seen on the upper side of this pipe. The flow divider located 4 meters ahead of the injection point shows significant incrustation (figure 5). The time taken for the liquor to travel from the injection point to this equipment is 6 seconds.

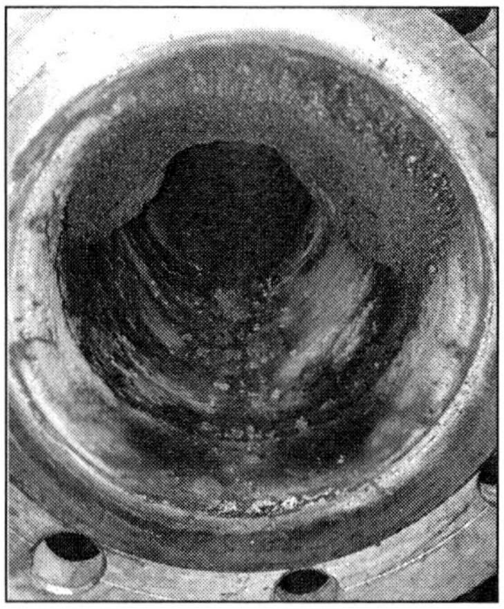

Figure 4. Incrustation in the pipe of CO_2 injection.

Figure 5. Incrustation in the flow divider to saturators.

The formation of incrustations in the pipes allowed us to conclude that the mixing of CO_2 with limed liquor was fast, and the formation of calcium carbonate followed within a few seconds, indicating that there is a good phase mixing. This has also been observed elsewhere [3] with static mixers and recirculation pipes, and working at lower pH.

The piping was modified to allow an easy disassembly for periodic cleaning. A set of spare pipes was manufactured, to quickly replace the pipes to cope with the occurrence of clogging.

The ejector was installed in the pipe as shown in figure 6. It was in perfect condition, showing no wearing or incrustation. The ejector was reused from a decommissioned gas motor, and is made of stainless steel.

Figure 6. Ejector of pure carbon dioxide.

Control

The automatic control of lime and liquor mixing, and CO_2 injection is diagrammed in figure 7. The software runs on a PLC, using standard PID control. The setup was made simpler, compared to the previous one, because no circulation pump is used. Details of the previous installation are given in reference 1.

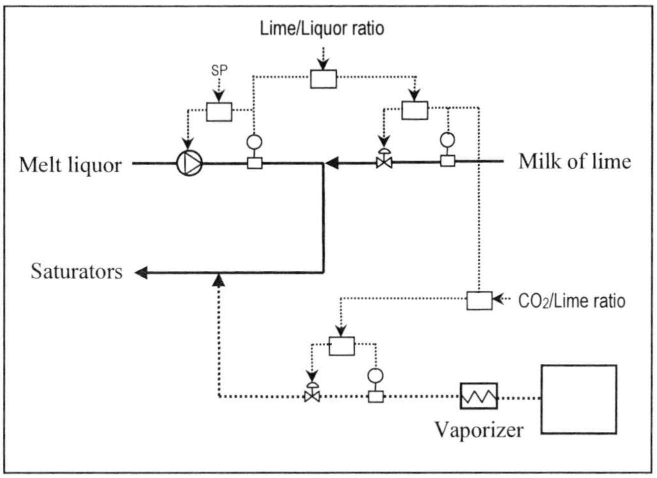

Figure 7. Flow control of liquor, lime, and CO_2.

Results

The modifications made to the previous installation brought the following advantages:

i) It has allowed the reduction of the pipe length subject to incrustation, and consequent cleaning procedures. A set of spare pipes is able to keep to the minimum any interruption of production due to clogging.

ii) The installation has less equipment, making it easier to operate and control. Maintenance costs are also lower since no circulation pump is used.

iii) A reduction of circa 25% in electricity, bringing the operating costs down to approximately 0.05 € per ton of raw sugar.

iv) Without circulation pumps and piping, the equipment costs of the new installation are 46% lower.

The new simplified pre-carbonatation setup in under operation since November 2016.

Conclusions

The modifications introduced in the pre-carbonatation setup proved to be a better solution to the injection of carbon dioxide in limed liquor. The new solution has a lower investment cost, and a lower operating cost. The control of the operation is simpler, and easy to accomplish with standard control equipment. Filterability of carbonated liquor has not been affected, as well as the performance of the whole carbonatation station.

References

1. Chorão, J., Alves, C., Araújo, A. (2016), "Pre-carbonatation with pure CO_2 in a cane sugar refinery", Paper #1151, 75th Annual Meeting of S.I.T., New York
2. Rein, P. (2017), "Cane Sugar Engineering", 2nd Ed., Verlag Dr Albert Bartens KG, Berlin
3. Varelius, E. P. (2014), "Optimizing the reaction efficiency of carbon dioxide in a sugar refinery carbonatation process", Department of Chemical Engineering, Lund University, Sweden (www.chemeng.lth.se/exjobb/E712.pdf)

TOTAL DISSOLVED SOLIDS MEASUREMENT FOR SUGAR LEAKAGE DETECTION IN CONDENSATE

K-PATENTS®
PROCESS INSTRUMENTS

Seppo Kilpinen, Eero Mielonen
K-Patents Oy, Finland.

S.I.T. Paper #1176

INTRODUCTION

Sugar contamination may cause serious damage to the steam and condensate systems in the sugar refineries and mills. Several methods have been used to monitor sugar in condensate, including conductivity, pH, flame photometer and Total Organic Carbon (TOC) tests (Table 1). A new system for detecting sugar in boiler feedwater has been developed based on modern, advanced refractive index technology. The system consists of a process refractometer installed in a temperature controlled condensate line to a drain/balance tank. The system provides instant, accurate and reliable information of sugar content in the condensates by means of Total Dissolved Solids (TDS) measurement. Moreover, the system operates in-line allowing sufficient time to react and eliminate any harmful feedwater to the boiler system.

Table 1. Comparison of methods for monitoring sugar content in condensatelines.

Method	Features					
	Sugar content	In-line	Measured in boiler feed	True alarms	Quick response	Low maintenance
TDS	✓	✓	✓	✓	✓	✓
pH	✓	✓		✓*		
Flame photometer		✓				
Conductivity		✓	✓		✓	
TOC	✓	✓	✓			
Auto analyzers - Colorimeter	✓	✓	✓			
Laboratory	✓		✓			✓

* Based on measurement *after* the boiler.

METHODS

The sample conditioning system has been installed in a sugar factory in Russia for continuous monitoring of the condensates.

The location was the condensates feed to the balance tank before the boiler. The refractometer's output was set to control the shutter valve to the drain, to discard the contaminated condensates and ensure only clean water enters the boiler (Figure 1). The system can also be installed in all the condensate lines, e.g. from the evaporators, vacuum pans, and settling tanks, for the earliest detection.

The sampling system comprises (Figure 2):

- a Sample cooler
- a Process refractometer (supplied by K-Patents, model PR-23-A)
- an Indicating transmitter (supplied by K-Patents, model DTR)

The solution involves the development of a special calibration method for accurate and reproducible determination of low sugar content (0.02 - 0.05 Brix) for detecting even small traces ofleakages.

The response of the system to sugar shots was evaluated by injecting periodically a sugar solution directly into the line before the refractometer.

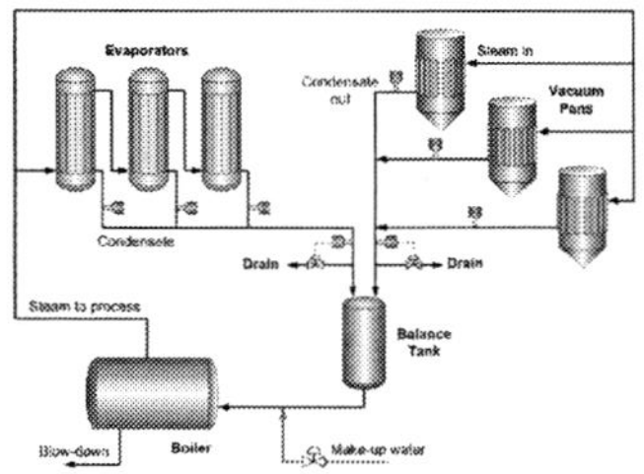

Figure 1. Recommended installation points for the refractometer through a sampling system. For early detection, the measurement points should be in the condensates lines from the evaporators, vacuum pans and settling tanks.

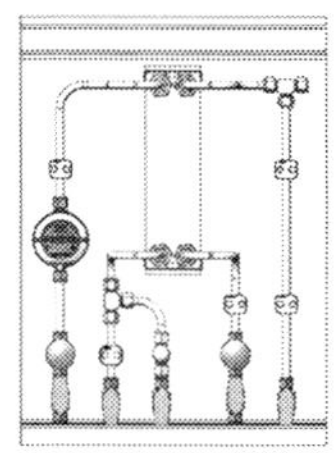

Figure 2. In the sample conditioning system, the cooled condensate is measured by the Process refractometer before it is sent to drain or balance tank. The transmitter displays the measurements and sends 4 – 20 mA output signal for real-time processcontrol.

RESULTS

It was demonstrated that the system can successfully detect sugar levels down to concentration of 200 ppm with a response time of less than 10 seconds (Figure 3). The measurement accuracy of 0.02 Brix was achieved.

The proposed system can detect the real harmful sugar content, which is above 200 ppm, while neglecting too low measurements caused by other components, including salts. False alarms can be eliminated, as the alarms will only be activated at harmful sugar levels.

The alarms control the shutter ensuring only quality condensate is fed to the boiler.

The condensate in the factory was monitored continuously and the results were satisfactory for the needs of the plant. The system works well after a year since its installation in the factory. The system was easy to install and start-up.

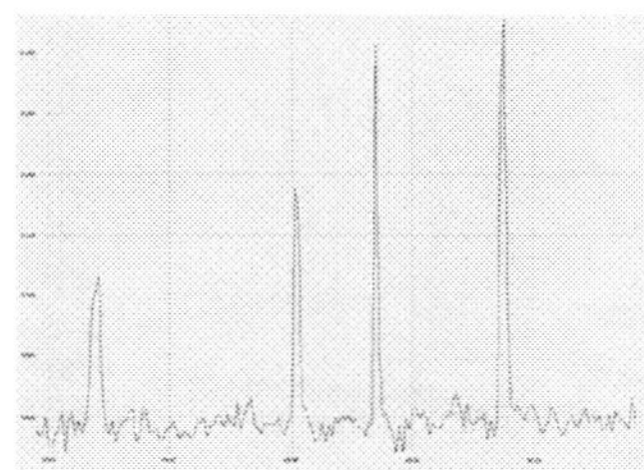

Figure 3. Real-time data of sugar levels (in Brix) measured by the refractometer. The refractometer immediately responded to the sugar shots injected to the feed line. These can be seen as high peaks.

CONCLUSIONS

TDS measurement by a refractometer proved to be a direct, reliable and fast method for detecting harmful levels of sugar in the condensate. Monitoring the TDS allows quick response to sugar leakages in the condensates and helps prevent system damages.

Continuous and accurate detection of condensate contamination:

- Prevents production and financial losses.
- Allows set-up of reliable and true alarms for detecting when the safe sugar limits are exceeded.
- Prevents carryover to boiler water and foaming in the boiler.
- Prevents corrosion and scaling inside the boilers, and consequently reduces equipment damage and maintenance.
- Increases equipment life.
- Reduces the use of chemicals in treatment of the boiler feedwater.

MORE INFORMATION

To ensure continuous support before and after purchase of our products, we offer local application consultation, training, maintenance and support expertise via our authorized sales representative network. Please refer to www.kpatents.com to contact your nearest representative.

K-PATENTS OY
PHONE: INT +358-207-291 570
INFO@KPATENTS.COM
WWW.KPATENTS.COM

S.I.T. Paper #1177

Molecular characterization of interspecific hybrids of sugarcane (*Saccharum* spp.) using SSR and RAPD markers

Jian-Cheng Chang[1], Li-Shyang Chen[2], Tso-Lin Cheng[2]

1 Taiwan Sugar Research Institute, Tainan, Taiwan
2 Taiwan Sugar Corporation, Tainan, Taiwan

Abstract

Sugarcane (*Saccharum* spp.) is considered one of the most important economic crops in tropical and subtropical regions of the world, primarily due to the high concentration of sucrose stored in the stem, which supplies the production of sugar. Sugarcane is also among the most suitable crops for biofuel production, as the waste product of sugar production can be recycled into the renewable fuel, and it thus plays an increasingly important role in circular agriculture. Hence, it is imperative to identify and protect critical sugarcane cultivars and varieties by use of molecular tools. A DNA barcode involves sequence variation within a standard DNA region as a tool to distinguish species. Simple sequence repeats (SSR) and random amplified polymorphic DNA (RAPD) are powerful and reliable PCR-based methods to detect DNA polymorphism. The current study analyzed genetic diversity in 12 TaiSugar commercial cultivars of sugarcane and a total of 13 SSR and 26 RAPD markers generated polymorphic bands. The constructed dendrogram matches the cross-breeding relationship. Sequences of DNA barcodes including *rbcL*, *matK*, *trnH-psbA* spacer, and ITS showed no variation among the cultivars, suggesting no natural selection occurred on the examined sugarcane populations. The developed molecular markers in this study would be useful for molecular characterization of sugarcanes, for future breeding programs to increase genetic diversity of sugarcane cultivars, and for providing objective and clear evidence for plant variety protection.

Introduction

Modern sugarcane cultivars (*Saccharum* spp.) are interspecific hybrids originating from crosses between *S. officinarum* (domesticated, $2n$=80) and *S. spontaneum* (wild species, $2n$=$5x$=40 to $16x$=128) (Arcenaux, 1967). The hybrid genome is a mixture of the both species with high and variable ploidy, $2n$=100–130 chromosomes (D'Hont *et al.*, 1996). Further gene incorporation can be made through crosses to other related species due to female restitution. The classification of sugarcane based on morphology is time and labor consuming; however, few attempts have been made to characterize sugarcane using molecular markers.

The aim of this study is to assess the genetic similarity among interspecific hybrids of TaiSugar sugarcane commercial cultivars by using chloroplast DNA barcodes and genomic SSR and RAPD markers and to investigate the potential of molecular marker identification.

Materials and Methods

Sugarcane leaf samples

Fresh and green sugarcane leaves were collected from the experimental field in Liuying Dist., Tainan City, Taiwan. A total of 12 TaiSugar commercial lines, ROC10, ROC16, ROC23, ROC28, 69-463, 76-12, 81-7049, 85-1389, 03-2026, 06-6043, 06-6111, and CP58-48 were examined in the present study. Leaf samples were preserved at -20°C for subsequent isolation of genomic DNA.

DNA extraction

Genomic DNA was isolated using Plant Genomic DNA Purification Kit (GeneMark Technolgy

Co., Ltd., Tainan, Taiwan) according to the manufacturer's protocol. DNA was quantified by absorbance measurement on a spectrophotometer at 260 nm, while the purity of the RNA was determined by the ratio of A_{260}/A_{280}.

PCR amplification

PCR amplification was conducted in a total reaction volume of 15 µl containing 5 ng/µL of genomic DNA, 1X GoTaq® Green Master Mix (Promega, Madison, WI, USA), and 0.4 µM of each primer, or 0.6 µM of a single RAPD primer. The PCR was carried out with an initial denaturation step for 5 min at 95°C; 30 cycles of 94°C for 30 s, the optimal annealing temperature of each primer set for 30 s, and 72°C for 1 min; and a final extension of 72°C for 5 min.

Primers

DNA barcodes

rbcLa_f	5'-ATGTCACCACAAACAGAGACTAAAGC-3'
rbcLa_r	5'-GTAAAATCAAGTCCACCRCG-3'
matK_JC_F	5'-GAAATCTCGTCAACACTACTTCGTC-3'
matK_JC_R	5'-TTCCCAGAAAAGATGATTCGTTCCA-3'
Sac_rpl2_F2	5'-CCCGTAGTAAGAGGGGTAGTTATG-3'
Sac_psbA_R2	5'-AATCTGTAGTTGATAGCCAAGGTC-3'
ITS-u1	5'-GGAAGKARAAGTCGTAACAAGG-3'
ITS-u4	5'-RGTTTCTTTTCCTCCGCTTA-3'

RAPD primers: One hundred primers for RAPD amplification were obtained from Operon Technologies (primer sets AA–AE).

SSR primers: Total 21 primer pairs from Pan's report (2010) were screened.

Electrophoresis:

PCR products were run on a 2% agarose gel at 100V for 50 minutes, then photographed under UV light.

Statistical analysis:

A dendrogram was constructed from band identification (presence=1; absence=0) using NTSYSpc version 2.2 based on UPGMA.

Results

DNA barcods

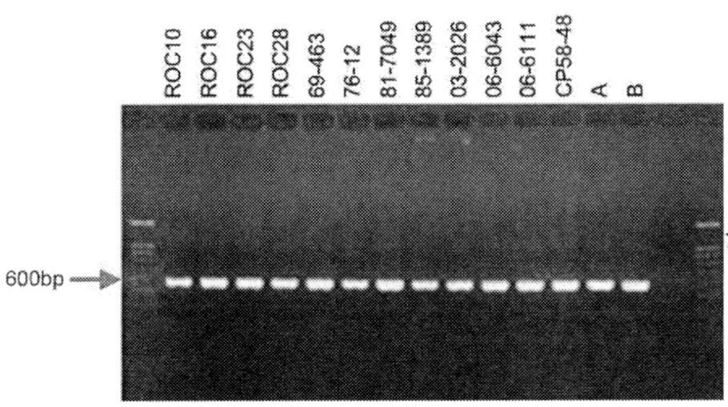

rbcL

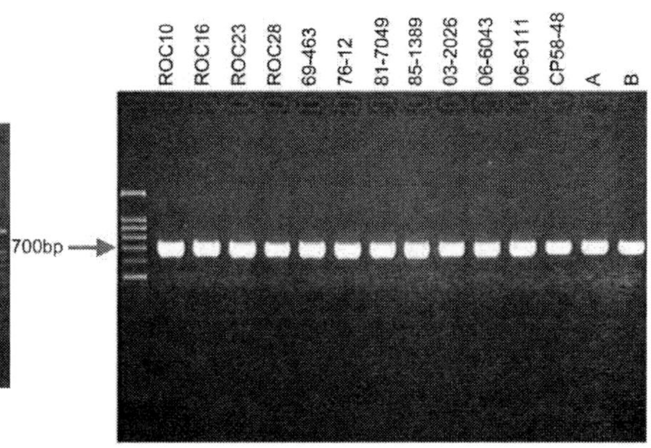

matK

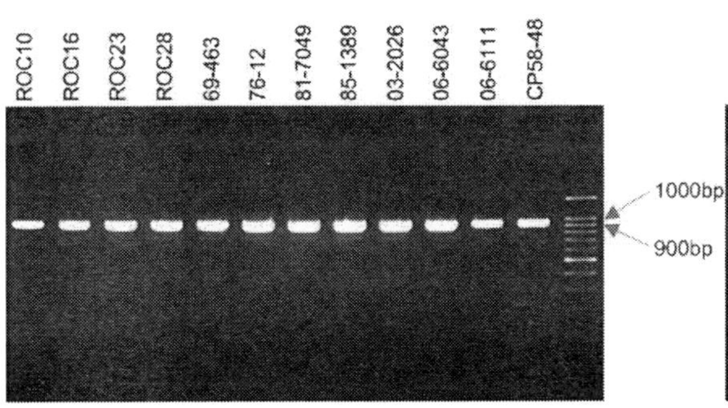

trnH-psbA

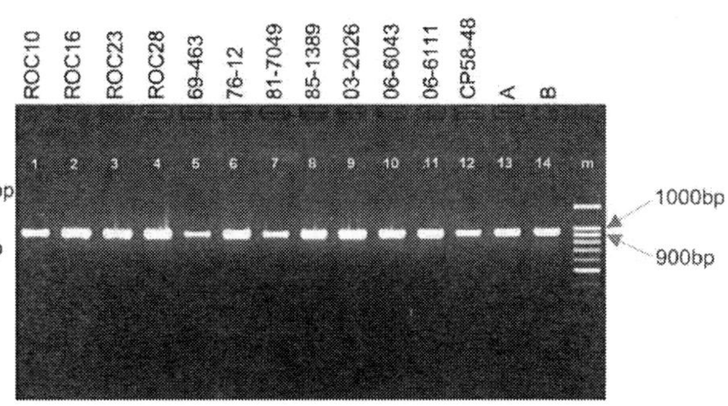

ITS

RAPD markers

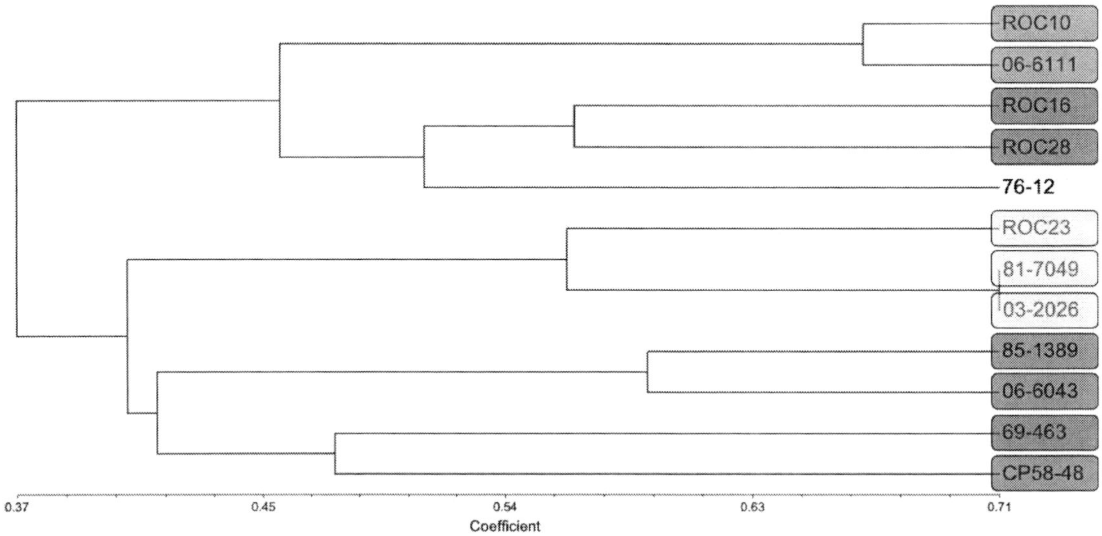

Phylogram of the evolutionary relationships of 12 TaiSugar commercial cultivars based on 26 RAPD markers. Nodes with the same color are parental and progeny lines.

SSR markers

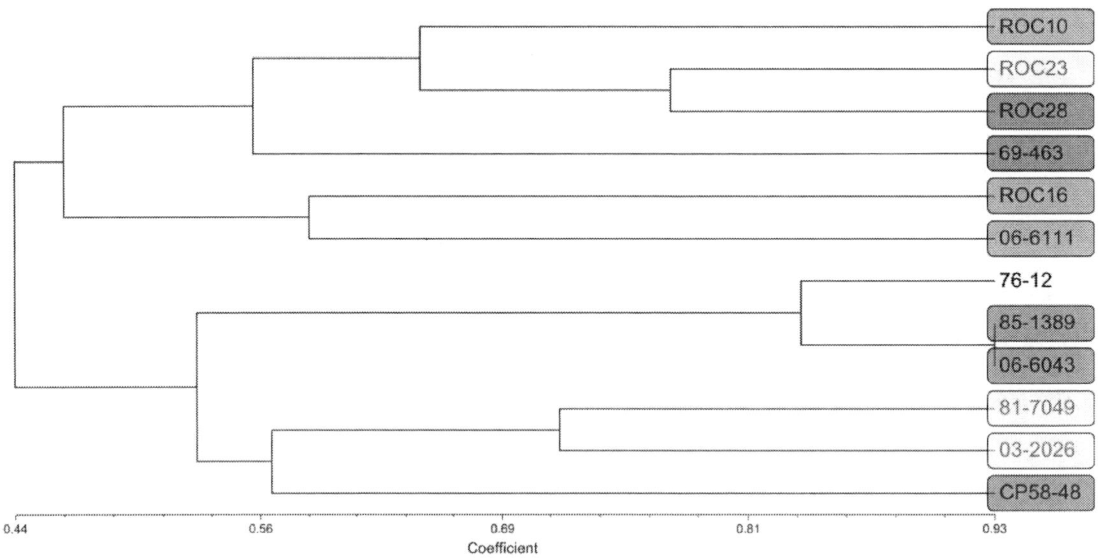

UPGMA dendrogram shows genetic diversity of 12 commercial lines generated from 13 SSR primers. Nodes with the same color represent parent-progeny relationship.

Progeny	Parental Lines (♀ x ♂)
ROC28	ROC16 x 69-463
03-2026	81-7049 x ROC23
06-6111	ROC10 x ROC16
06-6043	85-1389 x CP58-48

Conclusions

1. In the current study, no genetic diversity was observed amongst the 12 sugarcane lines examined using four different DNA barcodes, suggesting a lack of natural selection on the commercial lines developed in TaiSugar. Further hybridization breeding is urgently needed in order to expand the sugarcane gene pool.

2. The phylogenetic tree constructed with RAPD data matched the examined sugarcane cultivars parent and progeny relationship. Moreover, species identification can be achieved with the combination of different RAPD markers based on the presence and absence band patterns.

3. The cluster analysis based on SSR markers did not perfectly represent the phylogenetic relationship of studied commercial lines. Three out of the four progeny lines were found to be linked to only one of the parents. It is plausible that the allele dosage of the SSR markers led to a biased tree.

4. The developed molecular markers in this study would be useful for molecular characterization of sugarcanes, for future breeding programs to increase genetic diversity of sugarcane cultivars, and

for providing objective and clear evidence for plant variety protection.

References

Arceneaux, G., 1967, March. Cultivated sugarcanes of the world and their botanical derivation. In *Proc Int Soc Sugar Cane Technol* (Vol. 12, pp. 844–854).

D'Hont, A., Grivet, L., Feldmann, P., Glaszmann, J.C., Rao, S. and Berding, N., 1996. Characterisation of the double genome structure of modern sugarcane cultivars (*Saccharum* spp.) by molecular cytogenetics. *Molecular and General Genetics MGG, 250*(4), pp.405–413.

Pan, Y.B., 2010. Databasing molecular identities of sugarcane (*Saccharum* spp.) clones constructed with microsatellite (SSR) DNA markers. *Am J Plant Sci, 1*(02), p. 87.

RHEWUM®

S.I.T. Paper #1178. Survey of screen clothes – guideline to select the most suitable screen cloth for the specific screening task in the sugar industry

Prepared by Dipl.-Ing. Mario Gerards (RHEWUM GmbH)

Abstract

Sugar refining processes are well known for decades and screening of different types of sugar is an essential part of the process, whether it is involved at the preliminary production steps of screening sugar beet and cane, within the granulation process or at the very end of the production, for de-dusting the granulated sugar product before packing.

This poster focuses on screen clothes, which can be selected and installed within a screening machine. It gives an overview of the available screen meshes and outlines the pros and cons of different arrangements. Finally it can be

Types of Meshes

square mesh

oblong mesh (Elongated openings, usually in a ratio of 1:3 or 1:10)

Terminology

Mesh width (w) The opening size of the slots or squares on the screen panel through which the material passes.

Wire diameter (d) Diameter of the wire for a woven wire cloth, as measured before weaving.

Screen cut The particle size at which equal proportions of material report to the over size and undersize. The aim of the screening.

Pros & Cons of Thin and Thick Wires

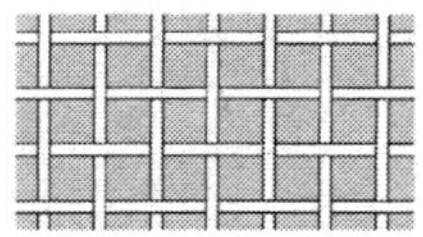
light screen cloths:

- high screening capacity
- thin wire
- big open screening area
- short operating life
- low tensioning force

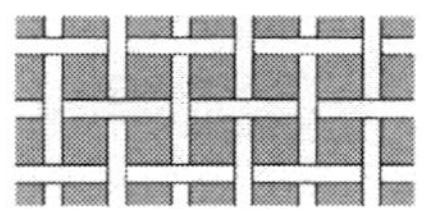
standard screen cloths:

- normal wire
- common screening area
- sufficient operating life

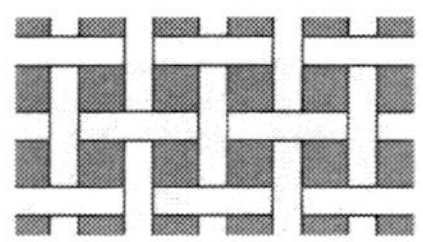
heavy screen cloth:

- mesh = wire diameter => minimum open screening area of 25%.
- low screening capacity
- thick wire
- reduced open screening

Open surface of square vs. oblong meshes

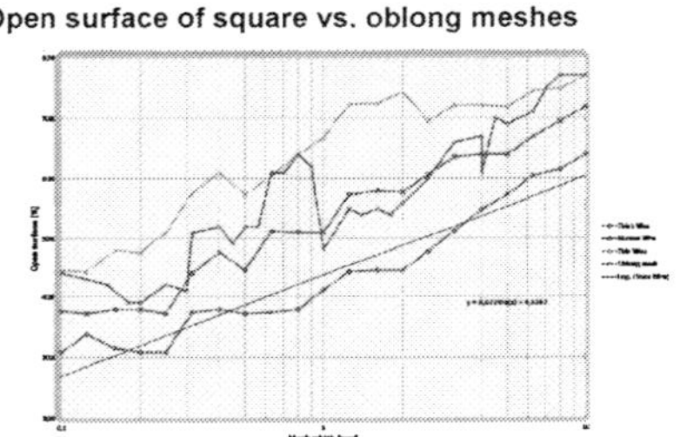

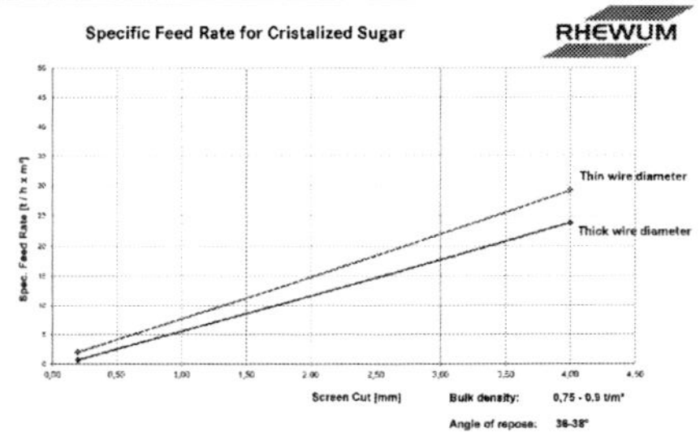

Conclusion

Various factors of the screen cloth can have a big impact on the efficiency. RHEWUM has been developing and manufacturing screening machines since the beginning of the 1950's and has extensive experience in choosing the ideal screen cloth for its customer's screening task. If desired, screening tests can be conducted at the in-

Open Screening Area

For woven wire cloth and wire screens the open area is defined as the ratio of the total area of the apertures to the total area of the cloth or the screen.

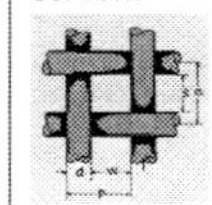

open screening area $A_0 = \varphi$

square mesh

$A_0 = \varphi = w^2 / (w + d)^2$

long mesh

$A_0 = \varphi = (w_q * w_l) / ((w_q + d_q) * (w_l + d_l))$

Case Study

RHEWUM WA 300 x 225/2 + WA 350 x 269/2 with rotary valve distribution

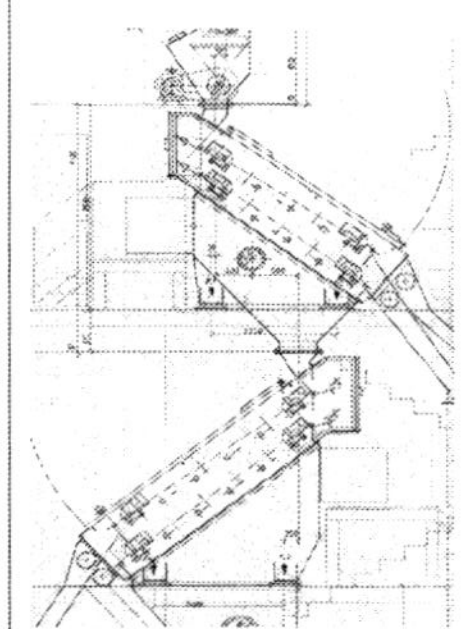

Feed material: White sugar

Capacity: 50–60 t/h

Bulk density: 0,9 – 1,0t/m³

Moisture content: 0,04 % H_2O

Temperature: max. 30°C

Cut point: 2,2 + 1,0 mm / 1,25 + 0,2 mm

Efficiency: + 95 %

Multivariate Methods to Quantify Key Chemical Traits of Sugar Crops

Sophie Minori Uchimiya, USDA Agricultural Research Service, Southern Regional Research Center, New Orleans, LA 70124

USDA United States Department of Agriculture

Agricultural Research Service

Introduction and Rationale

- Reversible electron-transfer and metal complexation of (poly)phenols underlie the antioxidant reactivity.
- Structure-reactivity relationships are necessary to understand the roles of polyphenolic chromophores beyond the traditional calorimetric methods for the tannin contents and free radical quenching ability.
- Rapid, easy, and inexpensive industrial methods are on demand for breeders, and for feedstock accounting/quality assurance at sorghum and sugarcane processing facilities for the food/biofuel production.

sample	λ_{max}	acid-butanol assay	Fe(II)	vanillin	DPPH
	210 nm	$\mu g\ g^{-1}$ delphinidin Cl	µM	[catechin]	[trolox]
	methanol	acetone-water	butanol	equiv. mg g^{-1}	equiv. µg g^{-1}
grape seed	0.97	33,165	37	320±27	542±22
Philippines	0.75	1,307	21	0	11±5
Taiwan	0.50	843	9	0	139±7

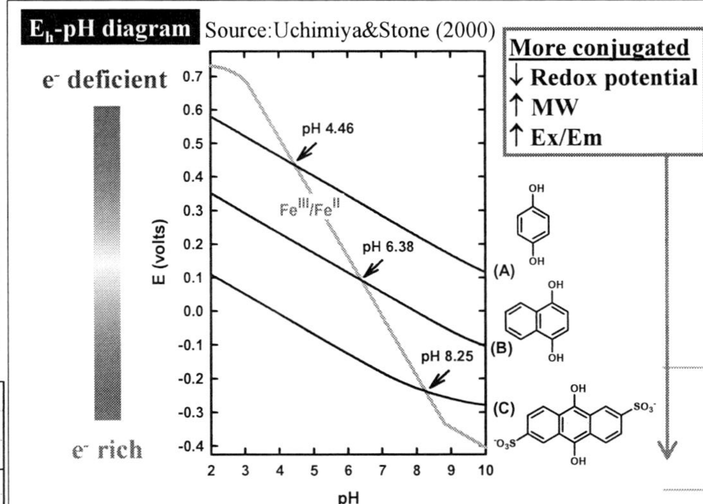

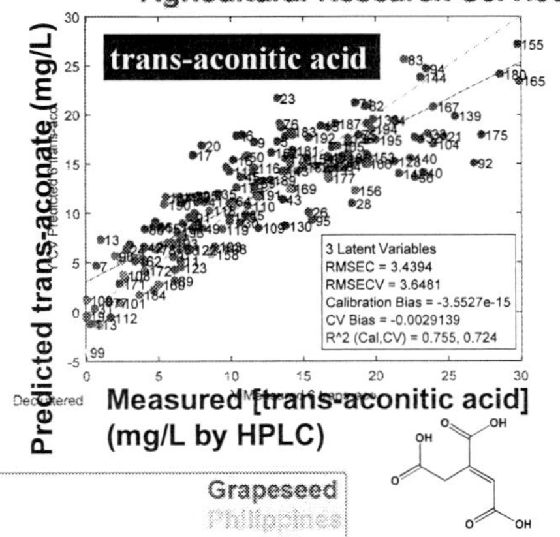

Materials and Methods

- Sugarcane extracts from Taiwan and Philippines were manufactured by heating cane juice at 90-100 °C for 40 min, and spray-drying at >90 °C. Grapeseed extracts were purchased off the market shelf, and contained dicalcium phosphate and cellulose bulking agent.
- Antioxidant activity by DPPH radical quenching (vs. Trolox); λ_{max} (UV/vis); condensed tannins by acid-butanol assay (vs. delphinidin); electron-donating capacity by Fe(III) reduction and ferrozine colorimetric method; flavonoids by vanillin assay (vs. catechin); and conjugated double bonds by fluorescence EEM.
- Cyclic voltammetry by screen-printed carbon electrodes (Ag/AgCl reference; 0.1M KCl, pH 5 by 20 mM phosphate buffer).

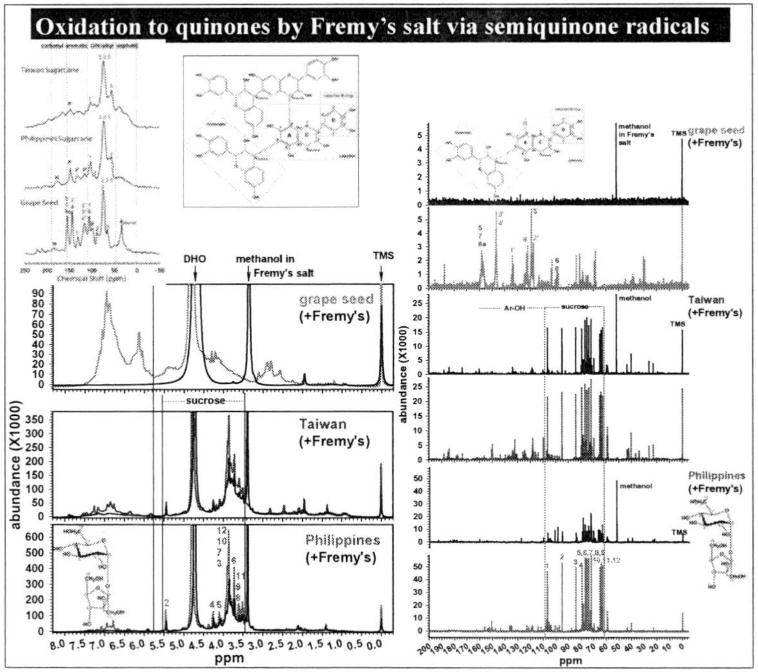

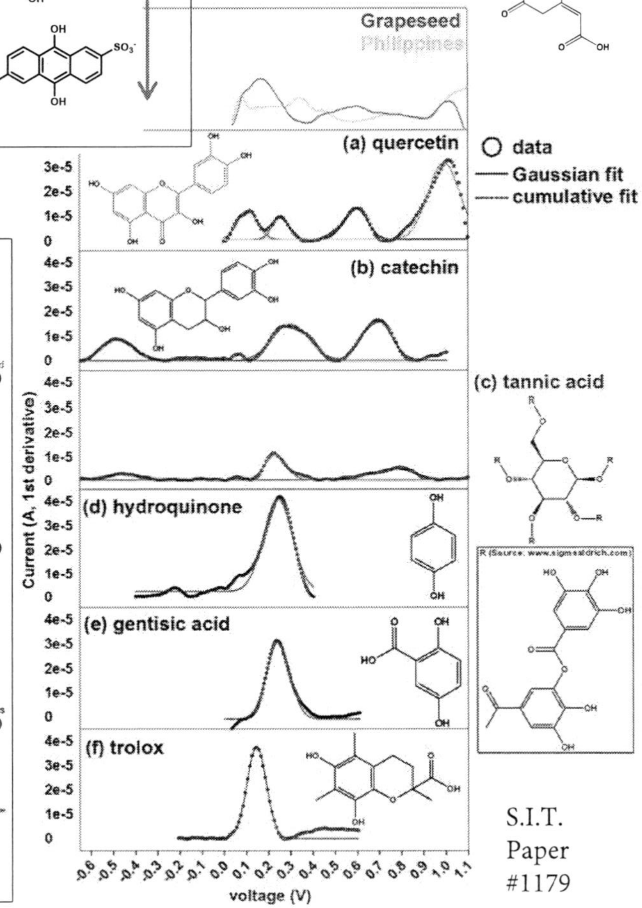

S.I.T. Paper #1179

ROSTER OF DELEGATES - ANNUAL MEETING 2017

Name	Affiliation	Address
Aguirre, Dr. Edgar	Sugar Industry Technologists, Inc.	Myrtle Beach, SC, U.S.A.
Babu, Roopa	ED&F Man Sugars Limited	London, ENGLAND
Barber, Kevin	Jord International Pty. Ltd.	Sydney, AUSTRALIA
Bouché, Catherine	ITECA Socadei	Aix in provence, FRANCE
Braun, Nickas	Nordic Sugar AB	Arlov, SWEDEN
Bruhns, Philipp	Sugar Industry/Zuckerindustrie	Berlin, GERMANY
Bryant, Chris	Jord International Pty. Ltd.	Sydney, AUSTRALIA
Burchell, Michael	ASR Group C&H Sugar Company, Inc.	Crockett, CA, U.S.A.
Carroll, Doug	Schenck Process, LLC	Whitewater, WI, U.S.A.
Chang, Jian-Cheng	Taiwan Sugar Corporation	Tainan, TAIWAN
Chen, Chad	Dr. Chou Technologies, Inc.	Austin, TX, U.S.A.
Chen, Pao-Sheng	Shun Ting Engineering Agency	Tainan, TAIWAN
Chen, Siou-Zheng	Lai Da Sustainable Development LED	Tainan, TAIWAN
Chen, Xiang-Rong	Gwofeng Technology Co., Ltd.	Tainan, TAIWAN
Chen, Yu-Fang	Fang Sheng Agency	Tainan, TAIWAN
Chen, Yu-Hsuan	Taiwan Sugar Corporation	Tainan, TAIWAN
Chettakul, Sarath	Wellman Co., Ltd.	Bangkok, THAILAND
Chettakul, Sayan	Wellman Co., Ltd.	Bangkok, THAILAND
Chorão, Jose Manuel	RAR - Rafinerias de A�úcar Reunidas	Porto, PORTUGAL
Chou, Dr. Chung Chi	Dr. Chou Technologies, Inc.	Austin, TX, U.S.A.
Chuasuwan, Kusumaln	United Farmer & Industry Co., Ltd	Nogrua, THAILAND
CLIFF	Taiwan Tang Xin Co., Ltd.	Tainan, TAIWAN
Cole, Dr. Marsha	ARS-USDA-SRRC	New Orleans, LA, U.S.A.
Dae Woo, Seo	HAITAI htb	Cheonan-si, KOREA
Dagallier, Adrien	Gaudfrin S.A.	Saint Garmain en Laye, FRANCE
DeBeer, Brian	Manildra Harwood Sugars	Harwood Island, AUSTRALIA
Demone, David P.	ASR Group - Domino Foods, Inc.	Iselin, NJ, U.S.A.
Dibella, Peter	Birkalla Sugar Technology Services Pty. Ltd.	Harwood, AUSTRALIA
Duc, Norbert	ITECA Socadei	Aix in provence, FRANCE
Eggleston, Dr. Gillian	ARS-USDA-SRRC	New Orleans, LA, U.S.A.
Fuchs, Philippe	pro-M-tec Theisen GmbH	Ettlingen, GERMANY
Fuhrmeister, Ron	Putsch GmbH & Co. KG	Hage, GERMANY
Gauchet, Sebastien	Novasep Asia	Shanghai, CHINA
Gerards, Mario	RHEWUM GmbH	Remscheid, GERMANY
Gupta, Ritesh	SPRAY Engineering Devices, Ltd.	PUNJAB, INDIA
Handa, Lalit	SPRAY Engineering Devices, Ltd.	PUNJAB, INDIA
Harada, Masakazu	Mitsui Sugar Co., Ltd.	Tokyo, JAPAN
Herrmann, Martin	Putsch GmbH & Co. KG	Hage, GERMANY
Hirata, Dr. Akio	Mitsui Sugar Co., Ltd	Tokyo, JAPAN
Ho, Wen-Hua	Fu Jiang Hydropower Agency	Tainan, TAIWAN
Ho, Zhi-Yuan	Dragon Enterprise Co., Ltd.	Tainan, TAIWAN
Huang, Charles	Taiwan Sugar Corporation	Tainan, TAIWAN
Huang, Jia-Lin	Jia Rui Co. Ltd.	Tainan, TAIWAN
Huang, Ming-Zhong	Taiwan Area Labor Agency	Tainan, TAIWAN
Huang, Zhi-Hao	Kenmec Mechanical Engineering Co., Ltd.	Tainan, TAIWAN
Inkson, Dr. Mike	Sugar Knowledge International Limited	Rushmore Hill, ENGLAND
Jung, Kwon Chae	CJ Cheiljedang Incheon Sugar Refinery	Seul, KOREA

ROSTER OF DELEGATES - ANNUAL MEETING 2017

Name	Affiliation	Address
Kaddoury, M.	Cosumar S. A.	Casablanca, MOROCCO
Kilpinen, Seppo	K-Patents Process Instruments	Vantaa, FINLAND
Kuan, Yu Hock	Asiatic Engineering Sdn. Bdh.	Penang, MALAYSIA
Kullman, Kjell	Silver Wibull Sweden AB	Hassleholm, SWEDEN
Lahlou, Dr. Mohamed	BMA AG	Braunschweig, GERMANY
Lai, Ming-The	Taiwan Sugar Corporation	Tainan, TAIWAN
Lai, Wei-Hsiang	Taiwan Sugar Corporation	Tainan, TAIWAN
Langhans, Bernd	Siemens	Nuremberg, GERMANY
LeBlanc, Dr. Jeffrey	Dr. Chou Technologies, Inc.	Austin, TX, U.S.A.
Lee, Jae Geun	CJ Cheiljedang Incheon Sugar Refinery	Seul, KOREA
Lee, Ming-Chun	Double Elle Indusrial Co., Ltd.	Tainan, TAIWAN
Lee, Xiao-Ping	Yi Cheng Engineering Agency	Tainan, TAIWAN
Lengen, Jerry	Graver Technologies	Glasgow, DE, U.S.A.
Lin, Chong-Xing	Star Crown Enterprise Co., Ltd.	Tainan, TAIWAN
Lin, Chun-Chi	Shye Ku Indiustry Co., Ltd.	Tainan, TAIWAN
Lin, Jian-Guo	Lumax International Corp., Ltd.	Tainan, TAIWAN
Lin, Tsung-Han	Taiwan Sugar Corporation	Tainan, TAIWAN
Liou, Jia-Jer	Taiwan Sugar Research Institute	Tainan, TAIWAN
Liu, Jie	Jord International Pty. Ltd.	Sydney, AUSTRALIA
Liu, Jinjun	BMA CHINA	Kunming, CHINA
Liu, Wei-Chun	Taiwan Sugar Corporation	Tainan, TAIWAN
Love, Dr. David	Tongaat Hulett Sugar	Glenashley, REPUBLIC OF SOUTH AFRICA
Mackenthun, Rüdinger	BWS Technologie GmbH	Grevenbroich, GERMANY
Maeda, Yasushi	Tsukishima Kikai Co., Ltd.	Tokyo, JAPAN
Mielonen, Eero	K-Patents Process Instruments	Vantaa, FINLAND
Milne, Timothy	Sugar Technology International	Plano, TX, U.S.A.
Muanme, Tawee	Sutech Engineering Co., Ltd.	Bangkok, THAILAND
Muneef, Abdulelah	United Sugar Company Saudi Arabia	Jeddah, SAUDI ARABIA
Myung Hoon, Seol	HAITAI htb	Cheonan-si, KOREA
Nakagawa, Atsushi	Mitsui Sugar Co., Ltd.	Tokyo, JAPAN
Nielsen, Bjarne Christian	Neltec Denmark A/S	Vojens, DENMARK
O'Donoghue, Alan	JM Canty International	Lockport, NY, U.S.A.
Odaira, Atsushi	Ebisu Kagaku & Co., Ltd.	Tokyo, JAPAN
Ohlson, Trevor	Lantic Inc. - Vancouver	Vancouver, B.C., CANADA
Okumura, Keigo	Mitsui Sugar Co., Ltd.	Tokyo, JAPAN
Paillat, Dominique	EURODIA	Pertuis, FRANCE
Parker, Craig	British Sugar PLC	Norfolk, ENGLAND
Peacock, Stephen	Amalgamated Research, LLC	Twin Falls, ID, U.S.A.
Pelham, David	U.S. Sugar Corporation	Clewiston, FL, U.S.A.
Pobi, Subhendu	PT. Jawamanis Rafinasi	Banten, INDONESIA
Prasobsom, Artit	United Farmer & Industry Co., Ltd	Nogrua, THAILAND
Pron, Frederic	Comessa S.A.	Strasbourg, FRANCE
Qian, Lai-Fu	Sunjuice Holdings Co., Ltd.	Tainan, TAIWAN
Rauch, Andreas	SternEnzym GmbH & Co. KG	Ahrensburg, GERMANY
Richard, Charley	Sugar Journal	New Orleans, LA, U.S.A.
Rohde, Alexander	SternEnzym GmbH & Co. KG	Ahrensburg, GERMANY
Rousset, Francois	Novasep Process	St. Maurice de Beynost, FRANCE

ROSTER OF DELEGATES - ANNUAL MEETING 2017

Name	Affiliation	Address
Saichua, Thiti	Sutech Engineering Co., Ltd.	Bangkok, THAILAND
Saleem, Altaf	Shakarganj Sugar Research Institute	Jhang Punjab, PAKISTAN
Sarir, Emmanuel	Carbo-Solutions International, LLC	Beverlywood, CA, U.S.A.
Schneider, Dirk	BMA AG	Braunschweig, GERMANY
Schorn, Paul	Tongaat Hulett Sugar South Africa, Ltd	Durban, REPUBLIC OF SOUTH AFRICA
Schulze, Dr. Bernd-Christoph	ESCON GmbH	Berlin, GERMANY
Schwanke, Uwe	BMA AG	Braunschweig, GERMANY
Sheen, Herng-Kuang	Taiwan Sugar Corporation	Tainan, TAIWAN
Shimasato, Itsuki	Toyo Hidoki Co., Ltd.	Tokyo, JAPAN
Silfversparre, Katarina	Nordic Sugar AB	Arlov, SWEDEN
Simpson, Peter	New Zealand Sugar Co., Ltd.	Auckland, NEW ZEALAND
Singh, Indrajit	Consultant	Hidd, BAHRAIN
Siriputtha, Saijai	United Farmer & Industry Co., Ltd.	Nogrua, THAILAND
Su, Dr. Cheng-Hao	Fusugar Industry Corporation	Taichung City, TAIWAN
Sugimoto, Akira	Mitsui Sugar Co., Ltd.	Tokyo, JAPAN
Suhr, Mark R.	Sugar Knowledge International Limited	Rushmore Hill, ENGLAND
Syu, Meng-Sheng	Chung Chan Hsing Ironworks Co.,Ltd.	Tainan, TAIWAN
Takei, Hiroyuki	Tsukishima Kikai Co., Ltd.	Tokyo, JAPAN
Takemori, Shinya	Tsukishima Kikai Co., Ltd.	Tokyo, JAPAN
Tanaka, Kazuho	Tsukishima Kikai Co., Ltd.	Tokyo, JAPAN
Tashiro, Daisuke	Mitsui Sugar Co., Ltd.	Tokyo, JAPAN
Thurapan, Amornsak	United Farmer & Industry Co., Ltd.	Nogrua, THAILAND
Tiwow, Hendrogiarto	PT. Jawamanis Rafinasi	Banten, INDONESIA
Tso, Hsi-Chun	Taiwan Sugar Corporation	Tainan, TAIWAN
Uchimiya, Dr.Sophie	USDA-ARS-SRRC	New Orleans, LA, U.S.A.
Udomsinca, Piyapong	Sutech Engineering Co., Ltd.	Bangkok, THAILAND
Urabe, Tetsuro	Ebisu Kagaku & Co., Ltd.	Tokyo, JAPAN
van Noord, Frank	Suiker Unie	Dinteloord, THE NETHERLANDS
Vawda, Ahmed	United Sugar Company Saudi Arabia	Jeddah, SAUDI ARABIA
Ventaka Ramana, Tamminidi	VRL Automation Engineering & Projects Pvt. Ltd.	Bangalore, INDIA
Visweswara Rao, T.K.	VRL Automation Engineering & Projects Pvt. Ltd.	Bangalore, INDIA
von Horn, Dennis	AMF-Bruns GmbH & Co. KG	Apen, GERMANY
Wang, Betty	Bonita Enterprise Co., Ltd.	Houston, TX, U.S.A.
Wang, Hong-Ming	Bonita Enterprise Co., Ltd.	Houston, TX, U.S.A.
Wang, Ming-Ching	Yong Teng Mechanical Industrial Co., Ltd.	Tainan, TAIWAN
Wassing, Paul	Veco International	Eerbeek, THE NETHERLANDS
Wei, Yu-Chi	Power Chain Enterprise Co., Ltd.	Tainan, TAIWAN
Wilson, Thomas	Imperial Sugar Company	Port Wentworth, GA, U.S.A.
Wittenberg, Dr. Arend	Suiker Unie	Dinteloord, THE NETHERLANDS
Wu, Bo-Jhin	Taiwan Sugar Corporation	Tainan, TAIWAN
Xiao, Chang-Geng	Taiwan Sugar Corporation	Tainan, TAIWAN
Xie, Guang-Yu	Yu Min Electromechanical Engineering Co., Ltd.	Tainan, TAIWAN
Xu, Shi-Han	Xin Jin Motor Transport Co. Ltd.	Tainan, TAIWAN
Yang, Sheng-Yun	Chiuan Sheng Mechanical Agency	Tainan, TAIWAN
Yang, Ya-Ting	Affiliate FUCC	Tainan, TAIWAN
Yatabe, Osamu	Mitsui Sugar Co., Ltd.	Tokyo, JAPAN
Yoshitomi, Akihiro	Mitsui Sugar Co., Ltd.	Tokyo, JAPAN

ROSTER OF DELEGATES - ANNUAL MEETING 2017

Name	Affiliation	Address
Zhang, Jia-ming	Cabot Norit Activated Carbon	Shanghai, CHINA
Zhang, Yang-Hong	Cabot Norit Activated Carbon	Shanghai, CHINA
Zheng, Hong-Dian	Taiwan Sugar Corporation	Tainan, TAIWAN
Zheng, Zuo-Lin	Taiwan Sugar Corporation	Tainan, TAIWAN
Zhou, Wei-Jia	Yu En Engineering Co., Ltd.	Tainan, TAIWAN

GUESTS OF DELEGATES - ANNUAL MEETING 2017

Name	Affiliation	Address
Aguirre, Margoth	Sugar Industry Technologists, Inc.	Myrtle Beach, FL, U.S.A.
Braun, Cecilia	Nordic Sugar AB	Arlov, Sweden
Chou, Yng	Emeritus	Austin, TX, U.S.A.
Dagallier, Yan	Gaudfrin S.A.	Saint Garmain en Laye, FRANCE
Demone, Kathy	ASR Group - Domino Foods, Inc.	Iselin, NJ, U.S.A.
Milne, Susie	Sugar Technology International	Plano, TX, U.S.A.
Pelham, Gaye	United States Sugar Corporation	Clewiston, FL, U.S.A.
Schwanke, Jean	BMA AG	Braunschweig, GERMANY
Silfversparre, Gustav	Nordic Sugar AB	Arlov, SWEDEN
Urabe, Mineko	Ebisu Kagaku & Co., Ltd.	Tokyo, JAPAN
van Noord-Ebshamp, Maartje	Suiker Unie	Dinteloord, THE NETHERLANDS
Vawda, Shereen	United Sugar Company Saudi Arabia	Jeddah, SAUDI ARABIA
Wittenberg, Claudia	Suiker Unie	Dinteloord, THE NETHERLANDS
Dibella, Dianne	Birkalla Sugar Technology Services	Harwood Island, AUSTRALIA